Multiferroic Materials

Properties, Techniques, and Applications

Series in Materials Science and Engineering

Other books in the series:

Computational Modeling of Inorganic Nanomaterials
S T Bromley, M A Zwijnenburg (Eds)

Automotive Engineering: Lightweight, Functional, and Novel Materials
B Cantor, P Grant, C Johnston

Strained-Si Heterostructure Field Effect Devices
C K Maiti, S Chattopadhyay, L K Bera

Spintronic Materials and Technology
Y B Xu, S M Thompson (Eds)

Fundamentals of Fibre Reinforced Composite Materials
A R Bunsell, J Renard

Novel Nanocrystalline Alloys and Magnetic Nanomaterials
B Cantor (Ed)

3-D Nanoelectronic Computer Architecture and Implementation
D Crawley, K Nikolic, M Forshaw (Eds)

Computer Modelling of Heat and Fluid Flow in Materials Processing
C P Hong

High-K Gate Dielectrics
M Houssa (Ed)

Metal and Ceramic Matrix Composites
B Cantor, F P E Dunne, I C Stone (Eds)

High Pressure Surface Science and Engineering
Y Gogotsi, V Domnich (Eds)

Physical Methods for Materials Characterisation, Second Edition
P E J Flewitt, R K Wild

Topics in the Theory of Solid Materials
J M Vail

Solidification and Casting
B Cantor, K O'Reilly (Eds)

Fundamentals of Ceramics
M W Barsoum

Aerospace Materials
B Cantor, H Assender, P Grant (Eds)

Series in Materials Science and Engineering

Multiferroic Materials
Properties, Techniques, and Applications

Edited by

Junling Wang

CRC Press
Taylor & Francis Group
Boca Raton London New York

CRC Press is an imprint of the
Taylor & Francis Group, an **informa** business

Cover Image: Reprinted with permission from A. F. Moreira dos Santos, et al. (2003). Epitaxial growth and properties of metastable $BiMnO_3$ thin films, Applied Physics Letters. 84:1. © 2003, AIP Publishing LLC. See Chapter 2 for detailed description.

CRC Press
Taylor & Francis Group
6000 Broken Sound Parkway NW, Suite 300
Boca Raton, FL 33487-2742

© 2017 by Taylor & Francis Group, LLC
CRC Press is an imprint of Taylor & Francis Group, an Informa business

Printed on acid-free paper
Version Date: 20160609

International Standard Book Number-13: 978-1-4822-5153-1 (Hardback)

Library of Congress Cataloging-in-Publication Data

Names: Wang, Junling, 1978- editor.
Title: Multiferroic materials : properties, techniques, and applications / edited by Junling Wang.
Other titles: Series in materials science and engineering.
Description: Boca Raton, FL : CRC Press, Taylor & Francis Group, [2017] | Series: Series in materials science and engineering | Includes bibliographical references and index.
Identifiers: LCCN 2016021831| ISBN 9781482251531 (hardcover ; alk. paper) | ISBN 1482251531 (hardcover ; alk. paper)
Subjects: LCSH: Ferromagnetic materials.
Classification: LCC QC761.5 .M85 2017 | DDC 620.1/7--dc23
LC record available at https://lccn.loc.gov/2016021831

Visit the Taylor & Francis Web site at
http://www.taylorandfrancis.com

and the CRC Press Web site at
http://www.crcpress.com

Printed and bound in the United States of America by Publishers Graphics, LLC on sustainably sourced paper.

Contents

Contents

Preface

MULTIFERROICS HAS EMERGED AS one of the hottest topics in solid state physics in this millennium. The coexistence of multiple ferroic/antiferroic properties makes them an ideal playground for fundamental physics study. Furthermore, the potentially strong couplings among the order parameters lead to tantalizing possibilities for applications. In particular, a strong coupling between magnetic spin and electric dipole, that is, the magnetoelectric coupling effect, may lead to electric-field control of magnetization and revolutionize memory technology as well as the next-generation spintronics industry.

Supported by advances in both materials synthesis and first-principles calculation techniques, the past decade has witnessed a renaissance of research in magnetoelectric and multiferroic materials, which has largely remained quiet after pioneering work back in the 1960s. Coincidentally, groundbreaking progress was made on both type-I ($BiFeO_3$, 2003) and type-II multiferroics ($TbMnO_3$ 2003 and $TbMn_2O_5$ 2004) at about the same time, which further supported the revival of this field. After more than a decade of intense research, we have achieved an unprecedented understanding about multiferroics and magnetoelectric coupling effect. It is time to pause, reflect on what we have understood, and identify future directions. With this intent, we wrote this book. This book contains recent progress in the field written by researchers who are active on the front lines. After the introductory chapter (Chapter 1), in which Professor Khomoskii has given a comprehensive overview of the topic with valuable historical background of the development of multiferroic materials, the rest of the book is divided into four sections. Section I (Chapters 2 through 5), Single-phase Multiferroics, covers single-phase multiferroics, both type-I and type-II, and their possible applications in spintronic devices. Section II (Chapters 6 through 8), Multiferroic Composites, reviews multiferroic composites (both bulk and nanocomposites). A comprehensive discussion of their various applications is also included in this section. In Section III (Chapters 9 through 11), Theoretical Approaches in Multiferroic Study, we present the theoretical approaches used in studying multiferroics, covering both phenomenological and first-principles calculations. The book ends with a final section (Chapters 12 through 14), Emerging Topics in the Field, on emerging topics in the field of multiferroics, focusing on domain walls and vortices.

Preface

I am especially grateful to all the authors who contributed chapters to this book. It took us almost two years to put this work together, but I am confident that the book will be a useful reference not only for students specializing in solid-state physics and materials science but also for scientists and engineers exploring multiferroic materials for fundamental physics and/or practical applications. I also would like to thank Dr. Lu You, Dr. Yang Zhou, and Mr. Lei Chang for their editorial assistance throughout this project. Last but certainly not least, I would like to express my gratitude to senior editor Luna Han of Taylor & Francis Group LLC. Without her patience and constant support, I would not have been able to complete this project.

Editor

Junling Wang earned a PhD at the University of Maryland, College Park, USA, in 2005. After one year of postdoctoral research at Pennsylvania State University, he joined the School of Materials Science and Engineering at Nanyang Technological University in 2006 and was promoted to associate professor with tenure in 2011. His research focuses on multiferroic thin films and their application in nonvolatile memories. He has extensive experience on (1) oxides heteroepitaxy by pulsed laser deposition, (2) ferroelectric and magnetic domain analysis, and (3) applications of these functional oxides in nanoelectronic and spintronic devices. He has published approximately 110 papers in top journals, including *Science, Nature Communications, NPG Asia Materials, Advanced Materials, ACS Nanos,* etc. His work has been cited more than 6700 times. His original work on multiferroic thin films and nanocomposites, both published in *Science,* have alone been cited over 3000 times.

Contributors

Marin Alexe
Department of Physics
University of Warwick
Coventry, United Kingdom

Manuel Angst
Peter Grünberg Institut and
  Jülich Centre for Neutron
  Science
Forschungszentrum Jülich GmbH
Jülich, Germany

and

Experimental Physics IVC
RWTH Aachen University
Aachen, Germany

Seung Chul Chae
Department of Physics Education
Seoul National University
Seoul, Korea

Aiping Chen
Center for Integrated
  Nanotechnologies (CINT)
Los Alamos National Laboratory
Los Alamos, New Mexico

Lang Chen
Department of Physics
South University of Science and
  Technology of China
Shenzhen, Guangdong, People's
  Republic of China

San-Wook Cheong
Rutgers Center for Emergent
  Materials and
Department of Physics and
  Astronomy
Rutgers University
Piscataway, New Jersey

Ying-Hao Chu
Department of Materials Science
  and Engineering and
Department of Electrophysics
National Chiao Tung
  University
Hsinchu, Taiwan

and

Institute of Physics
Academia Sinica
Taipei, Taiwan

Shuai Dong
Department of Physics
Southeast University
Nanjing, China

Shuxiang Dong
Department of Materials Science
  and Engineering
Peking University
Beijing, People's Republic of China

Contributors

Yue Hao
State Key Discipline Laboratory of
  Wide Band Gap Semiconductor
  Technology
School of Microelectronics
Xidian University
Xi'an, China

Jason T. Haraldsen
Department of Physics
University of North Florida
Jacksonville, Florida

Jia-Mian Hu
School of Materials Science and
  Engineering
and
State Key Lab of New Ceramics and
  Fine Processing
Tsinghua University
Beijing, China

Chuanwei Huang
Department of Physics
South University of Science and
Technology of China
Shenzhen, Guangdong, People's
  Republic of China

Yen-Lin Huang
Department of Materials Science
  and Engineering
National Chiao Tung University
Hsinchu, Taiwan

Quanxi Jia
Center for Integrated
  Nanotechnologies (CINT)
Los Alamos National Laboratory
Los Alamos, New Mexico

Daniel I. Khomskii
II Physikalisches Institut
Universitaet zu Koeln
Koeln, Germany

Jong-Woo Kim
Functional Ceramics Group
Korea Institute of Materials Science
  (KIMS)
Changwon, Korea

Heng Li
State Key Discipline Laboratory of
  Wide Band Gap Semiconductor
  Technology
School of Microelectronics
Xidian University
Xi'an, China

Xin Li
State Key Discipline Laboratory of
  Wide Band Gap Semiconductor
  Technology
School of Microelectronics
Xidian University
Xi'an, China

Jun-Ming Liu
National Laboratory of Solid State
  Microstructure
Nanjing University
Nanjing, China

Xiaoli Lu
State Key Discipline Laboratory of
  Wide Band Gap Semiconductor
  Technology
School of Microelectronics
Xidian University
Xi'an, China

Ce-Wen Nan
School of Materials Science and
  Engineering
and
State Key Lab of New Ceramics and
  Fine Processing
Tsinghua University
Beijing, China

Shashank Priya
Bio-Inspired Materials and Devices
    Laboratory
Center for Energy Harvesting
    Materials and Systems
Virginia Tech
Blacksburg, Virginia

Jungho Ryu
Functional Ceramics Group
Korea Institute of Materials Science
    (KIMS)
Changwon, Korea

Jan Seidel
School of Materials Science and
    Engineering
University of New South Wales
Sydney, Australia

Junling Wang
School of Materials Science and
    Engineering
Nanyang Technological University
Singapore

Jan-Chi Yang
Department of Materials Science
    and Engineering
National Chiao Tung University
Hsinchu, Taiwan

Lu You
School of Materials Science and
    Engineering
Nanyang Technological University
Singapore

Jincheng Zhang
State Key Discipline Laboratory of
    Wide Band Gap Semiconductor
    Technology
School of Microelectronics
Xidian University
Xi'an, Czhina

Jiwen Zhang
State Key Discipline Laboratory of
    Wide Band Gap Semiconductor
    Technology
School of Microelectronics
Xidian University
Xi'an, China

Yang Zhou
School of Materials Science and
    Engineering
Nanyang Technological University
Singapore

Yuan Zhou
Bio-Inspired Materials and Devices
    Laboratory
Center for Energy Harvesting
    Materials and Systems
Virginia Tech
Blacksburg, Virginia

Jian-Xin Zhu
Theoretical Division and
    Center for Integrated
    Nanotechnologies
Los Alamos National Laboratory
Los Alamos, New Mexico

1

Coupled Electricity and Magnetism in Solids
Multiferroics and Beyond

Daniel I. Khomskii

Contents

1.1 Introduction

The intrinsic coupling of electricity and magnetism is one of the corner-stones of modern physics. It goes back to the famous Maxwell equations, or even earlier, to Michael Faraday, and one can even find earlier reports pointing in that direction. This coupling plays crucial roles in all modern physics, and it is one of the foundations of modern technology, for example, in the generation of electricity in electric power stations, and in electric transformers. Recently this field acquired new life in spintronics—the idea of being able to use not only charge but also spin of electrons for electronic applications. In this field, one predominantly deals with the influence of magnetic field and/or magnetic ordering on transport properties of materials; for example, the well-known magnetoresistance or the work of magnetic tunnel junctions. But very interesting such effects can also exist in insulators. These are, for example, the (linear) magnetoelectric effect, or the coexistence and mutual influence of two types of ordering, magnetic and ferroelectric ordering in multiferroics. Such phenomena are very interesting in terms of its physical features, and are very promising for practical applications, such as for addressing magnetic memory electrically without the use of currents, or as very efficient magnetic sensors. These factors probably are reasons for creating such a significant interest in this field. It is now one of the hottest topics in condensed matter physics, and in addition to magnetoelectrics and multiferroics per se, the study of these has many spin-offs in the related fields of physics, such as the study of magnetoelectric effects in different magnetic textures (domain walls, magnetic vortices, skyrmions, etc.).

There are already several good reviews of this field [1–7], and there exists a very complete and useful collection of short reviews on multiferroics in the special issue of the *Journal of Physics of Condensed Matter* [8]. There is also a chapter on multiferroics in my recent book [9]. In this chapter, I will more or less follow the general outline of my short review on "Multiferroics for pedestrians," published in *Physics*: Trends [7]—of course with significant additions.

This book is mainly devoted to real multiferroics—materials with coexisting magnetic and ferroelectric orderings. These materials are extremely interesting as to their physics, and they promise many important practical applications. However, one has to realize that for many practical applications, such as attempts to write and read magnetic memory in hard discs electrically, using electric fields rather than currents (e.g., with gate voltage devices), one needs not so much multiferroics but rather materials with good magnetoelectric properties: one must be able *to modify* the magnetic state by a changing electric field. But the idea is that it is precisely multiferroic materials in which the change of magnetic state by electric field, or vice versa, may be especially strong. From this point of view, various textures in magnetic materials, which can have magnetoelectric response, such as certain domain walls of skyrmions, also attract now considerable attention. These topics are mentioned later in this chapter and extensively discussed in several chapters of this book.

1.2 Some Historical Notes

When describing the field of magnetoelectrics and multiferroics, the first reference one usually gives is the paper of Curie [10], who shortly noticed the possibility of having both magnetic and electric orderings in one material. But the real story began with a short remark in one of the famous books on theoretical physics by Landau and Lifshitz [11], who wrote in 1959:

> *Let us point out two more phenomena, which, in principle, could exist. One is piezomagnetism, which consists of linear coupling between a magnetic field in a solid and a deformation (analogous to piezoelectricity). The other is a linear coupling between magnetic and electric fields in a media, which would cause, for example, a magnetization proportional to an electric field. Both these phenomena could exist for certain classes of magnetocrystalline symmetry. We will not however discuss these phenomena in more detail because it seems that till present, presumably, they have not been observed in any substance.*

Indeed, at the moment of publication of that volume there were no known real examples of magnetoelectric or multiferroic systems. But already less than a year after its publication, the seminal paper by Dzyaloshinskii [12] appeared, who on symmetry grounds predicted that the well-known antiferromagnet Cr_2O_3 should exhibit the linear magnetoelectric effect. And the following year, this effect was indeed observed in Cr_2O_3 by Astrov [13]. After that a rapid development of this field followed, initially in the study of magnetoelectrics, see Reference [14]. But very soon the ideas of not only the magnetoelectric effect, but also those of real multiferroics were put forth. Soon the first multiferroic—a material in which (antiferro) magnetic and ferroelectric orderings are present simultaneously—was discovered by Ascher et al. [15]—the Ni-I boracite. (It was in fact Schmid [16] who later coined the very term "multiferroics" in connection with such materials.) An active program to synthesize such materials artificially was initiated, predominantly by two groups in the former Soviet Union: in the group of Smolenskii in Leningrad (present-day St. Petersburg) and in the group of Venevtsev in Moscow.

However, after considerable activity in the 1960s and 1970s, the interest in this field faded somewhat. A new surge of activity appeared at around 2000, and there were three factors that stimulated it.

The first was the realization of an interesting and challenging problem in the physics of magnetic and ferroelectric materials, mostly on the example of perovskites. There are quite a lot of magnetic perovskites, including the famous colossal magnetoresistance manganites, or the "two-dimensional perovskite" La_2CuO_4—the parent material of high-temperature superconducting cuprates. Extensive discussion of these materials, with many tables, is contained in the collection compiled by Goodenough and Longo in the well-known Landolt-Börnstein series [17]. Another, even more extensive collection of tables of ferroelectric perovskites, starting with the equally famous material $BaTiO_3$, was published by Japanese scientists [18]. And, surprisingly enough, a comparison of these extensive collections of tables, 100–300 pages each, demonstrates that there is practically no overlap between them: a perovskite is either magnetic or ferroelectric, but practically never both simultaneously (of course, it may be neither, as is the case with the prototype mineral perovskite $CaTiO_3$, which gave the name to this whole family). What is the reason for this mutual exclusion, and can one go around it? This problem was known already during the period 1970 to 1980, but was formulated only in 1999 during a workshop in Santa Barbara, and publicized after 2000 [19,20]; it attracted the attention of the scientific community. This problem is discussed in Section 1.6.

But, of course, the most important factors were two experimental break-throughs. One was the fabrication and study of very good films of, it seems, the best multiferroic material known, $BiFeO_3$, by Wang et al. [21]. This gave a possibility of studying the multiferroic effects, and immediately opened perspectives for very appealing practical applications. $BiFeO_3$ remains until today the favorite material for many investigations, both in basic research and in applied fields. The second achievement was the discovery by two groups, of Kimura et al. [22] and of Hur et al. [23], of a novel class of multi-ferroics. In multiferroics that were known previously, the ferroelectric and magnetic orderings occurred independently and were driven by different mechanisms; typically, although not always, ferroelectric ordering starts at a higher temperature. In the novel class of multiferroics discovered in [22,23], ferroelectricity is driven by a particular type of magnetic ordering, and occurs only in the magnetically ordered phase. One can call the first group "type-I multiferroics," and the second group "type-II multiferroics" [7]. I discuss this classification and the microscopic mechanisms in action in each of these classes in Section 1.5. Here, I only want to stress that these two experimental breakthroughs gave new life to the whole field and led to an enormous increase of activity in this field, the present book being a good example of that.

1.3 Magnetoelectric Effect; Symmetry Considerations

The specific feature of magnetoelectric materials is the possibility of generating electric polarization by magnetic field, and vice versa, magnetization by electric field. This can be described by the relations

$$P_i = \alpha_{ij} H_j + \beta_{ijk} H_j H_k + \dots \tag{1.1}$$

$$M_j = \alpha_{ji} E_i + \beta_{jik} E_i E_k + \dots \tag{1.2}$$

where we use the standard convention of summation over the repeated indices. Terms quadratic in E, H are typically more common and less interesting; the most interesting effect is the presence of the first, linear terms mentioned earlier. This is referred to as the linear magnetoelectric effect, or simply the magnetoelectric effect.

One can also describe the linear magnetoelectric effect by including in the expression for the free energy the term

$$F_{ME} = -\alpha_{ij} E_i H_j \tag{1.3}$$

As the polarization and the magnetization are given by $P = -\partial F/\partial E$ and $M = -\partial F/\partial H$, one immediately obtains from this expression for the free energy the first terms in Equations 1.1 and 1.2.

As we see in general, in a crystal, the magnetoelectric coefficient α is a tensor. It can have symmetric and antisymmetric components. The symmetric part of this tensor can always be transformed to a diagonal form

$$\alpha_{ij} = \alpha_j \delta_{ij} \tag{1.4}$$

where δ_{ij} is the Kronecker symbol. In this case, the polarization for the magnetic field along the main axes would be parallel to the magnetic field. But there can also be an antisymmetric part of the magnetoelectric tensor. It is known that such an antisymmetric tensor, with independent components α_{12}, α_{13}, and α_{23}, is equivalent to an axial vector, or pseudovector

$$T_i = \varepsilon_{ijk} \alpha_{jk} \tag{1.5}$$

where ε_{ijk} is the totally antisymmetric Levi-Civita symbol. This pseudovector T is called the toroidal moment. If the system has a nonzero toroidal moment, then, from Equations 1.1, 1.2, and 1.5 one can see that, for example, the polarization in an external magnetic field would be

$$P \sim T \times H \tag{1.6}$$

and magnetization would be

$$M \sim T \times E \tag{1.7}$$

that is, they would be perpendicular to the external fields.

Symmetry considerations play a very important role in the magnetoelectric effect. First of all, these refer to the symmetry with respect to spatial inversion, $\mathcal{J}$, and time reversal, T. Electric field, polarization and electric dipole moments are usual vectors, changing sign under spatial inversion, but remaining the same under time reversal:

$$\mathcal{J} \, P = -P \qquad \mathcal{J} \, E = -E$$
$$\mathcal{T} \, P = P \qquad \mathcal{T} \, E = E \tag{1.8}$$

On the other hand, magnetization and the magnetic field are axial vectors, or pseudovectors—odd with respect to time reversal, but even with respect to spatial inversion.

$$\mathcal{J} \, M = M \qquad \mathcal{J} \, H = H$$
$$\mathcal{T} \, M = -M \qquad \mathcal{T} \, H = -H \tag{1.9}$$

One can easily understand these rules when one recalls that magnetic field and magnetic moments are created by currents $J = e \, v = e \, dr/dt$. For example, for circular currents shown in Figure 1.1, we have $M \sim r \times J$. One sees from this expression that M is even for spatial inversion ($r \to -r$), but odd for time reversal ($t \to -t$).

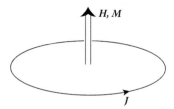

FIGURE 1.1 Magnetic field created by a current.

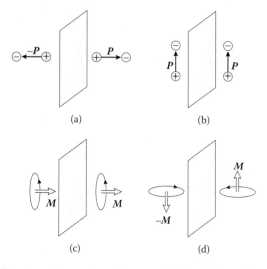

FIGURE 1.2 Transformation rules of magnetization (M) and polarization (P) under mirror reflection.

Using Figure 1.1, one can also obtain the transformation rules of, for instance, M or P under mirror reflections, illustrated in Figure 1.2 (see also [9]): under mirror reflections the components of P and M parallel and perpendicular to the mirror plane change as

$$P_\perp \rightarrow -P_\perp \qquad P_\parallel \rightarrow P_\parallel$$
$$M_\perp \rightarrow M_\perp \qquad M_\parallel \rightarrow -M_\parallel \tag{1.10}$$

From Equations 1.8 and 1.9 one can see that the linear magnetoelectric effect (linear terms in Equations 1.1, 1.2, or 1.3) can exist in a system only if both inversion and time reversal symmetries are simultaneously broken. The energy (Equation 1.3) should be a scalar, and thus, the magnetoelectric coefficient α should be both $\mathcal{J}$- and $\mathcal{T}$-odd. For that, first of all, the system should have some magnetic ordering, which breaks time reversal symmetry. In most cases, this is the standard magnetic ordering, for which the average spin at a site $\langle S_i \rangle \neq 0$. But one cannot exclude more complicated states such that not the magnetic dipole $\langle S \rangle$ is nonzero, but rather there exists some non-zero higher-order spin correlation function, containing an odd number of spins, for example, the magnetic octupole $\sim \langle S_1\, S_2\, S_3 \rangle$. And the spatial inversion symmetry should also be broken in order to have the linear magnetoelectric effect. Often this symmetry is broken just by a particular type of magnetic ordering.

In general, the free energy may also contain terms of higher order, not only those presented in Equation 1.3. For example, we may have terms of the type $\beta_{ijk}\, E_i\, H_j\, H_k$, or similar terms written as a function of order parameters P and M, for example, $\sim \beta_{ijk}\, P_i\, M_j\, M_k$, or terms $\sim P^2 M^2$. The conditions for their appearance are often not as stringent as those for linear magnetoelectric coupling. We will not, however, consider such terms below, and will concentrate on the linear magnetoelectric effect.

There is one general relation between the magnetoelectric response function α_{ij} and the usual dielectric and magnetic response, characterized by the dielectric constant ε_{ij} (or the corresponding electric susceptibility, or polarizability), and the magnetic response characterized by magnetic permeability μ or magnetic susceptibility χ, with $\mu = 1 + 4\pi\chi$. This constraint has the form [24]

$$\alpha^2 < \chi_e \chi_m \qquad (1.11)$$

where χ_e and χ_m are the electric and magnetic susceptibilities. We see that one can hope to obtain strong magnetoelectric coupling, for example, close to a ferroelectric or magnetic transition, in which (for II order transitions) χ_e or χ_m diverge, that is, χ_e or $\chi_m \to \infty$.

One more point is worth addressing here. We now know very well that the electric and magnetic responses are in general frequency- and momentum-dependent, $\varepsilon(\boldsymbol{q}, \omega)$, $\chi(\boldsymbol{q}, \omega)$. This dependence has very definite physical meaning. Thus, for example, the dielectric function contains terms such as

$$\varepsilon(\boldsymbol{q}, \omega) \sim \sum c_i / (\omega^2 - \omega_i^2(\boldsymbol{q})) \qquad (1.12)$$

That is, it has poles at the positions of dipole-active collective excitations $\omega_i(\boldsymbol{q})$, for example, optical phonons. These modes give definite signatures, for example in the optical properties of solids. Similarly, the structure of $\chi(\boldsymbol{q}, \omega)$, which can be measured, for example, by magnetic neutron scattering, tells us about magnetic excitations in the system, such as spin waves with their spectrum $\omega(\boldsymbol{q})$; and the existence of (strong) maximum of $\chi(\boldsymbol{q}, 0)$ at a certain q-value, $\boldsymbol{q}_0$, may be a signature of eventual magnetic instability of the system, such as the formation of spin density wave with momentum $\boldsymbol{q}_0$, and so on.

One should think that, similarly, the magnetoelectric response function α should also have both frequency and momentum dependence, $\alpha(\boldsymbol{q}, \omega)$. This question was not, to the best of my knowledge, yet studied in a general form for magnetoelectric materials. Apparently, the electromagnons [25] are related to this question—they should be the poles of both ε and α, similar to Equation 1.12. But what could be, for example, the q-dependence of α, what would be its significance, and how can one measure it, are still open questions. One could think that there should also be some general relations for $\alpha(\boldsymbol{q}, \omega)$ similar to the Kramers–Kronig relations or to the optical sum rule for $\varepsilon(\boldsymbol{q}, \omega)$; however, I am not aware of any such general treatment yet (possibly one could find some related results in the literature, but they are not formulated using this terminology).

1.4 Multiferroics

1.4.1 General Considerations

By multiferroics in a narrow sense we refer to materials having simultaneously both magnetic and ferroelectric orderings, that is, having two order parameters $M(r)$ and $P(r)$.* Magnetic ordering could be of different types: ferromagnetic, ferri- or antiferromagnetic, or it could be of a more complicated type. But for electric ordering one has in mind a real ferroelectric ordering, $\langle P \rangle \neq 0$. Sometimes in the field of ferroelectricity one also speaks about antiferroelectrics, but one has to realize that this notion has no strict physical meaning. Magnetic transitions, of any kind, always correspond to symmetry breaking: going from paramagnetic to magnetically ordered state, we break at least the time reversal symmetry (and maybe some spatial symmetries as well). Similarly, ferroelectric transition corresponds to a change of symmetry in the system from a centrosymmetric to a noncentrosymmetric one. However, the nominally antiferroelectric transitions do not necessarily break any symmetry: one can always formally consider any system as having electric dipoles, for example, inside a unit cell, pointing in opposite directions. In this sense, any structural transition in a solid is accompanied by some charge redistribution and could be formally called an antiferroelectric. Still, sometimes it can make sense to speak about an antiferroelectric transition, if this structural transition is accompanied by relatively strong anomalies in the dielectric constant ε. In any case, in the field of multiferroics we always have in mind the appearance of a real ferroelectric polarization, which is nonzero when averaged over the whole sample—although sometimes we also speak about *local polarization*.

The different characters of electric and magnetic orderings are reflected in one important aspect—the type and the meaning of the order parameter. For instance, the magnetization of a ferromagnet $\langle M \rangle$ or sublattice magnetization $L = \langle M_\uparrow - M_\downarrow \rangle$ of an antiferromagnet is a well-defined quantity, having absolute meaning. This, however, is not the case with electric polarization: it may depend, for example, on the choice of the unit cell, cf. Figure 1.3. It looks that with the choice of the unit cell, as shown in Figure 1.3a, the polarization points from left to right; however, in the same system but with a different choice of the unit cell, as shown in Figure 1.3b, it points from right to left. And indeed, accurate treatment shows, see for example, a very pedagogical explanation in [26], that the absolute value of polarization is not a uniquely defined quantity,

(a)   (b)

FIGURE 1.3 Local electrical polarization of a one-dimensional lattice.

---

* Sometimes one includes among multiferroics also systems with a third type of ordering—a ferroelastic one. But in this book, we will not consider it.

but *the change of polarization* with changing external conditions, such as temperature or electric field, is. This is also reflected in the fact that one has to use special theoretical methods (the so-called Berry phase methods) for *ab initio* calculation of polarization.

The very term "multiferroic" was proposed by Schmid [16]. In his review article [27], Schmid also presented a classification of different symmetry classes that allow for the simultaneous presence of both ferroelectric and magnetic orderings.

In speaking about multiferroics, symmetry considerations play a crucial role indeed. Both, time reversal T and spatial inversion symmetry $\mathcal{J}$, should be broken. And one also needs one unique vector P, which has to become $-P$ under inversion. One sees that these symmetry requirements are the same as those needed to get the linear magnetoelectric effect. An important question arises: in which cases would one get magnetoelectric, with polarization existing only in an external (magnetic) field, and when will we have real multiferroics, with spontaneous polarization existing without any external field? The answer is qualitatively the following: For magnetoelectric, both T and $\mathcal{J}$ should be broken, but the product $T\mathcal{J}$ is conserved: by consecutive application of time reversal and spatial inversion we would return to the initial state. For multiferroics, however, not only T and $\mathcal{J}$ but also the product $T\mathcal{J}$ should be broken. Thus, by looking at these symmetries one can understand whether a particular material with a given magnetic structure would be a real multiferroic or only a magnetoelectric. The examples mentioned here demonstrate this. The classical magnetoelectric material Cr_2O_3 has a crystal and magnetic structure with the main elements shown in Figure 1.4a (the crystal inversion

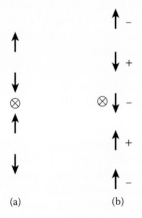

(a) (b)

FIGURE 1.4 (a) The main elements of crystal and magnetic structures of Cr_2O_3. The crystal inversion center is marked by the encircled cross $\otimes$. Magnetic ordering shown here breaks this inversion, but simultaneous inversion and time reversal (inversion of spin directions) return the state to the original one. (b) In a system with alternation of ions with different charges, for instance, + and −, and the magnetic structure ↑ ↑ ↓ ↓ (this is a schematic representation of a real situation in Ca_3CoMnO_6), not only T and $\mathcal{J}$, but also $T\mathcal{J}$ are broken. Time reversal following inversion leads to a state different from the initial one.

center is marked by the encircled cross $\otimes$). Of course, as in all magnetic states, time-reversal is broken right away. According to the rules formulated earlier, spatial inversion is also broken (it transforms spin ↑ to ↓). But simultaneous inversion and time reversal (inversion of spin directions) return the state to the original one, that is, $\mathcal{T}\,|\text{in}\rangle = -|\text{in}\rangle$, $\mathcal{J}\,|\text{in}\rangle = -|\text{in}\rangle$, but $\mathcal{T}\,\mathcal{J}\,|\text{in}\rangle = |\text{in}\rangle$. Therefore, this system is magnetoelectric but not multiferroic. On the other hand, for example, in the structure shown in Figure 1.4b, with alternation of ions with different charges, for example, + and −, and with the magnetic structure ↑ ↑ ↓ ↓ (this is a schematic representation of a real situation in Ca_3CoMnO_6 [28], see also [29] and Figure 1.10), not only $\mathcal{T}$ and $\mathcal{J}$, but also $\mathcal{T}\,\mathcal{J}$ are broken, time reversal following inversion leading to a state different from the initial one. And indeed this gives real multiferroics [28].

One still has to be slightly careful with this classification: to have real ferroelectricity in the usual sense, one has to be able to reverse polarization, $\boldsymbol{P} \to -\boldsymbol{P}$, by applying a proper electric field E (which gives the famous polarization loop, $P(E)$ hysteresis). In some systems, however, the electric field required for that is too strong to be realized in practice, so that the polarization is always pointing in one direction, and we cannot switch it. In this case, one speaks not about ferroelectrics, but rather about pyroelectrics [11]. There are such examples also among "multiferroic" compounds. $PbVO_3$ has a tetragonal structure of the same type as the famous ferroelectric $PbTiO_3$, but the distortion in it leading to the dipole moments is so strong that the polarization cannot be reversed [30,31] (maybe it would be possible to realize this in thin films). Thus, $PbVO_3$ should rather be called a magnetic pyroelectric, not a multiferroic. Still, symmetry considerations are extremely useful and actually crucial in the whole big field of multiferroics. One can find corresponding treatment, for example, in the review [27] or in [32]. We will not dwell on this topic any more, and will rather discuss more microscopic aspects of the physics of multiferroics.

1.5 Different Types of Multiferroics

Speaking of microscopic mechanisms, one can first of all say that, despite the huge variety of different types of magnetic ordering, most of all "strong" magnets are conceptually the same, see for example, [9]: due to strong electron–electron interaction or strong electron correlations, the state is formed with localized electrons (which for integer number of electrons per cite are the Mott, or Mott–Hubbard insulators)—the state with localized magnetic moments—localized spins. The exchange interaction between these localized moments leads to a certain magnetic ordering at low temperatures. Depending on specific details, such as electron configuration of respective ions, orbital occupation, detailed type of the lattice, and so on, we can have quite diverse types of magnetic ordering, but the general picture—the presence of localized electrons or localized spins with particular exchange interaction—remains the same.

The situation with ferroelectrics is more diverse and much more complicated. There exist many different microscopic mechanisms leading to ferroelectric behavior. And all the diversity of ferroelectrics and of eventual multiferroics is mainly connected just with this diversity of mechanisms of ferroelectricity. Thus, we can have systems in which there exist structural units, for example, molecules similar to HCl, each of which has a nonzero dipole moment just "by construction." And some ordered arrangement of such units could in principle make a material ferroelectric. It is known, for example, that among many forms of water ice, there is one that is ferroelectric.

Another mechanism is met in hydrogen-bonded ferroelectrics. To these belong some inorganic compounds, for example, KH_2PO_4 (KDP), but mostly organic systems, for example, the first ferroelectric discovered in nature—the Rochelle or Seignette's salt $NaKC_4H_4O_6 \cdot 4H_2O$ (this compound even gave the name to this very phenomenon in several languages, where ferroelectricity is called seignetoelectricity)—although the exact nature of ferroelectricity in this material is still a matter of debate. In both cases, there may exist some magnetic ions in "other parts" of the system, so that in effect such systems may become multiferroic.

For us, however, other types of multiferroics are of more importance. These are ferroelectrics or multiferroics with ferroelectricity driven by the covalency of transition metal ions with surrounding cations (ligands), for example, oxygen; "geometric" ferroelectrics; and ferroelectrics with lone pairs. We now proceed to a short description of these three classes of materials. However, before that, we will briefly discuss two general notions.

One can divide all multiferroics into two big groups, which we can call type-I and type-II multiferroics [7]. We have already shortly mentioned this classification in Section 1.2. The multiferroics we have mostly discussed until now, with the mechanisms of ferroelectric ordering listed earlier, have independent mechanisms of ferroelectric and magnetic orderings, occurring at different temperatures (usually with ferroelectric transition above the magnetic one, but not necessarily so). These are type-I multiferroics. Ferroelectricity often occurs in them at rather high temperatures—in $BiFeO_2$ $T_{FE} \sim 1100$ K, and in $YMnO_3$ $T_{FE} \sim 1000$ K. Magnetic ordering, occurring independently, can also be at a rather high transition temperature—in $BiFeO_3$ $T_N = 640$ K, so that it is a good multiferroic already at room temperature. In general, such systems can have quite large spontaneous polarization, which in $BiFeO_3$ reaches 80–100 $\mu C/cm^2$—larger that in $BaTiO_3$ ($\sim$60 $\mu C/cm^2$). Of course, there also exists a certain coupling between magnetism and ferroelectricity in these materials, but unfortunately it is usually not strong enough, although it was demonstrated, for example, that one can modify the magnetic structure of $BiFeO_3$ by electric field [33].

At the beginning of this century, the other, novel class of multiferroics was discovered [22,23]—the systems which we can call type-II multiferroics. These are the systems in which ferroelectricity exists and is generated only in certain magnetically ordered states. These materials attract now the main attention from the point of view of fundamental science. They are, however, at least as yet, less promising for practical applications than type-I systems such as $BiFeO_3$, or composite multiferroics consisting, for example, of layers of nonmagnetic ferroelectrics such as $(PbZr)TiO_3$ and adjacent layers of good magnets such as permalloy, with the coupling between these layers occurring via common strain (using magnetostriction of magnetic layers and piezoelectric response of ferroelectric ones). But from the physical point of view these type-II, or magnetically driven multiferroics, present special interest.*

1.6 Type-I Multiferroics

The varieties of type-I multiferroics differ first of all by the mechanisms leading to ferroelectricity. We have already briefly discussed two such types.

Systems having structural units with permanent dipoles. To such systems belong materials containing, for example, polar groups such as BO_3. If there are also magnetic ions in such compounds, these could be multiferroic. Examples of such materials are boracites, for example, the Ni–I boracite $Ni_3B_7O_{13}I$ [15], or iron borate $RFe_3(BO_3)_4$ [34].

Hydrogen-bonded ferroelectrics. As mentioned earlier, to these systems belong the first known ferroelectric—the Rochelle (or Seignette's) salt, first prepared in 1675. There are many materials in this class, but to the best of my knowledge there are no good multiferroics yet among those.

Transition metal perovskites. Probably the most important class of multiferroics is the perovskite family in which ferroelectricity is due to the "ferroelectric-active" transition metal ions. Such are, for example, the famous $BaTiO_3$. In classical physics, one usually describes ferroelectricity as a consequence of the so-called polarization catastrophe, at which the high polarizability of some constituent ions leads to the instability of the nonpolar state and thus the creation of ferroelectricity. However, there is another, more microscopic explanation of the appearance of ferroelectricity in systems such as $BaTiO_3$: the establishment of a strong covalent bond of the transition metal ion, here Ti, with one (or several) of the surrounding oxygen ions at the expense of weakening the bonds with the other ones (see Figure 1.5). One can show, see for example, [3], that the energy gained by the corresponding shift δu of Ti ions from the centers of O_6 octahedra is $\sim -g(\delta u)^2$. As the elastic energy loss is also quadratic in δu, $+B(\delta u)^2/2$, such phenomenon may occur only if the gain of covalency energy exceeds the elastic energy loss, for which one needs strong electron–lattice interaction (large coupling constant g above) and a not too stiff lattice (smaller bulk modulus B). These conditions are met not in all materials even with similar structures. That is why, for example,

---

* Sometimes one presents a different classification of multiferroics, mainly paying attention to whether polarization is a primary order parameter, or whether it is caused by the coupling to another one, for example, magnetic ordering due to coupling $\sim PM^2$ (but possibly also by coupling to some nonferroelectric structural distortion). One calls the first class of these systems proper ferroelectric, and the second one improper ferroelectric [35]. In the usual Landau theory of II order phase transitions, see for example[36,37], the primary order parameter η close to the critical temperature behaves as $\eta \sim \sqrt{(T_c - T)}$, but, for example, for the coupling of the type $P\eta^2$ and PM^2, polarization is linear in temperature, $P \sim T_c - T$.

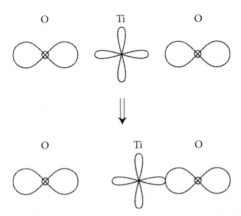

FIGURE 1.5 The establishment of a strong covalent bond of the transition metal ion, for example, Ti in $BaTiO_3$, with one (or several) of the surrounding oxygen ions at the expense of weakening the bonds with the other ones, leading to a spontaneous polarization and thus ferroelectricity.

$BaTiO_3$ is ferroelectric; $SrTiO_3$ is not but is "almost there" (it has a very high dielectric constant, and small perturbations such as uniaxial stress and even isotope substitution $^{16}O \rightarrow {}^{18}O$ make it ferroelectric); and $CaTiO_3$ is much further from the ferroelectric state.

However, $BaTiO_3$, a classical ferroelectric, is not magnetic; that is, it is not a multiferroic. The analysis of experimental data, for example, the comparison of extensive tables with the large amount of data on ferroelectric [17] and magnetic [18] perovskites shows that usually these two types of orderings in perovskite family are mutually exclusive. Ferroelectricity in "classical" systems is observed practically exclusively in perovskites ABO_3 with transition metal B-ions with empty d-shells, that is, with the occupation d^0 ($BaTi^{4+}O_3$; $LiNb^{5+}O_3$; etc.). On the other hand, magnetism requires partial occupation of d-levels. The realization of this dichotomy caused a long debate why this is so, and why none (or so few) of the magnetic perovskites with d^n ($n \neq 0$) shells are ferroelectric [3,19,20]. There are several physical factors proposed to explain this property.

One is simply that the ions with empty d-shells are usually smaller than those with d^n ($n \neq 0$) configurations, so that such small ions can easily shift from the center of a large O_6 octahedral cavity. This factor may play a certain role, see below, but it does not explain why, for example, $BaTiO_3$ is ferroelectric while $CaTiO_3$ is not. Even worse, $CaMnO_3$ is also not ferroelectric, although it contains Mn^{3+} ions that are smaller than Ti^{4+} in $BaTiO_3$ [38].

The other factor could be that, whereas the formation of a strong covalent bond with, say, one oxygen, see Figure 1.5, leads to a decrease of the electron energy (only the bonding orbital is occupied for d^0 transition metal ions, Figure 1.6a), in the presence of real d-electron(s) the antibonding orbital should also be filled, Figure 1.6b. Because of that we lose, in this simple example, half of the energy which we gained by shifting the d^0 transition metal ion. This effect, if it does not forbid ferroelectricity, at least makes it much less probable.

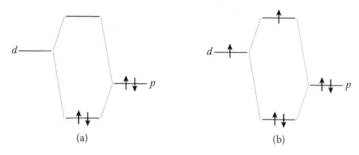

FIGURE 1.6 Difference between orbital occupation in d^0 and d^n ($n \neq 0$) transition metal ions.

One can think of yet another factor that could suppress the formation of transition metal–oxygen covalent bonds leading to ferroelectricity, or at least make it less favorable [3]. The covalent bond is a typical singlet chemical bond, for example, with the wave function of the type $(d\uparrow\, p\downarrow - d\downarrow\, p\uparrow)/\sqrt{2}$, as in the H_2 molecule in the Heitler–London approximation. However, in the presence of another localized d-electron (or several of those), as in Figure 1.6b, the Hund's rule exchange would destabilize the singlet covalent bond—similarly to the magnetic pair-breaking of singlet Cooper pairs in s-wave superconductors with magnetic impurities [39].

All these factors could be at work and make the coexistence of magnetism and ferroelectricity in these systems very unlikely. But one has to realize that this mutual exclusion is not a "theorem." It is rather a game of numbers, and the physical factors described earlier, though making this coexistence highly unlikely, do not forbid it. And indeed, it was predicted theoretically [40,41] and observed experimentally [42] that in the system $AMnO_3$ with magnetic Mn^{4+} (d^3, $S = 3/2$) the system may become ferroelectric when one increases the size of the A^{2+} ions (going from Ca to Sr to Ba).

Based on this discussion, the more natural route to create multiferroics in perovskites is the route first taken by several groups (see reviews in [43,44])—to make a mixed perovskite containing ferroelectric-active ions with the configuration d^0, and magnetic ions with configuration d^n, $n \neq 0$, in the hope that every species will do "what it wants." And indeed, on this route, several multiferroics of this series were synthesized, first by Smolenskii and his group [43]. Some of them were even ferro-, or rather ferrimagnetic, which is very favorable for practical applications. Unfortunately, the coupling between magnetic and ferroelectric degrees of freedom in these systems turned out to be rather weak.

There are other suggestions as well of how to make magnetic perovskites ferroelectric. One idea is to use the coupling between rotation and tilting of BO_6 octahedra often occurring in perovskite ABO_3 [45]. And indeed it was possible to create multiferroic on this route. This mechanism, however, belongs rather to another class of ferroelectric and multiferroic behavior, which we may call the "geometric" mechanism.

"Geometric" multiferroics. As mentioned earlier, there are many multiferroics with improper ferroelectricity, which appears as a secondary effect

(a "by-product") of a primary ordering, for example, of rotation and tilting of structural units in a crystal, such as MO_6 octahedra. The best known and the most important examples of systems with this mechanism of ferroelectricity are the hexagonal $RMnO_3$ (where R is a small rare earth ion) systems, for example, $YMnO_3$. These systems sometimes are called hexagonal perovskites, although they have not much in common with real perovskites except similar-looking chemical formulae. In these systems, the Mn^{3+} ions with fivefold coordination are located in the centers of trigonal bipyramids (two oxygen tetrahedra "glued" together by a common face, common oxygen triangle). They form layered structures with triangular Mn layers. Similar to conventional perovskites, here these building blocks also have a tendency toward tilting and rotation to guarantee the close packing of the lattice. With such distortions (in perovskites they are called $GdFeO_3$-distortion) there appear short AO pairs with a dipole moment. But in perovskites ABO_3 these dipoles in neighboring cells are oriented in opposite directions and cancel each other, and in the standard case they don't lead to net ferroelectricity (unless one uses special tricks to avoid this compensation). However, in systems such as $YMnO_3$ there is no such compensation. The geometric mechanism of ferroelectricity in these systems, which presents a good example of improper ferroelectrics, was established in [46]. These materials, with their interesting magnetic structure, nowadays present an important playground for studying, in particular, the characteristics of multiferroic domains and domain walls.

One extra comment must be made in connection with geometric ferroelectrics and multiferroics. We have explained its origin by the rotation and tilting of building blocks of a system, for example, MO_6 octahedra in perovskites or MO_5 trigonal bipyramids in systems such as $YMnO_3$. This one, in its turn, is usually explained by the tendency toward close packing of rigid ions, characterized, for perovskites BO_3, by the tolerance factor $t = (R_A + R_O)/\sqrt{2}(R_B + R_O)$: for $t \sim 1$ the system remains cubic, but for smaller values of t there occurs rotation and tilting of the BO_6 octahedra, see for example, [9]. But if we look more deeply, beyond the simplified picture of rigid ions, we realize that it is again a certain tendency of chemical bonds, in this case predominantly A–O bonds, which leads to such distortions. Thus, in effect it is always local chemistry that is responsible for the formation and stability of one or the other crystal structure, in particular, the ferroelectric one. Very often one can express this tendency in the language of the pseudo-Jahn–Teller, or second order Jahn–Teller effect [47].

Lone pair mechanism. Yet another "chemical" mechanism of ferroelectricity is provided by materials containing ions with the so-called lone pairs, or dangling bonds. These are usually materials containing Bi^{3+} or Pb^{2+}. Bi typically accepts valences 3+ and 5+. Bi^{5+} has the electronic structure $(Xe)4f^{14}5d^{10}$, and Bi^{3+} has two extra $6s$ electrons. In principle, these could become valence electrons and take part in the formation of chemical bonds (as they do for Bi^{5+}). However, in Bi^{3+} and similarly in Pb^{2+}, these two electrons do not participate in the formation of chemical bonds and are free to "rotate" in different directions in a crystal, which could lead to a particular orientation of dipole moments associated with them. (Of course these are not pure $6s$ electrons that are spherical, but they are usually hybridized with their own p-electrons, or with p-electrons of surrounding ligands, for example, oxygen.)

Ferroelectricity due to charge ordering. One more mechanism of ferroelectricity and of eventually multiferroic behavior is the possibility that the charge ordering, existing in some materials, can break inversion symmetry and lead to ferroelectricity. This topic is discussed in great detail in Chapter 3; thus, I will restrict myself to a few remarks illustrating the main idea of this phenomenon.

Suppose we have a structure consisting of dimers with equivalent sites, such as, for example, H_2 molecules, see Figure 1.7a. This structure is definitely centrosymmetric and is not ferroelectric (the inversion centers are marked in Figure 1.7 by small encircled crosses $\otimes$). If however, we now have an extra intradimer charge ordering, Figure 1.7b, making "left" and "right" ions in a dimer inequivalent, each such dimer (a "molecule") would have a dipole moment (double arrows in Figure 1.7b), and the entire system may become ferroelectric. This mechanism was first proposed in [48], and it is now detected in some materials. And if some constituting ions are magnetic, such systems would simultaneously become multiferroic.

Concluding this section, a few general remarks are in order. When considering electric polarization, one usually speaks about two contributions to it, the ionic and the electronic contributions. Ferroelectric transition is always a structural phase transition from the paraelectric phase with centrosymmetric crystal structure to the low-temperature state with broken lattice inversion symmetry, belonging to a pyroelectric class [11]. As with all structural transitions, besides a shift of ions there is also a redistribution of electron density (actually a change of chemical bonding). Consequently, one can speak about two contributions to the total polarization in ferroelectrics: an ionic contribution and an electronic one* (although, strictly speaking, it may be impossible

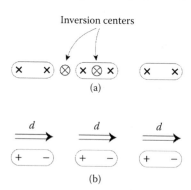

FIGURE 1.7 Illustration of how charge ordering can lead to ferroelectricity.

---

* Once again, if we look deeper, in effect all cohesion in solids, and consequently all structural transitions, are of electronic nature "deep inside." Nevertheless, it makes sense to separate electronic and ionic contribution to polarization—with the meaning, crudely speaking, that the ionic contribution is predominantly the contribution of ionic cores, electronic contribution being attributed to the change in the distribution of valence electrons.

to define them rigorously). By electronic contribution we have in mind, more specifically, the ferroelectricity caused by the change of electronic distribution at fixed positions of ions. Of course, if we then "release" the lattice, the ions would relax and the ionic positions would adjust to the change of electronic density. Still, the main driving mechanism could be predominantly electronic.

Usually these contributions are determined theoretically, for example, using *ab initio* calculations. And the outcome turns out to be not universal, and it strongly depends on the system. Thus, in perovskites $RMnO_3$ with small rare earth R = Er, ..., with the E-type magnetic structure, ionic and electronic contributions are comparable [49]. But in $TbMnO_3$, the ionic contribution dominates, electronic contribution being only about 10% of the ionic one (and in opposite direction) [50]. In principle, there can also exist situations with polarization of predominantly electronic character, which we discuss in detail in Section 1.7.

1.7 Type-II Multiferroics

By type-II multiferroics we refer to multiferroics in which ferroelectricity exists only in a magnetically ordered state, and is in fact driven by a particular type of magnetic ordering. It was the discovery of such multiferroics [22,23] that invigorated the entire field of multiferroics and made it such a hot topic. From a symmetry point of view, we are dealing here with materials in which the crystal structure has an inversion symmetry, that is, which in themselves are not ferroelectric, but in which a particular type of magnetic ordering breaks this inversion symmetry. As discussed in Section 1.4, if in such a situation both $\mathcal{T}$ and $\mathcal{J}$ invariance are broken, but the combined $\mathcal{J}\,\mathcal{T}$ invariance is preserved, the systems would be linear magnetoelectrics, but not multiferroics. If, however, the $\mathcal{J}\,\mathcal{T}$ invariance is also broken, the system is (or can become) multiferroic.

Microscopically one can also speak of several different groups of type-II multiferroics. In this introductory chapter, I will give only a short overview of these questions, paying attention to some more general or subtle points.

Type-II multiferroics with spiral magnetic structures. Probably the biggest group of type-II multiferroics discovered until now belongs to a class of materials with helicoidal or spiral magnetic structure. Such structures are often incommensurate with the underlying crystal lattice, and they present a subclass of spin density wave structure. We can have different types of spin density wave. They can be sinusoidal, with spins perpendicular (Figure 1.8a) or parallel (Figure 1.8b) to the wavevector of the spin density wave. Or they can be helicoidal (spiral) of two types: proper screw, (Figure 1.8c), with spins rotating in the plane perpendicular to the wavevector, or cycloidal, (Figure 1.8d), with the spin rotation plain containing the wavevector. There may also exist different types of conical structures, two of which are shown in Figure 1.8e and f. As is shown both experimentally [51] and theoretically [52–54], in most cases, ferroelectricity is produced by the cycloidal magnetic structures of Figure 1.8d. Such are, for example, multiferroic systems $TbMnO_3$ and

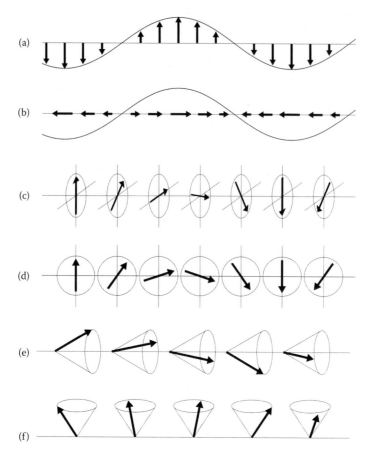

FIGURE 1.8 Different types of spin density wave that may give rise to ferroelectricity and eventually multiferroicity.

MnWO$_4$. The polarization of a pair of spins is given in this case by the following expression [52]:

$$P_{ij} \sim r_{ij} \times (S_i \times S_j) \tag{1.13}$$

where r_{ij} is the vector from site i to site j, and S_i and S_j are the spins at corresponding sites. For the cycloidal structure of Figure 1.8d, the polarizations (dipole moments) of consecutive bonds add, producing net ferroelectricity and multiferroic behavior. In this case, one can write down the expression for the total polarization, (1.13), in the equivalent form [2,54]

$$P \sim Q \times (S_i \times S_j) \sim Q \times e \tag{1.14}$$

where Q is the wavevector of the cycloid and e is the axis of rotation of spins in the cycloid. In other words, the polarization of the cycloid lies in the plane of rotating spins and perpendicular to the spiral axis. The same expression also gives the polarization of the conical spiral of Figure 1.8f: the "antiferromagnetic" component of the spins rotates in a cycloidal fashion and produces polarization given by the same expressions in Equations 1.13 and 1.14.

According to these expressions, the proper screw of Figure 1.8c or conic spiral of Figure 1.8e should not produce any polarization. This was indeed the accepted point of view for some time, until it was realized [55,56] that in certain cases, these magnetic structures can also lead to polarization, directed along the spiral direction Q. Indeed, in this case, all directions perpendicular to Q are equivalent, and polarization can be directed only along the spiral. But if a crystal has twofold symmetry axis C_2 perpendicular to Q (which was implicitly assumed in the derivation leading to expressions in Equations 1.13 and 1.14), then both directions, parallel and antiparallel to Q, are equivalent, that is, $C_2 P = -P$, which leads to $P = 0$. However, if such twofold axis in a crystal symmetry is absent, nonzero polarization can, in principle, exist. And indeed, several multiferroic materials with the proper screw magnetic structure were discovered experimentally [57].

From a microscopic point of view, the mechanism of ferroelectricity and multiferroicity in the most common case of a cycloidal structure is the inverse Dzyaloshinskii–Moriya effect [53]. The Dzyaloshinskii–Moriya interaction has the form

$$D_{ij} \cdot (S_i \times S_j) \qquad (1.15)$$

where the Dzyaloshinskii vector D_{ij} for a pair ij, in the case of systems with superexchange, for example via oxygen as shown in Figure 1.9, is proportional to the displacement δ of oxygen ion from the center of the i–j bond,

$$D_{ij} \sim \delta \times r_{ij} \qquad (1.16)$$

This interaction leads to the canting of otherwise collinear spins S_i, S_j. But, inversely, if by some reason (most often due to frustrations), the magnetic structure is of a cycloidal type with canted neighboring spins, it is favorable to shift the oxygen ions to gain the Dzyaloshinskii–Moriya energy (Equation 1.15). For a cycloidal structure of Figure 1.8d all such shifts would be in the same direction, which would produce an electric polarization given by the expressions in Equations 1.13 and 1.14.

There are also other microscopic mechanisms leading to multiferroicity in such cases, for example, for a proper screw [56]. I will not discuss them here. It is only important to notice that all of them rely on the presence of the (relativistic) spin–orbit interaction $\lambda l \cdot S$, and therefore typically, the ferroelectric

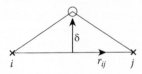

FIGURE 1.9 The Dzyaloshinskii vector D_{ij} for a pair ij, in the case of systems with superexchange, for example, via oxygen, is proportional to the displacement δ of the oxygen ion from the center of the i–j bond, see Equation 1.16.

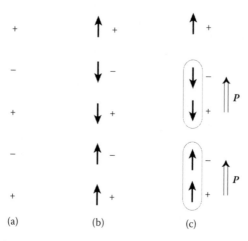

FIGURE 1.10 Schematic description of multiferroicity induced by magnetostriction.

polarization in such cases is rather small. There exists, however, another mechanism of multiferroicity, not relying on the real spin–orbit interaction and acting also for collinear magnetic structures.

Magnetostriction mechanism of type-II multiferroics. Another mechanism of multiferroic behavior is the standard magnetostriction due to the dependence of the exchange integral J_{ij} on the distance between sites i and j, and often also on the angle M_i–O–M_j for the superexchange mechanism shown in Figure 1.9. Due to this magnetostriction, some ions shift in the magnetically ordered phase, which can lead to electric polarization.

The simplest (and actually realistic) example is shown in Figure 1.10 (cf. Figure 1.7), with which it has much in common (see also the discussion at the end of Section 1.4). Suppose we have a lattice made of ions with different charges, which in itself is centrosymmetric, (Figure 1.10a; each ion is an inversion center). If, however, the magnetic structure is of the type ↑ ↑ ↓ ↓, (Figure 1.10b), then the inversion symmetry would be broken by this magnetic structure and the material could become ferroelectric. And indeed, due to magnetostriction, the ferromagnetic and antiferromagnetic bonds would become inequivalent, and if, for example, the ferromagnetic bond would become shorter, we would get the situation shown in Figure 1.10c, with polarization shown by the double arrow. We see that this mechanism resembles very much that shown in Figure 1.7, where we started with equivalent bonds and obtained ferroelectricity by putting charge ordering on top. Here we consider an opposite situation: with inequivalent sites, bonds becoming inequivalent due to magnetostriction in a particular magnetic structure. Note that this mechanism also works for a collinear spin structure, and it does not require any relativistic spin–orbit interaction. Therefore, one could expect larger values of ferroelectric polarization in such systems, and indeed this is the case: for typical cycloidal multiferroics, the polarization is usually ~10^{-2} μC/cm^2; in multiferroics with the magnetostriction mechanism it can reach several μC/cm^2.

Thus, theoretical considerations predict for Mn perovskites with E-type magnetic structure (resembling somewhat what is shown in Figure 1.10b) a polarization of ~2 µC/cm² [49]. The first measurement [58] demonstrated a much smaller value, but the improved sample quality led to values almost equal to the theoretical prediction.

Experimentally there exist many multiferroics with this mechanism of ferroelectricity: for example, one of the very first type-II multiferroics, $TbMn_2O_5$ [23,59]; Ca_3CoMnO_6 [28], which is well described by the structure depicted in Figure 1.10; CdV_2O_4 [60] and a few others.

Electronic mechanism of ferroelectricity in frustrated magnets. In the examples of type-II multiferroics considered earlier the ionic displacements played a crucial role, and the ionic contribution to polarization was significant, sometimes dominant [50] (although the electronic contribution was also important). There exist however also the possibility of a purely electronic mechanism of multiferroic behavior. Such mechanism, proposed in [61,62], can operate in frustrated magnets with a particular magnetic structure. One can show that in a magnetic triangle with one electron per site, described by the strongly interacting Hubbard model

$$\mathcal{H} = -t \sum_{\langle ij \rangle} c_{i\sigma}^{\dagger} c_{j\sigma} + U \sum_{i} n_{i\uparrow} n_{i\downarrow}, \tag{1.17}$$

there would appear, for certain magnetic structures, a charge redistribution, given by the expression

$$n_i = 1 + (32t^3 / U^2)[\boldsymbol{S}_1 \cdot (\boldsymbol{S}_2 + \boldsymbol{S}_3) - 2\boldsymbol{S}_2 \cdot \boldsymbol{S}_3] \tag{1.18}$$

(see Figure 1.11). If the spin correlation function entering (18) is nonzero, as is the case for the structure of Figure 1.11, charge redistribution would occur, and the triangle would acquire a dipole moment (double arrow in Figure 1.11). In principle, this could give net polarization in a bulk solid consisting of such triangles, see [62], in which case the material would be multiferroic. (Of course, if we then release the lattice, it would also distort somewhat, thereby contributing to the polarization. But the main mechanism and the driving force of multiferroic behavior is in this case purely electronic.)

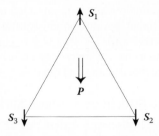

FIGURE 1.11 Schematic illustration of spin frustration leading to charge redistribution and electric polarization.

1.8 Beyond Multiferroics

It is a very interesting development that, after learning many things while studying multiferroics, we can go back to some more traditional magnetic systems, and predict (and observe) many very nontrivial effects connected with the coupling between magnetic and electric phenomena using the "multiferroic know-how." I will briefly discuss a few of these below. More detailed treatments can be found in the other chapters of this book.

1.8.1 Electric Activity of Magnetic Domain Walls

The specific features of domain wall in multiferroics such as $BiFeO_3$ or $YMnO_3$ is a very important and well-studied field. Here I want to draw attention to a slightly different aspect. It was first pointed out by Mostovoy [54] that there should exist nontrivial magnetoelectric and multiferroic effects in some domain walls even in ordinary insulating ferromagnets.

Typically, there exist two types of domain walls in ferromagnets: Bloch walls shown in Figure 1.12a, in which spins rotate in the plane of the wall, perpendicular to direction from one domain to the other; and Néel walls shown in Figure 1.12b, in which spins in the center of the domain wall point alongside the normal to this wall. We immediately see that the Bloch wall presents a part of the proper screw of Figure 1.8c, whereas the Néel wall has a "cycloidal" structure as shown in Figure 1.8d. According to Equations 1.13 and 1.14, one should then expect the appearance of electric polarization at every Néel domain wall.

One can then propose a beautiful experiment, which was indeed carried out [63]. It is well known that if we put magnetic dipoles (magnetic needles) in an inhomogeneous magnetic field, such needles would be attracted (or repelled, depending on the direction of the dipoles) to the region of the stronger field. The same of course is also true for electric dipoles in the gradient of an electric field. The group of Logginov et al. [63] carried out such experiment not with the usual magnetic dipoles but with insulating ferromagnets with Néel domain walls, which, according to the arguments presented earlier, should carry electric dipoles. They used a film of a well-known such material, the iron garnet $(Bi, Lu)_3(Fe, Ga)_5O_{12}$, with $T_c \sim 450$ K, approached it by a sharpened copper wire and applied a voltage pulse to the wire. This produced an inhomogeneous electric field in the film as shown in Figure 1.13, and it was observed that the Néel domain walls were attracted to the region of stronger electric field. This experiment, besides demonstrating the appearance of electric dipoles on Néel domain walls, opens a way to control such domain wall by the electric field, which may be extremely useful in manufacturing new electrically controlled memory media and devices.

1.8.2 Spiral Magnetic Structures on Metal Surfaces

Yet another experimental observation that can be easily explained if we invoke the physics described earlier is the detection of helicoidal magnetic structures in thin films (monolayer or bilayer) of magnetic metals on nonmagnetic substrates [64–66]. The first such results were obtained on a

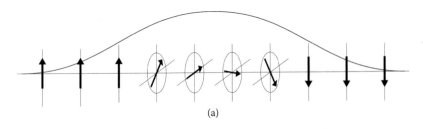

(a)

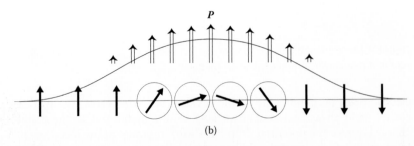

(b)

FIGURE 1.12 Different magnetic domain walls: (a) Bloch wall and (b) Néel wall.

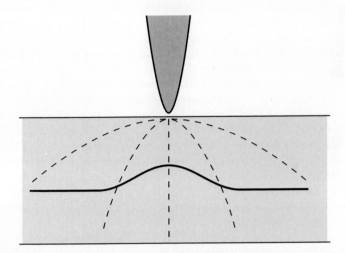

FIGURE 1.13 Schematic illustration of how a Néel domain wall can be attracted to the region of stronger electric field, demonstrating the appearance of electric dipoles on such domain walls.

monolayer of manganese on tungsten [64](see Figure 1.14). It was observed that instead of forming a collinear magnetic structure, a cycloidal spiral was formed (Figure 1.14 is a simplified schematic picture, which shows the situation if the Mn layer were ferromagnetic; see the original publication for the actual structure). From what we have learned in Section 1.7, it is clear that the cycloidal spiral would produce a polarization and a corresponding

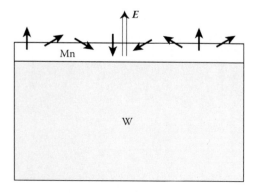

FIGURE 1.14 Schematic illustration of the formation of cycloidal magnetic structure in a monolayer of manganese on tungsten.

electric field, perpendicular to the surface. But, vice versa, the intrinsic electric field always existing at the surface of a metal (due to the double-charge layer, or the potential drop—the work function of the metal), which is normal to the surface, would have the tendency to create a cycloidal spiral from the initially collinear spin structure. And as this intrinsic electric field in the double layer always has the same sign on the entire surface, all the spirals in such monolayer would have the same sense of rotation, for example, clockwise.

This phenomenon should be quite general, since such potential drop exists at the surface of every metal. But this tendency to rotate the spins would act against the magnetic anisotropy that is always present at the surface, and in order to be observable it must be strong enough to overcome this anisotropy. This is why such cycloidal structures are found not on all magnetic layers on top of any metal. One needs at least a strong spin–orbit coupling at the surface (we remind that the mechanism leading to Equations 1.13 and 1.14, is in fact the spin–orbit interaction), for which it is better to use a heavy metal as a substrate; and it is also desirable to have magnetic metals without very strong single-site anisotropy.

The authors of the original papers themselves proposed an explanation that relies on the Dzyaloshinskii–Moriya interaction at the surface [66] (which always exists in this case, since the inversion symmetry is broken by the surface itself). In fact, physically this is the same explanation as the one described earlier (as mentioned previously, the microscopic mechanism of the relations in Equations 1.13 and 1.14 is in fact the same Dzyaloshinskii–Moriya interaction). However, the picture presented in Figure 1.14 is simpler and more transparent, even if it is conceptually the same. By the way, the same coupling of magnetic and electric degrees of freedom at the surface, which lies at the core of our picture, is also very similar to the recent proposal [67] explaining the large magnitude of Rashba spin–orbit coupling at the surface by the important role of electric dipoles at the surface layer.

1.8.3 Magnetoelectric Effects in Magnetic Vortices and Skyrmions

There may exist different magnetic textures in normal magnets: not only domain walls, but also, for example, magnetic vortices, Figure 1.15a and b, or skyrmions, Figure 1.15c.

One can show that all such objects would exhibit a linear magnetoelectric effect [68]. The "head to tail" vortices of Figure 1.15a are classical examples of systems with toroidal moment $T \sim \sum_i r_i \times S_i$, shown in Figure 1.15a by a double arrow. This means that here we would have a multiferroic of the type described in Section 1.6, with for example, the polarization P perpendicular to the magnetic field H, $P \sim T \times H$, and similarly with magnetization M induced by electric field E, $M \sim T \times E$. On the other hand, the "radial" vortex of the type of Figure 1.15b would have a linear magnetoelectric effect with a symmetric tensor α_{ij}; that is, the polarization would be parallel to the external magnetic field (along the principal directions in which α_{ij} is diagonal).

Skyrmions ("magnetic hedgehogs") can be of two types. Most often, in the "middle part" of a skyrmion, the spins are rotating as in Figure 1.15a. In this case (Figure 1.15c), the skyrmions would also exhibit the transverse magnetoelectric effect. But in the case of "radial" skyrmions (real "hedgehogs") the

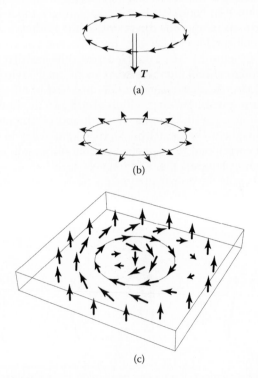

T

(a)

(b)

(c)

FIGURE 1.15 Schematic illustrations of (a, b) magnetic vortices and (c) skyrmion in magnetic materials.

magnetoelectric would be longitudinal. The magnetoelectric effect in skyrmions was recently experimentally observed in [69]. Interestingly enough, skyrmions can also be created on magnetic layers on the surface of a metal [70,71], apparently due to the mechanism described in Section 1.8.2.

1.8.4 Electric Activity of Spin Waves

From what we have learned in studying the cycloidal type-II multiferroics, we can predict yet another interesting effect—electric activity of spin waves [7]. As is well known [9], a quasiclassical picture of magnons in a ferromagnet is the precession of spins along the direction of magnetization as shown in Figure 1.16. As we see, in this case the instantaneous picture (a "snapshot") is that in which the perpendicular (xy) component of magnetization forms a cycloid in the xy-plane. Consequently, according to our understanding reached in studying type-II multiferroics, Equations 1.13 and 1.14, such spin wave should carry, besides magnetization, also an electric dipole perpendicular both to magnetization and to the magnon wavevector Q (double arrow in Figure 1.16).

One can also give this effect a completely classical interpretation. It is well known that there exists a circular magnetic field around a wire carrying a current as shown in Figure 1.17a. Correspondingly, the field of two such wires with currents in opposite directions would produce a magnetic field, or magnetic moment, located between the wires and pointing in the direction marked on Figure 1.17b by a double arrow. (The fields far away from the double wire would cancel.) But there exists a well-known symmetry between magnetic and electric phenomena, exemplified best of all by the Maxwell equations. Correspondingly, the motion of a magnetic dipole, which can be represented as a parallel motion of south and north poles, or positive and negative magnetic charges (although such monopoles do not exist separately), would correspond to magnetic currents running in opposite directions, as shown in Figure 1.17b, and in effect such "currents" created by a magnon—classically the propagating magnetic dipole—would be accompanied by an electric field, or electric dipole as shown in Figures 1.16 and 1.17c.

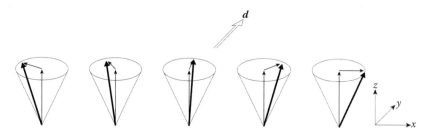

FIGURE 1.16 Precession of spins along the direction of magnetization in a spin wave. At any given moment, the projection of magnetic moments on the xy-plane forms a cycloid and induces an electric dipole indicated by the double arrow.

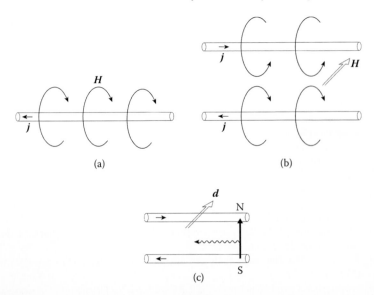

FIGURE 1.17 A completely classical interpretation of the relationship between a propagating magnetic dipole, that is, a magnon, and an electric dipole. (a) magnetic field H around the wire with electric current j; (b) magnetic field in the vicinity of two wires with opposite currents; and (c) electric field, or electric dipole d in the vicinity of a moving magnetic dipole, which can be represented as "currents" created by moving positive and negative magnetic poles S and N, by analogy with the case (b).

1.9 Conclusions

In this introductory chapter, I have tried to give an overview of some basic notions and phenomena that we meet in studying this interesting class of materials—multiferroics. A more detailed exposition of particular topics will be given in the following chapters of this book. Concluding this short overview I only want to stress once again that not only multiferroics themselves present significant interest, both from a general point of view and that of practical applications, but also the study of other interesting phenomena in other types of solids benefits strongly from the experience we get in studying multiferroics.

References

1. Fiebig, M. 2005. Revival of the magnetoelectric effect. *Journal of Physics D: Applied Physics* 38:R123.
2. Cheong, S.-W., and M. Mostovoy. 2007. Multiferroics: A magnetic twist for ferroelectricity. *Nature Materials* 6:13–20.
3. Khomskii, D. 2006. Multiferroics: Different ways to combine magnetism and ferroelectricity. *Journal of Magnetism and Magnetic Materials* 306:1–8.

4. Eerenstein, W., N. Mathur, and J. F. Scott. 2006. Multiferroic and magnetoelectric materials. *Nature* 442:759–765.

5. Wang, K., J.-M. Liu, and Z. Ren. 2009. Multiferroicity: The coupling between magnetic and polarization orders. *Advances in Physics* 58:321–448.

6. Tokura, Y. 2007. Multiferroics-toward strong coupling between magnetization and polarization in a solid. *Journal of Magnetism and Magnetic Materials* 310:1145–1150.

7. Khomskii, D. 2009. Trend: Classifying multiferroics: Mechanisms and effects. *Physics* 2:20.

8. *Journal of Physics: Condensed Matter, 2008,* 20:434201–434221.

9. Khomksii, D. 2014. *Transition Metal Compounds.* Cambridge: Cambridge University Press.

10. Curie, P. 1894. Sur la symétrie dans les phénomenes physiques, symétrie d'un champ électrique et d'un champ magnétique. *Journal of Physics* 3:393–415.

11. Landau, L. D., and E. M. Lifshitz. 1960. *Electrodynamics of Contintuous Media.* Oxford: Pergamon Press.

12. Dzyaloshinskii, I. 1959. On the magneto-electrical effect in antiferromagnets. *Soviet Physics JETP* 10:628–629.

13. Astrov, D. 1960. The magnetoelectric effect in antiferromagnetics. *Soviet Physics JETP* 11:708–709.

14. Freeman, A. J., and H. Schmid. 1975. *Magnetoelectric Interaction Phenomena in Crystals.* London: Gordon & Breach Science.

15. Ascher, E., H. Rieder, H. Schmid, and H. Stössel. 1966. Some properties of ferromagnetoelectric nickel-iodine boracite, $Ni_3B_7O_{13}I$. *Journal of Applied Physics* 37:1404–1405.

16. Schmid, H. 1994. Multi-ferroic magnetoelectrics. *Ferroelectrics* 162:317–338.

17. Goodenough, J. B., and J. M. Longo. 1970. Magnetic and other properties of oxides and related compounds. In *Landolt-Börnstein, Numerical Data and Functional Relations in Science and Technology,* New Series Vol. III. 4. Berlin, Germany: Springer.

18. Mitsui, T., and S. Nomura. 1981. Ferroelectrics and related substances, In *Landolt-Börnstein, Numerical Data and Functional Relations in Science and Technology,* New Series Vol. 16 (1). Berlin, Germany: Springer.

19. Khomskii, D. 2001. Magnetism and ferroelectricity: Why do they so seldom coexist? *Bulletin of the American Physical Society C* 21:002.

20. Hill, N. A. 2000. Why are there so few magnetic ferroelectrics? *The Journal of Physical Chemistry B* 104:6694–6709.

21. Wang, J., J. Neaton, H. Zheng, V. Nagarajan, S. Ogale, B. Liu, D. Viehland, V. Vaithyanathan, D. Schlom, and U. Waghmare. 2003. Epitaxial $BiFeO_3$ multiferroic thin film heterostructures. *Science* 299:1719–1722.

22. Kimura, T., T. Goto, H. Shintani, K. Ishizaka, T. Arima, and Y. Tokura. 2003. Magnetic control of ferroelectric polarization. *Nature* 426:55–58.

23. Hur, N., S. Park, P. Sharma, J. Ahn, S. Guha, and S. Cheong. 2004. Electric polarization reversal and memory in a multiferroic material induced by magnetic fields. *Nature* 429:392–395.

24. Brown Jr, W., R. Hornreich, and S. Shtrikman. 1968. Upper bound on the magnetoelectric susceptibility. *Physical Review* 168:574–577.

25. Pimenov, A., A. Mukhin, V. Y. Ivanov, V. Travkin, A. Balbashov, and A. Loidl. 2006. Possible evidence for electromagnons in multiferroic manganites. *Nature Physics* 2:97–100.

26. Spaldin, N. A. 2012. A beginner's guide to the modern theory of polarization. *Journal of Solid State Chemistry* 195:2–10.

27. Schmid, H. 2008. Some symmetry aspects of ferroics and single phase multiferroics. *Journal of Physics: Condensed Matter* 20:434201.

28. Choi, Y., H. Yi, S. Lee, Q. Huang, V. Kiryukhin, and S.-W. Cheong. 2008. Ferroelectricity in an Ising chain magnet. *Physical Review Letters* 100:047601.

29. Van Den Brink, J., and D. I. Khomskii. 2008. Multiferroicity due to charge ordering. *Journal of Physics: Condensed Matter* 20:434217.

30. Shpanchenko, R. V., V. V. Chernaya, A. A. Tsirlin, P. S. Chizhov, D. E. Sklovsky, E. V. Antipov, E. P. Khlybov, V. Pomjakushin, A. M. Balagurov, and J. E. Medvedeva. 2004. Synthesis, structure, and properties of new perovskite $PbVO_3$. *Chemistry of Materials* 16:3267–3273.

31. Belik, A. A., M. Azuma, T. Saito, Y. Shimakawa, and M. Takano. 2005. Crystallographic features and tetragonal phase stability of $PbVO_3$, a new member of $PbTiO_3$ family. *Chemistry of Materials* 17:269–273.

32. Harris, A. B., A. Aharony, and O. Entin-Wohlman. 2008. Order parameters and phase diagrams of multiferroics. *Journal of Physics-Condensed Matter* 20:434202.

33. Lebeugle, D., D. Colson, A. Forget, M. Viret, A. Bataille, and A. Gukasov. 2008. Electric-field-induced spin flop in $BiFeO_3$ single crystals at room temperature. *Physical Review Letters* 100:227602.

34. Zvezdin, A. K., G. P. Vorob'ev, A. M. Kadomtseva, Y. F. Popov, A. P. Pyatakov, L. N. Bezmaternykh, A. Kuvardin, and E. Popova. 2006. Magnetoelectric and magnetoelastic interactions in $NdFe_3(BO_3)_4$ multiferroics. *JETP Letters* 83:509–514.

35. Levanyuk, A., and D. G. e. Sannikov. 1974. Improper ferroelectrics. *Physics-Uspekhi* 17:199–214.

36. Landau, L. D., and E. M. Lifshitz. 1969. *Statistical Physics*. Reading, MA: Addison-Weslsy; or 1980. 3rd edition, Part 1, Oxford, United Kingdom: Butterworth-Heinemann.

37. Khomskii, D. I. 2010. *Basic Aspects of the Quantum Theory of Solids: Order and Elementary Excitations*: Cambridge: Cambridge University Press.

38. Shannon, R. T. 1976. Revised effective ionic radii and systematic studies of interatomic distances in halides and chalcogenides. *Acta Crystallographica Section A* 32:751–767.

39. Abrikosov, A. A., and L. P. Gorkov. 1961. Contribution to the theory of superconducting alloys with paramagnetic impurities. *Soviet Physics JETP* 12:1243–1253.

40. Bhattacharjee, S., E. Bousquet, and P. Ghosez. 2009. Engineering multiferroism in $CaMnO_3$. *Physical Review Letters* 102:117602.

41. Rondinelli, J. M., A. S. Eidelson, and N. A. Spaldin. 2009. Non-d^0 Mn-driven ferroelectricity in antiferromagnetic $BaMnO_3$. *Physical Review B* 79:205119.

42. Sakai, H., J. Fujioka, T. Fukuda, D. Okuyama, D. Hashizume, F. Kagawa, H. Nakao, Y. Murakami, T. Arima, and A. Baron. 2011. Displacement-type ferroelectricity with off-center magnetic ions in perovskite $Sr_{1-x}Ba_xMnO_3$. *Physical Review Letters* 107:137601.

43. Smolenskii, G. A., and I. E. Chupis. 1971. *Segnetoelectrics and Antisegnetoelectrics*, Leningrad: Nauka Publishers; Smolenskiĭ, G., and I. E. Chupis. 1982. Ferroelectro-magnets. *Soviet Physics Uspekhi* 25:475–493.

44. Venevtsev, Y. N., and V. Gagulin. 1994. Search, design and investigation of seignettomagnetic oxides. *Ferroelectrics* 162:23–31.

45. Rondinelli, J. M., and C. J. Fennie. 2012. Octahedral rotation-induced ferroelectricity in cation ordered perovskites. *Advanced Materials* 24:1961–1968.

46. Van Aken, B. B., T. T. Palstra, A. Filippetti, and N. A. Spaldin. 2004. The origin of ferroelectricity in magnetoelectric $YMnO_3$. *Nature Materials* 3:164–170.

47. Bersuker, I. B. 2006. *The Jahn-Teller effect*: Cambridge: Cambridge University Press.

48. Efremov, D. V., J. Van Den Brink, and D. I. Khomskii. 2004. Bond-versus site-centred ordering and possible ferroelectricity in manganites. *Nature Materials* 3:853–856.

49. Picozzi, S., K. Yamauchi, G. Bihlmayer, and S. Blügel. 2006. First-principles stabilization of an unconventional collinear magnetic ordering in distorted manganites. *Physical Review B* 74:094402.

50. Malashevich, A., and D. Vanderbilt. 2008. First principles study of improper ferroelectricity in TbMnO$_3$. *Physical Review Letters* 101:037210.

51. Kenzelmann, M., G. Lawes, A. Harris, G. Gasparovic, C. Broholm, A. Ramirez, G. Jorge, M. Jaime, S. Park, and Q. Huang. 2007. Direct transition from a disordered to a multiferroic phase on a triangular lattice. *Physical Review Letters* 98:267205.

52. Katsura, H., N. Nagaosa, and A. V. Balatsky. 2005. Spin current and magnetoelectric effect in noncollinear magnets. *Physical Review Letters* 95:057205.

53. Sergienko, I. A., C. Şen, and E. Dagotto. 2006. Ferroelectricity in the magnetic E-phase of orthorhombic perovskites. *Physical Review Letters* 97:227204.

54. Mostovoy, M. 2006. Ferroelectricity in spiral magnets. *Physical Review Letters* 96:067601.

55. Arima, T.-H. 2007. Ferroelectricity induced by proper-screw type magnetic order. *Journal of the Physical Society of Japan* 76:073702.

56. Jia, C., S. Onoda, N. Nagaosa, and J. H. Han. 2007. Microscopic theory of spin-polarization coupling in multiferroic transition metal oxides. *Physical Review B* 76:144424.

57. Kimura, T., J. Lashley, and A. Ramirez. 2006. Inversion-symmetry breaking in the noncollinear magnetic phase of the triangular-lattice antiferromagnet CuFeO$_2$. *Physical Review B* 73:220401; Seki, S., Y. Onose, and Y. Tokura. 2008. Spin-driven ferroelectricity in triangular lattice antiferromagnets ACrO$_2$ (A= Cu, Ag, Li, or Na). *Physical Review Letters* 101:067204; Kimura, K., H. Nakamura, K. Ohgushi, and T. Kimura. 2008. Magnetoelectric control of spin-chiral ferroelectric domains in a triangular lattice antiferromagnet. *Physical Review B* 78:140401.

58. Lorenz, B., Y.-Q. Wang, and C.-W. Chu. 2007. Ferroelectricity in perovskite HoMnO$_3$ and YMnO$_3$. *Physical Review B* 76:104405.

59. Radaelli, P., and L. Chapon. 2008. A neutron diffraction study of RMn$_2$O$_5$ multiferroics. *Journal of Physics: Condensed Matter* 20:434213.

60. Giovannetti, G., A. Stroppa, S. Picozzi, D. Baldomir, V. Pardo, S. Blanco-Canosa, F. Rivadulla, S. Jodlauk, D. Niermann, and J. Rohrkamp. 2011. Dielectric properties and magnetostriction of the collinear multiferroic spinel CdV$_2$O$_4$. *Physical Review B* 83:060402.

61. Bulaevskii, L., C. Batista, M. Mostovoy, and D. Khomskii. 2008. Electronic orbital currents and polarization in Mott insulators. *Physical Review B* 78:024402.

62. Khomskii, D. 2010. Spin chirality and nontrivial charge dynamics in frustrated Mott insulators: Spontaneous currents and charge redistribution. *Journal of Physics: Condensed Matter* 22:164209.

63. Logginov, A. S., G. Meshkov, A. Nikolaev, and A. P. Pyatakov. 2007. Magnetoelectric control of domain walls in a ferrite garnet film. *JETP Letters* 86: 115–118; Logginov, A., G. Meshkov, A. Nikolaev, E. Nikolaeva, A. Pyatakov, and A. Zvezdin. 2008. Room temperature magnetoelectric control of micromagnetic structure in iron garnet films. *Applied Physics Letters* 93:182510.

64. Bode, M., M. Heide, K. Von Bergmann, P. Ferriani, S. Heinze, G. Bihlmayer, A. Kubetzka, O. Pietzsch, S. Blügel, and R. Wiesendanger. 2007. Chiral magnetic order at surfaces driven by inversion asymmetry. *Nature* 447:190–193; Ferriani, P., K. Von Bergmann, E. Vedmedenko, S. Heinze, M. Bode, M. Heide, G. Bihlmayer, S. Blügel, and R. Wiesendanger. 2008. Atomic-scale spin spiral with a unique rotational sense: Mn monolayer on W (001). *Physical Review Letters* 101:027201.

65. Kubetzka, A., M. Bode, O. Pietzsch, and R. Wiesendanger. 2002. Spin-polarized scanning tunneling microscopy with antiferromagnetic probe tips. *Physical Review Letters* 88:057201; Vedmedenko, E., A. Kubetzka, K. Von Bergmann, O. Pietzsch, M. Bode, J. Kirschner, H. Oepen, and R. Wiesendanger. 2004. Domain wall orientation in magnetic nanowires. *Physical Review Letters* 92:077207.

66. Heide, M., G. Bihlmayer, and S. Blügel. 2008. Dzyaloshinskii-Moriya interaction accounting for the orientation of magnetic domains in ultrathin films: Fe/W (110). *Physical Review B* 78:140403.

67. Park, S. R., C. H. Kim, J. Yu, J. H. Han, and C. Kim. 2011. Orbital-angular-momentum based origin of Rashba-type surface band splitting. *Physical Review Letters* 107:156803.

68. Delaney, K. T., M. Mostovoy, and N. A. Spaldin. 2009. Superexchange-driven magnetoelectricity in magnetic vortices. *Physical Review Letters* 102:157203.

69. Okamura, Y., F. Kagawa, M. Mochizuki, M. Kubota, S. Seki, S. Ishiwata, M. Kawasaki, Y. Onose, and Y. Tokura. 2013. Microwave magnetoelectric effect via skyrmion resonance modes in a helimagnetic multiferroic. *Nature Communications* 4:2391.

70. Heinze, S., K. von Bergmann, M. Menzel, J. Brede, A. Kubetzka, R. Wiesendanger, G. Bihlmayer, and S. Blügel. 2011. Spontaneous atomic-scale magnetic skyrmion lattice in two dimensions. *Nature Physics* 7:713–718.

71. Romming, N., C. Hanneken, M. Menzel, J. E. Bickel, B. Wolter, K. von Bergmann, A. Kubetzka, and R. Wiesendanger. 2013. Writing and deleting single magnetic skyrmions. *Science* 341:636–639.

SECTION I

Single-Phase Multiferroics

2

Single-Phase Type-I Multiferroics

Resolving the Seemingly Contradictive Requirements for Ferroelectricity and Magnetism

Jan-Chi Yang, Yen-Lin Huang, and Ying-Hao Chu

Contents

Among the type-I multiferroics, the most important group belongs to the transition metal pervoskite oxide family. They possess intriguing properties and strong coupling of order parameters, serving as unique platforms for fundamental study and offering tantalizing opportunities for practical applications. In this chapter, we will review the current status of research that has been done on this group of multiferroics.

As mentioned in Chapter 1, there are seemingly contradictive requirements for ferroelectricity and magnetism that prevent them from coexisting in the same perovskite oxide family [1]. Empirical observations have shown that conventional ferroelectric systems containing transition metal elements usually have ions with d^0 configuration, that is, empty d orbitals, while magnetism is driven by partially filled d orbitals. Furthermore, ferroelectric materials usually are insulating in nature since the conduction of free carriers may suppress ferroelectricity even if the structure is noncentrosymmetric. Otherwise, an external electric filed would induce current flow instead of polarization reversal. On the other hand, ferromagnets generally show the property of conduction due to the presence of partially filled d orbitals. To circumvent these contradicting requirements, various strategies have been employed. There are mainly three groups for perovskite oxide multiferroics, namely, (1) materials with a double-perovskite structure, where two different B-site ions give rise to magnetism; (2) perovskite oxides, where the B-site ions are responsible for magnetic order. In both cases, the A-site ions are responsible for ferroelectricity; and (3) the so-called geometric ferroelectric, where structural distortion drives ferroelectricity.

2.1 Multiferroics with a Double-Perovskite Structure

Early studies on double-perovskite oxides were based on the prediction of the Goodenough–Kanamori rules [2], which address the resulting magnetic order of superexchange coupling between two next-to-nearest magnetic ions through a nonmagnetic anion. Researchers have attempted to design and synthesize new ferromagnetic materials based on these rules in the past decades. In general, Pauli's exclusion principle dictates that between two magnetic ions with half-occupied orbitals coupled through a nonmagnetic ion (e.g., O^{2-}), the superexchange interaction will be strongly antiferromagnetic. On the other hand, the coupling between an ion with filled orbitals and one with half-filled orbitals can be ferromagnetic.

One typical example of the double-perovskite multiferroics is $LaLuMnNiO_6$. First-principle calculations predicted that it is multiferroic with ferromagnetism originating from B-site ions and ferroelectricity due to an A-site disorder [3]. However, there is still no experimental confirmation of this prediction and it is rather difficult to synthesize the material due to phase separation [4]. Note that the prediction of Goodenough–Kanamori rules may not always hold in double-perovskite systems. For example, in $ALaMnBO_6$ systems, where A = Ca, Sr, Ba and B = Fe, Ru, antiferromagnetism is observed instead of ferromagnetism [5].

Another example is perovskites containing Fe^{3+} and Cr^{3+} at the B-site [7]. It is expected that the ordering of Fe^{3+} and Cr^{3+} leads to ferromagnetism. Ueda et al. [6] have fabricated $LaCrO_3$–$LaFeO_3$ superlattices on $SrTiO_3$ (111) single crystal substrate using pulsed laser deposition assisted with reflection high-energy electron-diffraction (RHEED). The B-site-ordered structure results in ferro-magnetism with a Curie temperature of 375 K (Figure 2.1). Inspired by this success, Baettig and Spaldin [8] proposed a potential multiferroic material, Bi_2CrFeO_6, in which the Bi ion with $6s^2$ long pair electrons drives the fer-roelectricity. Calculations suggest that the net magnetic moment is 2 μB per Fe–Cr pair, which corresponds to a magnetization of ~160 emu/cm³ and the ferroelectric polarization is 80 μC/cm². Despite the challenges of synthesizing Bi_2CrFeO_6 with well-ordered B-site ions due to the same valence state and similar ionic radius of Fe^{3+} and Cr^{3+}, several works have indeed revealed the multiferroic properties at room temperature [9–11]. However, most Bi_2CrFeO_6 samples investigated so far suffer from drawbacks like high leakage and the formation of secondary phases. One way to combat these issues is to replace one of the B-site ions with ions of different valence, such as Mn^{4+} or Ni^{2+}, to reduce phase separation [12].

The concept of double-perovskite multiferroics can also be achieved by magnetism-driven ferroelectricity. Breaking the magnetic ordering leads to noncentrosymmetric lattice distortion and results in the development of ferroelectricity. In the double-perovskite system, R_2CoMnO_6 (R = Y or rare earth metals), the special collinear magnetic structure ↑↑↓↓ is formed from the identical spins in Co–Co and Mn–Mn ions. The competition between fer-romagnetic and antiferromagnetic exchange results in an asymmetric ionic lattice structure, thereby giving rise to different centers of negative and posi-tive charges. For example, Y_2CoMnO_6 has been estimated with saturation

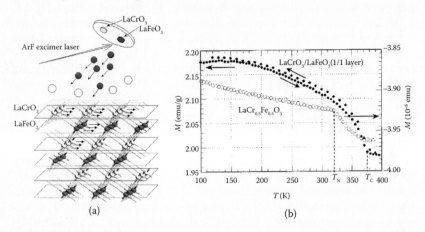

(a)

(b)

FIGURE 2.1 With advanced thin-film fabrication technology, artificial superlat-tice of complex oxides can be realized. (a) The schematic shows $LaCrO_3$–$LaFeO_3$ superlattice along the [111] direction fabricated by laser molecular beam epi-taxy (LMBE). In this periodical FeO–CrO structure, the Fe^{3+} (d^5)–O–Cr^{3+} (d^3) chain gives rise to the ferromagnetism, which can be described by Goodenough–Kanamori's rules. (b) In contrast to solid solution $LaCr_{0.5}Fe_{0.5}O_3$ showing the typi-cal antiferromagnetic behavior, the artificial $LaCrO_3$–$LaFeO_3$ superlattice exhibits a ferromagnetic character. (From Ueda, K. et al., *Science*, 280, 1064–1066, 1998. Reprinted with permission of AAAS.)

polarization ~65 µC/cm² and magnetization 6.2 µB per formula [13,14]. Another similar double-perovskite system, Lu_2MnCoO_6 shows a magnetic order below 43 K and ferroelectricity below 35 K [15]. Such a magnetostriction-driven ferroelectricity induced by long-range magnetic structure can also be found in a similar double-perovskite system, Y_2NiMnO_6, in which ferroelectricity is provoked by breaking the spatial-inversion symmetry [16]. Details on this type of multiferroics are provided in Chapter 4.

2.2 Lone-Pair Electron Multiferroic Systems

In the previous section, we discussed multiferroics with a double-perovskite structure in which the ordered B-site magnetic ions are expected to generate a robust ferromagnetic order while various mechanisms can lead to ferroelectricity. Among the possible origins of ferroelectricity in perovskite oxides, the stereochemical activity of the A-site cations plays a significant role. Combined with a B-site ion that has partially filled d-orbitals, it is possible to obtain multiferrocity without the need for a double-perovskite structure. Such multiferroic materials are usually referred to as lone-pair electron systems.

In typical ferroelectric perovskite oxides such as $PbTiO_3$ and $BiFeO_3$, the Pb^{2+} and Bi^{2+} ions both possess lone-pair electrons, that is, two valence electrons that do not participate in the chemical bonding. The presence of such long-pair electrons gives rise to high polarizability and drives the off-center distortion of the crystal structure, generating robust ferroelectricity [17–19]. If the B-site ions possess magnetic moments, the materials could also have a long-range magnetic order, though they are usually antiferromagnetic based on the Goodenough–Kanamori rules. To date, tremendous efforts have been devoted to three materials in this group, namely, $BiFeO_3$, $BiMnO_3$, and Pb-based mixed perovskites.

2.2.1 Lone-Pair Electron Multiferroic Systems: $BiFeO_3$

$BiFeO_3$ is the most studied single-phase multiferroic material. This is because of the coexistence of ferroelectric and magnetic orders at room temperature, making it particularly appealing for applications in the manufacture of non-volatile memories and nanoelectronic products. It possesses a large ferroelectric polarization (~100 µC/cm²) with a high Curie temperature of ~1103 K, and G-type antiferromagnetism with high Néel temperature of ~673 K. It belongs to the *R3c* space group, and spontaneous polarization takes place along the pseudocubic <111> directions. (Note that the pseudocubic index will be used in this chapter unless otherwise specified.)

The crystal and magnetic structures of $BiFeO_3$ are shown in Figure 2.2. The two oxygen octahedra rotate along the [111] direction in an opposite direction by ±13.8°. The presence of lone-pair electrons leads to a large displacement of the Bi-ions relative to the FeO_6 octahedra. Combined with the shift of Fe ion with respect to the center of the octahedra, they give rise to the spontaneous polarization along the <111> direction [20,22]. The reported spontaneous

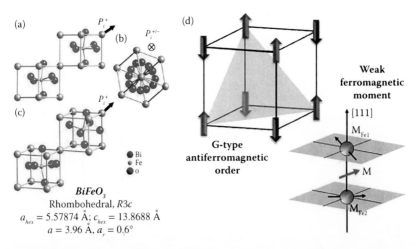

FIGURE 2.2 Characteristics of multiferroic $BiFeO_3$. Structure of $BiFeO_3$ on a view along the $[110]_{pc}$ (a) and $[111]_{pc}$ (b) directions. (c) A general three-dimensional view of $BiFeO_3$. (d) G-type antiferromagnetic ordering in $BiFeO_3$ and the formation of the weak ferromagnetic moment resulting from Dzyaloshinskii–Moriya interaction. ((a–c) Reprinted from *Journal of Physics and Chemistry of Solids*, 32, Moreau, J.-M. et al., Ferroelectric $BiFeO_3$ X-ray and neutron diffraction study, 1315–1320, Copyright 1971, with permission from Elsevier. (d) The source of the material Martin, L. et al., *Journal of Physics: Condensed Matter*, 20, 434220, copyright 2008, IOP Publishing is acknowledged.)

polarization was relatively small (~6.1 μC cm^{-2}) in earlier studies, which could be due to the presence of impurities and measurement limitations [23]. Recent studies based on both thin-film and single crystal samples have confirmed that the spontaneous polarization of $BiFeO_3$ is ~100 μC cm^{-2} [24,25].

$BiFeO_3$ is a G-type antiferromagnet (where the Fe spins are coupled ferromagnetically within the same (111) planes and antiferromagnetically between adjacent (111) planes, Figure 2.2d). The symmetry of $BiFeO_3$ allows a small canting of the Fe magnetic moment originated from the Dzyaloshinskii–Moriya interaction [26,27], which results in a weak ferromagnetic moment of ~0.05 μB per unit cell.

A critical motivation for the study on multiferroics is to obtain strong coupling between the order parameters, and to achieve this objective, much effort has been devoted [21,28–30]. In 2006, Zhao et al. [28] revealed that the antiferromagnetic and ferroelectric orders in $BiFeO_3$ are highly coupled. By combining photoemission electron microscopy (PEEM) and piezoresponse force microscopy (PFM) at the same location of the thin-film samples, they demonstrated that switching the ferroelectric polarization along certain paths induces the rotation of the antiferromagnetic easy axes. Figure 2.3 illustrates the correlation between the two order parameters, in which the magnetic easy plane is always perpendicular to the ferroelectric polarization. The strong coupling has enabled a direct modulation of magnetization through an electric field,

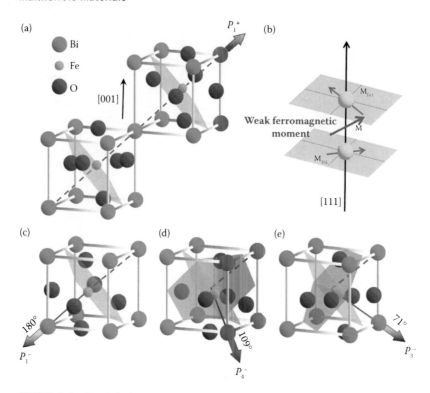

FIGURE 2.3 Crystal structure and correlated ferroelectricity and antiferromagnetism in BiFeO₃. (a) (001)-oriented BiFeO₃ crystal structure with an upward out-of-plane polarization component, in which the ferroelectric polarization is indicated by bold arrows, while corresponding antiferromagnetic plane is represented by shaded planes. (b) Formation of a weak ferromagnetic moment resulting from Dzyaloshinskii–Moriya interaction in BFO. Reorientation of the antiferromagnetic plane with regard to 180° (c), 109° (d), and 71° (e) polarization switching achieved with the application of an external electrical field (the antiferromagnetic plane changes from the orange plane to the green and blue planes on 109° and 71° polarization switching, respectively). ((a), (c)–(e) Reprinted by permission from Macmillan Publishers Ltd. *Nat. Mater.*, Zhao, T. et al. 5, 823–829, copyright 2006. (b) Reprinted with permission from Ederer, C. and Spaldin, N.A., *Physical Review B*, 71, 060401. Copyright 2005 by the American Physical Society.)

which is essential for applications in spintronics. (A more detailed discussion on applications of BiFeO₃ in spintronic devices can be found in Chapter 5.)

BiFeO₃ has been investigated in various forms, including single crystals, ceramic particles, nanoparticles, and thin films. Yet the recent flurry of research on BiFeO₃ has been made possible mainly by the development of thin-film growth techniques, such as pulsed laser deposition, chemical solution methods, sputtering, metal–organic chemical vapor deposition, molecular beam epitaxy, and so on. Furthermore, thin films also possess many advantages from the application point of view. We thus focus on the recent development in studies on $BiFeO_3$ thin films in this section. For a detailed review on the properties of $BiFeO_3$, readers are referred to Reference [31].

For an epitaxial thin film grown on a single crystal substrate, strain due to lattice mismatch is inevitable. The resulting changes in crystal symmetry would directly affect the physical properties of material. For $BiFeO_3$, theoretical calculations using phenomenological approach and density function theory, as well as experimental studies, have revealed a rich-phase diagram with different symmetries tunable by epitaxial strain [30,32–40]. Thin films of orthorhombic [41–44], strained rhombohedral (R) [24,45–49], and tetragonal (T) [50–56] structures have been found and studied. For example, when $BiFeO_3$ film is grown on the $LaAlO_3$ substrate, the large compressive strain stabilizes a tetragonal phase with a very large c/a ratio. Upon increasing the film thickness, the epitaxial strain is gradually released and T and R phases can coexist as shown in Figure 2.4 [57–62]. The detailed atomic structure of the R-T mixed-phase $BiFeO_3$ films has been studied by high-resolution transmission electron microscopy, which revealed the presence of a high-quality interface without any defects between the two phases [30,63]. Furthermore, external electric fields can induce a conversion between the two phases, leading to a very large piezoelectric response similar to that of a chemical doping-induced morphotropic phase boundary, suggesting a new approach to engineer piezoelectrics.

2.2.1.1 Domain Engineering in BiFeO$_3$ Thin Films

Besides affecting the crystal symmetry, epitaxial strain can also have a strong influence on domain structures. In ferroic systems (ferroelectric, ferromagnetic, and ferroelastic materials), domains form as a result of minimizing the total free energy. Where adjacent domains meet, it forms domain walls. In addition, domain structures affect the overall properties of the film, but domain walls themselves often possess unique physical properties that do not exist in bulk materials. This is especially true in $BiFeO_3$, and great effort has been devoted to its domain engineering.

Theoretical work on domain structures in ferroelectric materials can be traced back to the 1990s. In 1998, Streiffer et al. [64,65] proposed the generic domain structures of rhombohedrally distorted perovskite ferroelectrics. The domain configuration and domain walls are usually described by the angle between polarization directions in the adjacent domains. For $BiFeO_3$, this can be 71°, 109°, or 180° since the spontaneous polarization is along the pseudocubic <111> direction (Figure 2.5a through c) [66].

Domain engineering can be achieved by designing the elastic and/or electrostatic boundary conditions of the film. For example, if substrates with different in-plane lattice constants, for example, $DyScO_3$, are used, the anisotropic epitaxial strains would eliminate two of the structural variants shown in Figure 2.5d, leaving two structural variants with four possible polarization directions. The degeneracy between equivalent polarization states can be further broken by controlling the polar discontinuity (valence state mismatch) at the $BiFeO_3$/bottom electrode interface. Controlling the valence

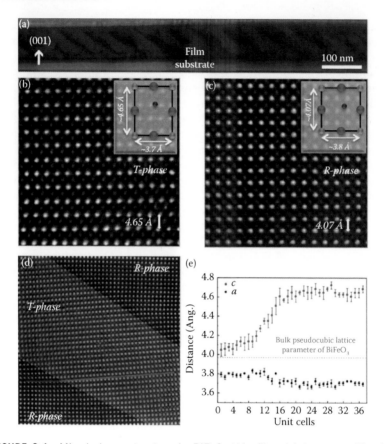

FIGURE 2.4 Mixed-phase structure in BiFeO₃ thin film. (a) Low-magnification transmission electron microscopy (TEM) image on the mixed-phase portion. (b) (c) Scanning-TEM (STEM) images of tetragonal (T)-like (b) and rhombohedral (R)-like (c) phase in mixed phase region, respectively. (d) High-magnification TEM image of morphotropic phase boundaries (T/R phase boundaries). (e) In-plane (in black) and out-of-plane (in red) lattice parameters measured on the line shown in (d). (From Zeches, R. et al., *Science*, 326, 977–980, 2009. Reprinted with permission of AAAS.)

state mismatch results in different carrier accumulating at the interface and consequently reaches a minimal energy to ferroelectric polarization preference. With a careful control of the boundary conditions, periodic 71° or 109° domains can be obtained, as shown in Figure 2.5e and f for BiFeO₃ films grown on DyScO₃ substrate. Note that more energy is required to form 180° domains in BiFeO₃ films and it is rarely observed [67].

Besides controlling the domain structure using boundary conditions, one can also create certain types of domain walls in BiFeO₃ by applying external electric field using a scanning probe microscope (SPM) tip. Traditionally, PFM based on SPM is often used to characterize the local polarization direction in a ferroelectric sample. By combining the out-of-plane and in-plane PFM

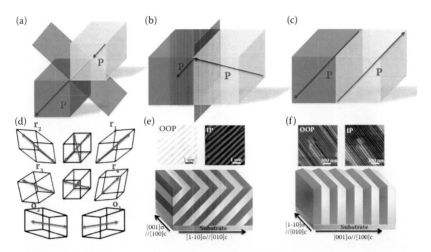

FIGURE 2.5 Domain patterns and domain engineering of BiFeO$_3$ [66]. Schematic of (a) 71°, (b) 109°, and (c) 180° domain configurations in BiFeO$_3$. The yellow and green bricks, respectively, represent individual domains with their well-defined polarizations (blue arrows). The red planes represent the domain walls planes in corresponding domain configurations. (d) The schematics of structural (r_1–r_4) and corresponding polarization (red arrows) variants. Schematics (lower panels) and corresponding out-of-plane and in-plane (upper panels) PFM images of (e) 71°, and (f) 109° periodic domain patterns grown on DyScO$_3$ single crystal substrates. (Reprinted with permission from Yang, J.C. et al., *J. Appl. Phys.* 116, 066801. Copyright 2014, American Institute of Physics.)

images, one can easily identify the type of domain walls present in the sample (Figure 2.6a) [68]. In order to artificially create certain types of domain walls, various parameters such as the voltage applied to the tip, scan rate, and scan direction have to be optimized. Using this method, Cruz et al. [69] have successfully created all three types of domain walls in BiFeO$_3$ as shown in Figure 2.6d through f. It was noticed that smaller electric fields favor a 180° polarization switching, while higher electric fields lead to a 109° switching.

2.2.1.2 Properties of Domain Walls in BiFeO$_3$

The motivation of domain engineering in BiFeO$_3$ is to acquire the unique properties and functionalities of domain walls, which are absent in bulk, that have been discovered over the past decade [71–74]. Seidel et al. [70] reported the enhanced conduction of certain domain walls in BiFeO$_3$. By using SPM, different types of domain walls were written in (110)-oriented BiFeO$_3$ films (Figure 2.7a). And it was observed that only 109° and 180° walls exhibited enhanced conduction but not the ones with 71° (Figure 2.7b). Theoretical calculation indicated that the change of the polarization across the 109° and 180° domain walls can reduce the energy bandgap at domain walls, resulting in increased conductivity. In Figure 2.7c, it was shown that the domain walls in a planar capacitor can be created or erased using an SPM tip, leading

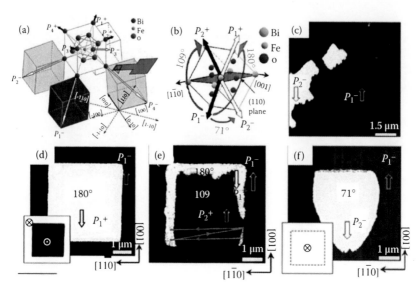

FIGURE 2.6 Advanced control of artificially created domain patterns in BiFeO$_3$. (a) Schematic of the four different structural variants along with a PFM cantilever oriented along [110] direction. Pseudocubic with different colors represent for the contrasts observed in PFM images, which resulting from the relative deformation of between the polarization and cantilever. (b) Schematic of the rotation types of the polarizations. (c) In-plane PFM image revealing two domain variants in the as-grown BFO (110) thin film. In-plane PFM images after electrical poling at (d) lower negative voltage, (e) higher negative voltage, and (f) higher positive voltage. The corresponding out-of-plane PFM images are shown in the insets. ((a) Adapted with permission from Zavaliche, F. et al., *Phase Transitions*, 79, 991–1017, 2006. Copyright 2006, Taylor & Francis Group. (b)–(f) Reprinted with permission from Cruz, M. et al., *Phys. Rev. Lett.*, 99, 217601. Copyright 2007, American Physical Society.)

to a well-controlled reversible change in the current. Various nanoelectronic devices can be envisioned based on this demonstration.

Of all the unique properties of the domain walls found in BiFeO$_3$, one in particular is very interesting for application in spintronic devices. In a study by Martin et al. [75], it was shown that certain types of domain walls may possess net magnetization much larger than those of the domain bulk. As shown in Figure 2.8, two distinctly different types of domain architectures, a stripelike BiFeO$_3$ film with mainly 71° domain walls and a mosaiclike BiFeO$_3$ film with mixed domain walls, are used to reveal the inherent magnetism of different domain walls. X-ray magnetic circular dichroism (XMCD) provides an element-sensitive approach for revealing surface spin structure as well as magnetic properties in the thin films. The difference in the absorption spectrum for left- and right-circularly polarized (LCP and RCP) light is essentially a measure of uncompensated spins in the material, while the asymmetry and spectra difference is proportional to the magnetic moment. Figure 2.8e and f

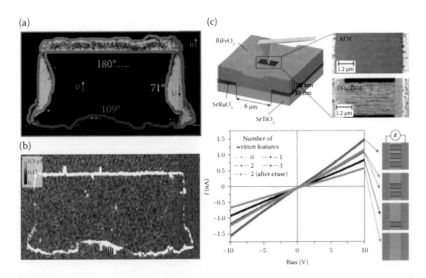

FIGURE 2.7 Multifunctionalities of BiFeO₃ domain walls. A (110)-oriented BiFeO₃/SrRuO₃/SrTiO₃ sample was first written by PFM for creating a region with coexistence of different types of domain walls (a). The types of domain walls are depicted in green (180°), blue (71°), and red (109°) colors, respectively. (b) Corresponding conductive-AFM image indicating the higher conductance at both 180° and 109° domain walls, while the conductance of 71° was absent in the experiment. (c) The nanoelectronic prototype was constructed by the BiFeO₃ ferroelectric domain wall. Conduction currents are incrementally controlled through creating or erasing the number of conductive domain walls. (Reprinted by permission from Macmillan Publishers Ltd. *Nat. Mater.*, Seidel, J. et al., 8, 229–234, copyright 2009.)

shows X-ray absorption data collected from BiFeO₃ films with stripelike and mosaiclike domain structures, respectively. The stripelike BiFeO₃ films exhibit almost no measurable asymmetry (Figure 2.8e). In contrast, the mosaiclike BiFeO₃ films consistently exhibit normalized asymmetries of between 0.5% and 1% (at zero applied field; Figure 2.8f). The presence of circular dichroism in mosaic films, even at a zero applied field, suggests a significant enhancement of magnetism at a certain type of domain walls in BiFeO₃. Further experimental work has confirmed the role of the domain-wall types in determining the exchange bias interaction, which takes place between pinned and uncompensated spins occurring at domain walls in BiFeO₃ and spins in the CoFe layer. It has been revealed that the 109° domain walls exhibit significantly enhanced exchange bias; furthermore, an exchange enhancement interaction has arisen from an interaction of the spins in the ferromagnet and the fully compensated (001) surface of the BiFeO₃ domain, which results in an enhancement of the coercive field of the ferromagnetic layer. Other interesting properties of the domain walls in BiFeO₃ have also been discovered. He et al. [76] reported intriguing magnetoresistance behaviors associated with the domain walls.

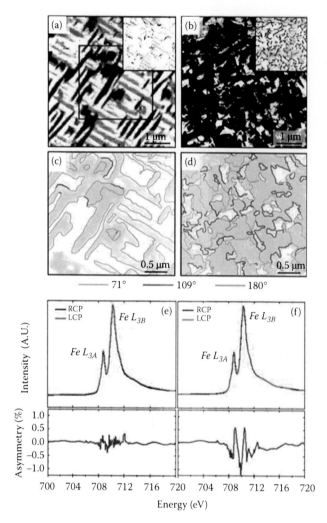

FIGURE 2.8 Magnetic study on various domain structures. In-plane and out-of-plane (inset) PFM images of (a) stripelike and (b) mosaiclike $BiFeO_3$ films, respectively. Correspondent domain-wall distribution for (e) stripelike and (f) mosaiclike BFO thin films. X-ray magnetic circular dichroism measurements on (e) stripelike and (f) mosaiclike domain structures (upper panel) as well as the respective asymmetry spectra for each measurement (lower panel). (Reprinted with permission from Martin, L.W. et al., *Nano Letters*, 8, 2050–2055. Copyright 2008, American Chemical Society.)

Yang et al. [45] observed a photovoltage wall above the band gap of $BiFeO_3$ in thin films with stripe domains, pointing to a fundamentally different mechanism for photovoltaic charge separation. A detailed discussion regarding the application of $BiFeO_3$ in various devices can be found in References [77,78] and in Chapter 5 of this book.

2.2.2 Lone-Pair Electron Multiferroic Systems: $BiMnO_3$

$BiMnO_3$ is another example of the lone-pair electron multiferroic system. It is expected to possess both ferroelectricity and ferromagnetism (Figure 2.9) at low temperatures [79,80]. In bulk $BiMnO_3$, the magnetic moment is 3.6 µB per unit cell at low temperature [81], which is close to the theoretical prediction [82]. The first $BiMnO_3$ sample was synthesized under high pressure at high temperatures in 1968 [83]. Early studies suggested the C2 symmetry based on electron and neutron powder diffraction [80]. Atou et al. [80] also revealed that the bond angle of Mn–O–Mn, between 140° and 160°, is significantly smaller than the ideal 180° for strong superexchange coupling. It was suggested that Bi $6s^2$ lone pair electrons trigger a three-dimensional orbital ordering of Mn d_{x2-z2} orbitals along the c-axis. As a result, the ferromagnetic ordering is favored over the antiferromagnetic ones below the critical temperature.

Theoretical calculations carried by Hill [1] suggested that $BiMnO_3$ can possess both ferroelectric and ferromagnetic orders. The origin of ferroelectricity of $BiMnO_3$ is undoubtedly related to the localized Bi $6s^2$ lone-pair electrons, which hybridize with O 2p electrons. A detailed comparison between $BiMnO_3$ (ferroelectric) and $LaMnO_3$ (non-ferroelectric) can be used to further elucidate the key role played by lone-pair electrons. $BiMnO_3$ shows ferromagnetism below 100 K [83,84], while $LaMnO_3$ exhibits A-type antiferromagnetic ordering [85]. The A-site ions of $BiMnO_3$ and $LaMnO_3$ have similar ionic radius (Bi^{3+} = 1.24 Å, La^{3+} = 1.22 Å). By the visualization of electronic bonding and lone pair in real space (Figure 2.10a), we can see the detailed functions of electron localization on different lattice planes. Comparing Mn–O plane in $BiMnO_3$ with Mn–O plane in $LaMnO_3$, both of them show similar spin arrangement and electron bonding state (Figure 2.10a). Different ferroelectric ordering in these two compounds is mainly caused by a different degree of electron localization on the La–O and Bi–O planes, while differences in the electronic structures of these two compounds are attributed to Bi $6s^2$ lone pair and the off-centered MnO_6 octahedra in $BiMnO_3$. Structure

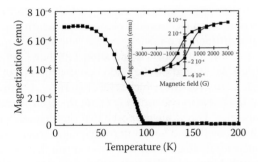

FIGURE 2.9 Magnetization versus temperature curve of a $BiMnO_3$ film shows a clear ferromagnetic transition at 105 K. The inset shows the ferromagnetic hysteresis loop at 5 K. (Reprinted with permission from Moreira Dos Santos, A.F. et al., *Appl. Phys. Lett.*, 84, 91. Copyright 2004, American Institute of Physics.)

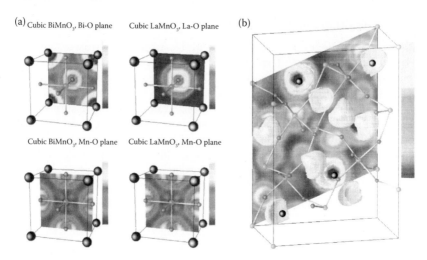

(a) Cubic BiMnO₃, Bi-O plane   Cubic LaMnO₃, La-O plane   (b)

Cubic BiMnO₃, Mn-O plane   Cubic LaMnO₃, Mn-O plane

FIGURE 2.10 Electronic structures of BiMnO₃ and LaMnO₃. (a) Valence electron localization function projected onto different planes. The light color of the scale bar represents a complete localized electron, while the dark color of the scale bar represents almost no electron localization. (b) Electron localization function representation of the isosurface of valence electrons in BiMnO₃ with monoclinic structure, projected within one unit cell. (Reprinted with permission from Seshadri, R. and Hill, N.A., *Chem. Mater.*, 13, 2892–2899. Copyright 2001 American Chemical Society.)

distortion including Mn–O octahedral distortion induced by $6s^2$ lone pair in Bi^{3+} is responsible for the magnetic properties in BiMnO₃. Figure 2.10b shows the structure of monoclinic BiMnO₃ with black spheres representing Bi and cyan spheres representing Mn. The visible regions in the isosurface correspond to the lobelike Bi lone pair electron, which is permitted by the distorted geometry of the monoclinic structure. The stereochemically active Bi lone-pair electrons give rise to the different structure distortions around the Mn–O octahedral, leading to the differences between BiMnO₃ and LaMnO₃.

Later on, further experimental results including electron diffraction [86–88], convergent-beam electron diffraction [86], neutron powder diffraction [81], electron diffraction on single-crystal [89], Raman spectroscopy on single-crystal [89], and theoretical calculation [90] have indicated that the centrosymmetric C2/c should be the real crystal structure of BiMnO₃. The centrosymmetric C2/c structure, which is incompatible with any ferroelectricity, implies that the ferroelectric ordering is not the ground state of BiMnO₃. I. V. Solovyev and Z. V. Pchelkina proposed that the ferroelectric polarization of BiMnO₃ is driven by the hidden antiferromagnetic order, which breaks the inversion symmetry [91]. To date, the true mechanism that drives the intrinsic ferroelectricity in BiMnO₃ is still a matter of debate. More experimental and theoretical approaches are required and are underway.

To facilitate the growth of BiMnO₃ under normal conditions, the chemical pressure method based on various dopants has been introduced. For example, the substitution of La^{3+} for Bi^{3+} has led to a smaller structure distortion and made the $BiMnO_3$ films easier to grow at relatively low temperature. Solid solutions of $LaMnO_3$ and $BiMnO_3$ perovskites ($La_xBi_{1-x}MnO_3$) have been studied [92–94] (Figure 2.11a through d). The phase diagram of $La_xBi_{1-x}MnO_3$ indicates that a mixture of ferromagnetic and antiferromagnetic state can be obtained with higher Bi concentrations ($x < 0.3$), while the compounds are weak ferromagnets when $x > 0.35$. Bulk $La_{0.1}Bi_{0.9}MnO_3$ ($x = 0.1$) shows a T_C of ~95 K and is magnetically and structurally similar to $BiMnO_3$ [93]. Gajek et al. [92] reported that ultrathin film (2 nm) of $La_{0.1}Bi_{0.9}MnO_3$ retains both ferromagnetic and ferroelectric properties at 10 K. A multiferroic tunnel junction was fabricated using the ultrathin $La_{0.1}Bi_{0.9}MnO_3$ film, and four tunable resistance states by electric and magnetic fields demonstrated. The result demonstrated the possibilities of encoding quaternary information by ferromagnetic and ferroelectric order parameters, and to read the stored information nondestructively through a resistance measurement, as shown in Figure 2.11e. In $BiMnO_3$ films with 20% La substitution of Bi, a much larger magnetodielectric effect compared with pure $BiMnO_3$ was observed; this suggested an enhanced coupling between magnetic and electric orders [94]. However, an increase of La content is accompanied by a decrease in ferroelectric Curie temperature. Therefore, a balance should be achieved between the doping concentration and functionalities in pursuit. To maintain or even increase the ferromagnetic Curie temperature of $BiMnO_3$, smaller A-site (Bi in $BiMnO_3$) substitution ions have been studied [95]. $SmMnO_3$ has a large Jahn-Teller distortion than $BiMnO_3$. And, it was reported that a 10% substitution of Sm (1.13 Å) for Bi increases the ferromagnetic Curie temperature of $BiMnO_3$ from 100 to 140 K, while an effective piezoelectric response (10 pm/V) was maintained at room temperature.

2.2.3 Solid Solution Lone-Pair Systems

Even though single-phase multiferroics is rare, perovskite oxides that possess strong ferroelectricity or ferromagnetism are abundant. It is thus straightforward to consider solid solutions of such oxides as an alternative strategy for developing new multiferroics or to enhance the properties of existing ones.

$Pb(Fe_{0.66}W_{0.33})O_3$ was investigated in the late 1950s. It possesses a ferroelectric relaxor transition at a temperature close to 180 K and an antiferromagnetic phase transition above room temperature (~343 K) [97]. Relaxor ferroelectrics possess short-range polarization ordering and frequency-dependent dielectric properties. They are characterized by strong frequency dependence of dielectric constant, and a broadened phase transition peak [98,99]. To improve the multiferroic and magnetoelectric coupling, PFW has been widely combined by classic ferroelectric material, lead zirconium titanate ($Pb(Zr_{1-x},Ti_x)O_3$, PZT ($0 < x < 1$)), forming a new branch of solid-solution and single-phase multiferroics (PFW–PZT). PZT with a composition of Zr:Ti = 0.53:0.47 shows a morphotropic phase boundary (MPB) composed by Zr-rich rhombohedral and Ti-rich tetragonal region phases. PZT with such

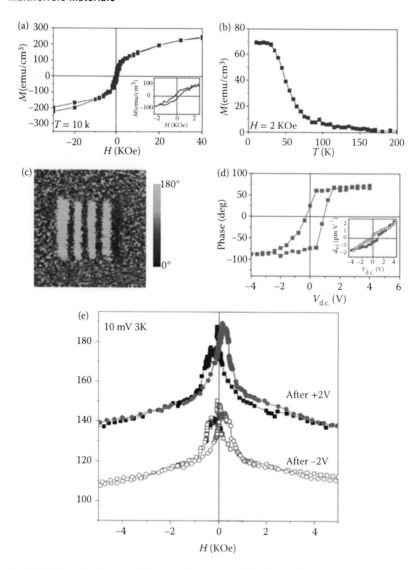

FIGURE 2.11 Multiferroic BiMnO$_3$ thin film. (a) Field and (b) temperature dependence of the magnetization of a 30 nm La$_{0.1}$Bi$_{0.9}$MnO$_3$ film. (c) Piezoresponse force microscopy phase image with electrically written stripes on La$_{0.1}$Bi$_{0.9}$MnO$_3$ film grown on a La$_{2/3}$Sr$_{1/3}$MnO$_3$ bottom electrode and variation of the piezoresponse phase as a function of applied voltage. (d) Variation of the piezoresponse phase with the applied voltage. (e) Tunnel magnetoresistance curves in a La$_{2/3}$Sr$_{1/3}$MnO$_3$/La$_{0.1}$Bi$_{0.9}$MnO$_3$/Au junction after applying voltage of +2V and −2V, in which the combination of the tunnel magnetoresistance and electroresistance effect gives rise to a four-resistance-state logic system. (Reprinted by permission from Macmillan Publishers Ltd. Gajek, M. et al., *Nat. Mater.*, 6, 296–302, 2007, Copyright 2007.)

a composition exhibits extraordinarily high pyroelectric, dielectric, piezo-electric and ferroelectric properties as well as a ferroelectric–paraelectric/tetragonal-cubic phase transition at ~660 K. During the prior investigation, properties of PFW–PZT as a function of PFW mixture have been examined (20%, 30%, and 40% PFW) [96]. These samples are all ferroelectric and weakly ferromagnetic. Among them, the PFW–PZT sample with 20% PFW and 80% PZT ($PbZr_{0.53}Ti_{0.47}O_3$) shows the most interesting properties at room temperature (Figure 2.12). A magnetic switching between the normal ferroelectric state and a magnetically quenched ferroelectric state was further reported [96]. PFW–PZT with this composition gives rise to a large polarization, low dielectric loss (<1%), high resistivity (~10^8–10^9 Ω cm), as well as a very large magnetoelectric effect. The ferroelectric polarization of the sample can be switched from P_r to zero with an applied magnetic field of less than 1 T (Figure 2.12).

In addition to the PFW–PZT solid solution system, $Pb(Fe_{0.5}Ta_{0.5})O_3$ (PFT) is another end member that is frequently used. In PFT, the magnetic (Fe^{3+} with $3d^5$ configuration) and nonmagnetic (Ta^{5+} with $3d^0$ configuration) B-site ions distribute randomly in the oxygen octahedra, leading to an antiferromangetic order with a Néel temperature of ~150 K. However, when combined with PZT ($PbZr_{0.53}Ti_{0.47}O_3$), the PFT–PZT solid solution clearly shows a magnetic hysteresis loop, a typical feature of a ferromagnetic/ferrimagnetic material, as

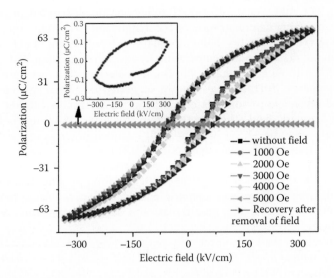

FIGURE 2.12 Multiferroic characteristics of Pb-based solid solution multiferro-ics. The three-state logic switching (+Pr, 0, −Pr) in 0.2 PFW–0.8 PZT using an application of magnetic field. Polarization-electric field (*P–E*) hysteresis loop indicates an application of external magnetic field (from 0 to 0.5 T), which gives rise to a polarization change from from |Pr| to zero. The inset shows the *P* = 0 relaxor state [96]. (The source of the material Kumar, A. et al., *J. Phys.: Condens. Matter*, 21, 382204. Copyright 2009, IOP Publishing is acknowledged.)

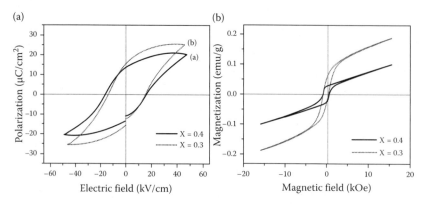

FIGURE 2.13 Multiferroic characteristics of Pb-based solid solution multiferroics. (a) Room temperature P-E hysteresis loop of PZT–PFT (PZT_{1-x}–PFT_x) with different compositions. (b) Room temperature ferromagnetic hysteresis loop of PZT–PFT with different compositions [100]. (Reprinted with permission from Sanchez, D.A. et al., *AIP Adv.*, 1, 042169. Copyright 2011, American Institute of Physics.)

shown in Figure 2.13a and b. Furthermore, electric-field control of the remnant magnetization was also reported, revealing the strong coupling between the two order parameters [100].

A further study on the PFT–PZT solid solution system led to an even more astonishing demonstration of the magnetoelectric coupling. A clear change in the ferroelectric domain structures (revealed by PFM) induced by magnetic field at room temperature was reported, as shown in Figure 2.14 [101]. Although the changes in the ferroelectric domains were found to be stochastic, further investigation suggested that they are reproducible and reversible to some extent, depending on the orientation of the applied magnetic field. The effective magnetoelectric coupling coefficient was measured to be $\sim 1.3 \times 10^{-7}$ s m^{-1} [101], defined in terms of the change in polarization under a given magnetic field. This value is much larger than the values found in magnetoelectric heterostructures [102]. Although the origin and mechanism of magnetoelectric coupling in solid-solution systems still remain unresolved currently, enough evidences exist to show that solid solution systems offer an alternative to obtain multiferroics at room temperature.

Our discussion here is by no means a complete coverage of solid-solution multiferroic systems. Other examples include $(1-x)Pb(Fe_{0.33}W_{0.67})O_3$–$xPbTiO_3$, $(1-x)Pb(Fe_{0.5}Nb_{0.5})O_3(PFN)$–$xPb(Zr_{0.2}Ti_{0.8})O_3$, and $Pb(Fe_{0.5}Nb_{0.5})O_3$–$Pb(Zn_{1/3}Nb_{2/3})O_3$ [103–105]. Related systems such as $[(BiFeO_3)_x$–$(BiMg_{0.5}Ti_{0.5}O_3)_{1-x}]_{0.55}$ $[PbTiO_3]_{0.45}$, $0.45(Bi_{1-x}La_x)FeO_3$–$0.55PbTiO_3$, and $Pb(Fe_{0.67}W_{0.33})O_3$–$BiFeO_3$ also show separate or combined ferroelectric and ferromagnetic properties [106,107]. For a detailed discussion on this topic, the readers are directed to refer to listed references and review papers [103–105,108,109].

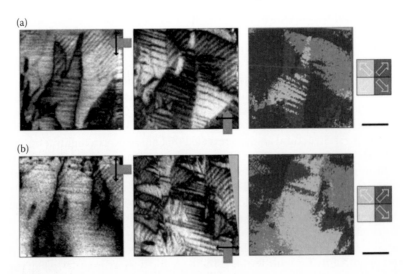

FIGURE 2.14 Ferroelectric domain modulation in PZT–PFT with an applied magnetic field. Lateral piezoresponse force microscopy images of PZT–PFT samples before (a) and after (b) the application of an 18 kOe magnetic field. The images captured at the same area indicate an obvious change in ferroelectric domains. The amplitude images are shown in the left and middle panels of (a) and (b), revealing variations in the local in-plane piezoelectric component, where the relative orientation of scanning cantilever is illustrated by the blue tips. The right panels of (a) and (b) show the sample in-plane component maps of the local polarization direction [101]. (Reprinted by permission from Macmillan Publishers Ltd. *Nat. Commun.*, Evans, D. et al., 4, 1534, copyright 2013.)

2.2.4 Other Single-Phase Lone-Pair Multiferroic Systems

In addition to the ones we have discussed, several other single-phase multiferroics have been predicted and investigated. For example, Hill et al. [110] suggested that $BiCrO_3$ possesses antiferromagnetic and antiferroelectric properties based on first-principles calculations. In 2006, $BiCrO_3$ thin films were grown on various substrates by pulsed laser deposition. A weak ferromagnetism with a Curie temperature of 120 K was reported [111]. $BiCoO_3$ is another example. It is a pyroelectric insulator with the C-type antiferromagnetic order. The Co ion is pyramidally coordinated by five oxygen ions, resulting from the noncentrosymmetric coordination for Bi ions with $6s$ lone pair electrons. Neutron diffraction revealed that the Co ions with d^6 occupancy typically shows high-spin configuration ($S = 2$). However, a transition from a high-spin state to a low-spin state ($S = 0$) can be induced by high pressure [112]. Recently, theoretical works on bulk samples suggested a giant polarization of more than 150 $\mu C\ cm^{-2}$ for $BiCoO_3$ [113]; these works inspire more researchers to conduct further studies. First-principles calculations also indicated that a morphotropic transition might be present in $BiFe_{1-x}Co_xO_3$ solid solution ($BiFe_{1-x}Co_xO_3$), and electric-field-driven phase changes could be expected [114].

PbVO$_3$ has been proposed to have large ferroelectric polarization with antiferromagnetic ordering. Early studies focused on high-temperature and high-pressure single crystal growth. However, experimental studies, especially electrical measurements, are limited by the low resistivity of the samples. In 2007, Martin et al. [115] successfully prepared PbVO$_3$ thin films on various substrates ($LaAlO_3$, $SrTiO_3$, ($La_{0.18}Sr_{0.82}$; $Al_{0.59}Ta_{0.41})O_3$, NdGaO$_3$, and LaAlO$_3$/Si substrates). Structural analysis revealed that it possesses a highly tetragonal phase with a c/a ratio of 1.32. A further study by Kumar et al. [116] using X-ray dichroism and second harmonic generation suggested that PbVO$_3$ is polar and piezoelectric, and likely ferroelectric. The simultaneous presence of piezoelectricity and magnetic order at low temperatures (<130 K) was also confirmed. These examples together with the well-established BiFeO$_3$, BiMnO$_3$, and the Pb-based multiferroic systems clearly establish that perovskite oxides with lone-pair electrons on the A-site ions and magnetic ions on the B-site offer the best opportunity for multiferroics. There are likely more multiferroic materials in this group to be explored.

2.3 Other Single-Phase Multiferroic Systems: Geometric Ferroelectricity

Hexagonal manganites, RMnO$_3$ (R, rare earth element: Ho, Er, Tm, Yb, Lu, Sc, and Y), represent a very interesting family of type-I multiferroics. They show relative high ferroelectric Curie temperatures, but much lower magnetic ordering temperatures [117,118]. Note that manganites in the orthorhombic phases are also multiferroic. But the ferroelectricity is driven by frustrated magnetic orders, leading to low ferroelectric ordering temperatures [119,120]. A detailed discussion on type-II multiferroics can be found in Chapter 4 of this book. Here we only briefly discuss the hexagonal manganites, where the ferroelectricity is driven by the tilting of the rigid MnO$_5$ bipyramid [121], focusing on YMnO$_3$. A review with great detail on the hexagonal manganite systems can be referred to Reference [122].

Compared with other rare earth manganite systems, YMnO$_3$ (Figure 2.15) is considered as a good prototype to investigate the magnetic structure and magnetoelectric coupling in hexagonal manganites [123,124]. Early work on YMnO$_3$ in the 1960s indicated that it possesses both ferroelectric with space group P6$_3$cm and A-type antiferromagnetic ordering with noncollinear spins of Mn in a triangular arrangement [117,120,125]. The origin of ferroelectricity in YMnO$_3$, especially the origin of hexagonal manganites in general, has been a matter under debate. They neither fulfill the "d^0-ness" requirement that leads to the off-center displacement in conventional perovskite ferroelectrics nor possess the "lone-pair electrons" that drive ferroelectricity in typical multiferroics such as BiFeO$_3$. For YMnO$_3$, it was initially suggested that the ferroelectricity arises from an off-center distortion of Mn ions toward one of the apexes of the MnO$_5$ bipyramids. However, a detailed structural study revealed that this is

(a) (b)

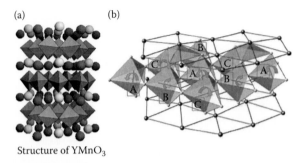

Structure of $YMnO_3$

FIGURE 2.15 Schematic of crystal structure and cooperative bipyramid rotation of multiferroic $YMnO_3$. (a) Crystal structure of $YMnO_3$. (Reprinted with permission from Muñoz, J. A. et al., *Phys. Rev. B*, 62, 9498. Copyright 2000, American Physical Society.) (b) Ferroelectricity results from the cooperative rotation of the bipyramids in $YMnO_3$, as indicated by the axes and by the arrows. The Y ions are shown in blue, O ions in yellow, and Mn ions in red. (Reprinted with permission from Kreisel, J. and Kenselmann, N., *Europhys. News*, 40, 17, 2009. Copyright 2009, EDP Sciences.)

not the case [121], suggesting that the off-center shift of the Mn^{3+} ions is not the main mechanism for ferroelectricity in the system. Based on first-principle calculations, Van Aken et al. [121] proposed that electrostatic and size effects are responsible for the structural distortion and ferroelectricity. The rotation of MnO_5 bipyramids shifts the O atoms from their centrosymmetric positions and leads to the displacement of Y atoms along the *c*-axis [121]. However, a later investigation by Cho et al. [126] using the polarization-dependent X-ray absorption spectroscopy (XAS) revealed strong anisotropic hybridization not only for Mn *3d*–O *2p* but also for Y *4d*–O *2p* bonds. This result indicates that the off-center Y–O ions contribute large anomalies in Born effective charges through rehybridization [126]. They concluded that the ferroelectricity originates from the Y d^0-ness with rehybridization together with the structural phonon instability, and the scenario is applicable for other hexagonal manganites.

The coupling between ferroelectric and magnetic orders in $YMnO_3$ was investigated by Fiebig et al. [127]. Using optical second harmonic generation imaging, they revealed that the antiferromagnetic and ferroelectric domains are strongly coupled together. The sign of the product *Pl* must be conserved upon crossing a ferroelectric domain wall, where *P* and *l* are the independent ferroelectric and antiferromagnetic order parameters, respectively (Figure 2.16). The strong coupling not only plays an important role in fundamental physics study but also offers possibilities for applications in next-generation electronics [128].

Thin films of manganites, including $YMnO_3$, have been prepared using various techniques, including radiofrequency magnetron sputtering and chemical vapor deposition on different substrates [129–131]. By tuning the growth conditions and selecting suitable substrates, various phases can be stabilized, demonstrating the advantages of thin-film studies. Clear ferroelectric hysteresis loops have been observed in $YMnO_3$ thin films, as shown in Figure 2.17 [132].

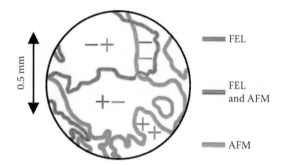

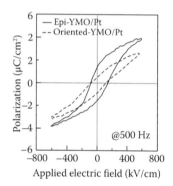

FIGURE 2.16 Four types of 180° domain walls in YMnO$_3$ are expressed as (+P, +I), (−P, −I), (+P, −I), and (−P, +I). P is the independent component of the ferroelectric order parameter and I is the independent component of the antiferromagnetic order parameter. Ferroelectric walls accompany antiferromagnetic walls, and therefore, there is no sole antiferromagnetic wall. (Reprinted by permission from Macmillan Publishers Ltd. *Nature*, Fiebig, M. et al., 419, 818–820, Copyright 2002.)

FIGURE 2.17 *P–E* hysteresis of the epitaxial YMnO$_3$/Pt thin film and that of the oriented YMnO$_3$/Pt. (Reprinted with permission from Ito, D. et al., *J. Appl. Phys.*, 93, 5563–5567. Copyright 2003, American Institute of Physics.)

In the thin-film form, YMnO$_3$-based field-effect transistors exhibit a long retention time of 10^4 s, suggesting that these films are promising for use in applications [132]. However, the growth temperatures for YMnO$_3$ are relatively high (800°C–850°C [132–134]). Chemical substitution with an element with a low melting temperature such as Bi has been investigated to decrease the growth temperature [134]; yet this is usually accompanied by a degradation of multiferroic properties.

In order to enhance multiferroic properties and to solve the critical drawbacks of YMnO$_3$, chemical alloying has been employed. Like many other manganites, A-site doping could have strong effects on the properties of YMnO$_3$ [133,135]. For example, doping with Li and Mg has led to an enhancement of carrier density in YMnO$_3$. Li-doped samples exhibit weak ferromagnetism behavior, for which the small canting of the Mn spins is thought to be the origin. Substitution of B-site cations in $YMnO_3$ is also exploited to tailor its physical properties such as electrical conductivity [136] and magnetoelectric coupling [137,138]. Twenty percent Ti substitution of Mn in $YMnO_3$ can result in a structural phase transition from P6$_3$cm to R3c and give rise to the tilting of MnO_5 bipyramids toward the center oxygen, which effectively dominates the magneto-dielectric coupling [139]. The findings support the conclusion of the work done by Aikawa et al., stating that Mn trimerization plays an important role to enhance the magnetoelectric coupling [137]. In addition, Ga doping (up to 50% of Mn and the same hexagonal structure remains) on the B-site has been shown to enhance the magnetoelectric coupling in the form of changes in the magnetocapacitance by two orders of magnitude [138]. Over the last few years, a wide range of hexagonal-$RMnO_3$ materials have been grown and investigated. Despite the progress, researchers are yet to find a hexagonal-$RMnO_3$ compound that exhibits both magnetism and ferroelectricity at room temperature. However, it remains a system with intriguing scientific implications for multiferroic materials.

2.4 Conclusions and Outlook

The fascinating properties and functionalities of type-I multiferroics have shown great potentials for both scientific research and practical applications. However, despite the tremendous progress that has been made, two issues remain unresolved: one is to look for multiferroic materials that are intrinsically ferromagnetic and ferroelectric at room temperature; the other is to realize electric-field control of magnetism at room temperature. Development of single-crystal and thin-film growth techniques and progress in theoretical calculation methodologies have led to the identification of many new multiferroics, resulting in a flurry of activities in this field. With this chapter, we try to bring to the readers a general view on type-I multiferroics and hope to inspire more researchers to join this exciting field.

References

1. Hill, N. A. 2000. Why are there so few magnetic ferroelectrics? *The Journal of Physical Chemistry B* 104:6694–6709.
2. Goodenough, J. B. 1955. Theory of the role of covalence in the perovskite-type manganites [La, M(II)] MnO_3. *Physical Review* 100:564.
3. Singh, D., and C. H. Park. 2008. Polar behavior in a magnetic perovskite from A-site size disorder: A density functional study. *Physical Review Letters* 100:087601.
4. Park, S., N. Hur, S. Guha, and S.-W. Cheong. 2004. Percolative conduction in the half-metallic-ferromagnetic and ferroelectric mixture of (La, Lu, Sr)MnO_3. *Physical Review Letters* 92:167206.
5. Ramesha, K., V. Thangadurai, D. Sutar, S. Subramanyam, G. Subbanna, and J. Gopalakrishnan. 2000. ALaMnBO$_6$ (A = Ca, Sr, Ba; B = Fe, Ru) double perovskites. *Materials Research Bulletin* 35:559–565.

6. Ueda, K., H. Tabata, and T. Kawai. 1998. Ferromagnetism in $LaFeO_3$–$LaCrO_3$ superlattices. *Science* 280:1064–1066.
7. Blasse, G. 1965. Ferromagnetic interactions in non-metallic perovskites. *Journal of Physics and Chemistry of Solids* 26:1969–1971.
8. Baettig, P., and N. A. Spaldin. 2005. Ab initio prediction of a multiferroic with large polarization and magnetization. *Applied Physics Letters* 86:012505.
9. Nechache, R., C. Harnagea, A. Pignolet, F. Normandin, T. Veres, L.-P. Carignan, and D. Ménard. 2006. Growth, structure and properties of epitaxial thin films of first principles predicted multiferroic Bi_2FeCrO_6. *Applied Physics Letters* 89:102902.
10. Nechache, R., C. Harnagea, and A. Pignolet. 2012. Multiferroic properties—structure relationships in epitaxial Bi_2FeCrO_6 thin films: Recent developments. *Journal of Physics: Condensed Matter* 24:096001.
11. Nechache, R., C. V. Cojocaru, C. Harnagea, C. Nauenheim, M. Nicklaus, A. Ruediger, F. Rosei, and A. Pignolet. 2011. Epitaxial patterning of Bi_2FeCrO_6 double perovskite nanostructures: Multiferroic at room temperature. *Advanced Materials* 23:1724–1729.
12. Azuma, M., K. Takata, T. Saito, S. Ishiwata, Y. Shimakawa, and M. Takano. 2005. Designed ferromagnetic, ferroelectric Bi_2NiMnO_6. *Journal of the American Chemical Society* 127:8889–8892.
13. Sharma, G., J. Saha, S. Kaushik, V. Siruguri, and S. Patnaik. 2013. Magnetism driven ferroelectricity above liquid nitrogen temperature in Y_2CoMnO_6. *Applied Physics Letters* 103:012903.
14. Nair, H. S., R. Pradheesh, Y. Xiao, D. Cherian, S. Elizabeth, T. Hansen, T. Chatterji, and T. Brückel. 2014. Magnetization-steps in Y_2CoMnO_6 double perovskite: The role of antisite disorder. *Journal of Applied Physics* 116:123907.
15. Yanez-Vilar, S., E. Mun, V. Zapf, B. Ueland, J. Gardner, J. Thompson, J. Singleton, M. Sánchez-Andújar, J. Mira, and N. Biskup. 2011. Multiferroic behavior in the double-perovskite Lu_2MnCoO_6. *Physical Review B* 84:134427.
16. Kumar, S., G. Giovannetti, J. van der Brink, and S. Picozzi. 2009. Theoretical prediction of multiferroicity in double perovskite Y_2NiMnO_6. *Physical Review B* 82:134429.
17. Trinquier, G., and R. Hoffmann. 1984. Lead monoxide. Electronic structure and bonding. *The Journal of Physical Chemistry* 88:6696–6711.
18. Watson, G., S. Parker, and G. Kresse. 1999. Ab initio calculation of the origin of the distortion of α-PbO. *Physical Review B* 59:8481.
19. Seshadri, R., and N. A. Hill. 2001. Visualizing the role of Bi 6s "lone pairs" in the off-center distortion in ferromagnetic $BiMnO_3$. *Chemistry of Materials* 13:2892–2899.
20. Moreau, J.-M., C. Michel, R. Gerson, and W. J. James. 1971. Ferroelectric $BiFeO_3$ X-ray and neutron diffraction study. *Journal of Physics and Chemistry of Solids* 32:1315–1320.
21. Martin, L., S. Crane, Y. Chu, M. Holcomb, M. Gajek, M. Huijben, C. Yang, N. Balke, and R. Ramesh. 2008. Multiferroics and magnetoelectrics: Thin films and nanostructures. *Journal of Physics: Condensed Matter* 20:434220.
22. Michel, C., J.-M. Moreau, G. D. Achenbach, R. Gerson, and W. J. James. 1969. The atomic structure of $BiFeO_3$. *Solid State Communications* 7:701–704.
23. Teague, J. R., R. Gerson, and W. J. James. 1970. Dielectric hysteresis in single crystal $BiFeO_3$. *Solid State Communications* 8:1073–1074.
24. Wang, J., J. Neaton, H. Zheng, V. Nagarajan, S. Ogale, B. Liu, D. Viehland, V. Vaithyanathan, D. Schlom, and U. Waghmare. 2003. Epitaxial $BiFeO_3$ multiferroic thin film heterostructures. *Science* 299:1719–1722.

25. Lebeugle, D., D. Colson, A. Forget, and M. Viret. 2007. Very large spontaneous electric polarization in BiFeO$_3$ single crystals at room temperature and its evolution under cycling fields. *Applied Physics Letters* 91:022907.

26. Dzyaloshinsky, I. 1958. A thermodynamic theory of "weak" ferromagnetism of antiferromagnetics. *Journal of Physics and Chemistry of Solids* 4:241–255.

27. Moriya, T. 1960. Anisotropic superexchange interaction and weak ferromagnetism. *Physical Review* 120:91.

28. Zhao, T., A. Scholl, F. Zavaliche, K. Lee, M. Barry, A. Doran, M. Cruz, Y. Chu, C. Ederer, and N. Spaldin. 2006. Electrical control of antiferromagnetic domains in multiferroic BiFeO$_3$ films at room temperature. *Nature Materials* 5:823–829.

29. Ederer, C., and N. A. Spaldin. 2005. Weak ferromagnetism and magnetoelectric coupling in bismuth ferrite. *Physical Review B* 71:060401.

30. Zeches, R., M. Rossell, J. Zhang, A. Hatt, Q. He, C.-H. Yang, A. Kumar, C. Wang, A. Melville, and C. Adamo. 2009. A strain-driven morphotropic phase boundary in BiFeO$_3$. *Science* 326:977–980.

31. Catalan, G., and J. F. Scott. 2009. Physics and applications of bismuth ferrite. *Advanced Materials* 21:2463–2485.

32. Infante, I. C., S. Lisenkov, B. Dupé, M. Bibes, S. Fusil, E. Jacquet, G. Geneste, S. Petit, A. Courtial, and J. Juraszek. 2010. Bridging multiferroic phase transitions by epitaxial strain in BiFeO$_3$. *Physical Review Letters* 105:057601.

33. Dupé, B., S. Prosandeev, G. Geneste, B. Dkhil, and L. Bellaiche. 2011. BiFeO$_3$ films under tensile epitaxial strain from first principles. *Physical Review Letters* 106:237601.

34. Lubk, A., S. Gemming, and N. Spaldin. 2009. First-principles study of ferroelectric domain walls in multiferroic bismuth ferrite. *Physical Review B* 80:104110.

35. Shang, S., G. Sheng, Y. Wang, L. Chen, and Z. Liu. 2009. Elastic properties of cubic and rhombohedral BiFeO$_3$ from first-principles calculations. *Physical Review B* 80:052102.

36. Biegalski, M. D., D. H. Kim, S. Choudhury, L. Chen, H. Christen, and K. Dörr. 2011. Strong strain dependence of ferroelectric coercivity in a BiFeO$_3$ film. *Applied Physics Letters* 98:142902.

37. Ravindran, P., R. Vidya, A. Kjekshus, H. Fjellvåg, and O. Eriksson. 2006. Theoretical investigation of magnetoelectric behavior in BiFeO$_3$. *Physical Review B* 74:224412.

38. Sando, D., A. Agbelele, D. Rahmedov, J. Liu, P. Rovillain, C. Toulouse, I. Infante, A. Pyatakov, S. Fusil, and E. Jacquet. 2013. Crafting the magnonic and spintronic response of BiFeO$_3$ films by epitaxial strain. *Nature Materials* 12:641–646.

39. Ederer, C., and N. A. Spaldin. 2005. Effect of epitaxial strain on the spontaneous polarization of thin film ferroelectrics. *Physical Review Letters* 95:257601.

40. Hatt, A. J., N. A. Spaldin, and C. Ederer. 2010. Strain-induced isosymmetric phase transition in BiFeO$_3$. *Physical Review B* 81:054109.

41. Arnold, D. C., K. S. Knight, F. D. Morrison, and P. Lightfoot. 2009. Ferroelectric–paraelectric transition in BiFeO$_3$: Crystal structure of the orthorhombic β phase. *Physical Review Letters* 102:027602.

42. Khomchenko, V., D. Karpinsky, A. Kholkin, N. Sobolev, G. Kakazei, J. Araujo, I. Troyanchuk, B. Costa, and J. Paixão. 2010. Rhombohedral-to-orthorhombic transition and multiferroic properties of Dy-substituted BiFeO$_3$. *Journal of Applied Physics* 108:074109.

43. Yang, J., C. Yeh, Y. Chen, S. Liao, R. Huang, H. Liu, C. Hung, S. Chen, S. Wu, and C. Lai. 2014. Conduction control at ferroic domain walls via external stimuli. *Nanoscale* 6:10524–10529.

44. Yang, J., Q. He, S. Suresha, C. Kuo, C. Peng, R. Haislmaier, M. Motyka, G. Sheng, C. Adamo, and H. Lin. 2012. Orthorhombic BiFeO$_3$. *Physical Review Letters* 109:247606.

45. Yang, S., J. Seidel, S. Byrnes, P. Shafer, C.-H. Yang, M. Rossell, P. Yu, Y.-H. Chu, J. Scott, and J. Ager. 2010. Above-bandgap voltages from ferroelectric photovoltaic devices. *Nature Nanotechnology* 5:143–147.

46. Bhatnagar, A., A. R. Chaudhuri, Y. H. Kim, D. Hesse, and M. Alexe. 2013. Role of domain walls in the abnormal photovoltaic effect in BiFeO$_3$. *Nature Communications* 4:2835.

47. Chen, Z., Y. Qi, L. You, P. Yang, C. Huang, J. Wang, T. Sritharan, and L. Chen. 2013. Large tensile-strain-induced monoclinic M B phase in BiFeO$_3$ epitaxial thin films on a PrScO$_3$ substrate. *Physical Review B* 88:054114.

48. Ren, W., Y. Yang, O. Diéguez, J. Íñiguez, N. Choudhury, and L. Bellaiche. 2013. Ferroelectric domains in multiferroic BiFeO$_3$ films under epitaxial strains. *Physical Review Letters* 110:187601.

49. Li, J., J. Wang, M. Wuttig, R. Ramesh, N. Wang, B. Ruette, A. Pyatakov, A. Zvezdin, and D. Viehland. 2004. Dramatically enhanced polarization in (001),(101) and (111) BiFeO$_3$ thin films due to epitaxial-induced transitions. *Applied Physics Letters* 84:5261.

50. Mazumdar, D., V. Shelke, M. Iliev, S. Jesse, A. Kumar, S. V. Kalinin, A. P. Baddorf, and A. Gupta. 2010. Nanoscale switching characteristics of nearly tetragonal BiFeO$_3$ thin films. *Nano Letters* 10:2555–2561.

51. Singh, M. K., S. Ryu, and H. M. Jang. 2005. Polarized Raman scattering of multiferroic BiFeO$_3$ thin films with pseudo-tetragonal symmetry. *Physical Review B* 72:132101.

52. Christen, H. M., J. H. Nam, H. S. Kim, A. J. Hatt, and N. A. Spaldin. 2011. Stress-induced R–MA–MC–T symmetry changes in BiFeO$_3$ films. *Physical Review B* 83:144107.

53. Vasudevan, R. K., Y. Liu, J. Li, W.-I. Liang, A. Kumar, S. Jesse, Y.-C. Chen, Y.-H. Chu, V. Nagarajan, and S. V. Kalinin. 2011. Nanoscale control of phase variants in strain-engineered BiFeO$_3$. *Nano Letters* 11:3346–3354.

54. Béa, H., B. Dupé, S. Fusil, R. Mattana, E. Jacquet, B. Warot-Fonrose, F. Wilhelm, A. Rogalev, S. Petit, and V. Cros. 2009. Evidence for room-temperature multiferroicity in a compound with a giant axial ratio. *Physical Review Letters* 102:217603.

55. Pailloux, F., M. Couillard, S. Fusil, F. Bruno, W. Saidi, V. Garcia, C. Carrétéro, E. Jacquet, M. Bibes, and A. Barthélémy. 2014. Atomic structure and microstructures of supertetragonal multiferroic BiFeO$_3$ thin films. *Physical Review B* 89:104106.

56. Infante, I., J. Juraszek, S. Fusil, B. Dupé, P. Gemeiner, O. Diéguez, F. Pailloux, S. Jouen, E. Jacquet, and G. Geneste. 2011. Multiferroic phase transition near room temperature in BiFeO$_3$ films. *Physical Review Letters* 107:237601.

57. Zhang, J., Q. He, M. Trassin, W. Luo, D. Yi, M. Rossell, P. Yu, L. You, C. Wang, and C. Kuo. 2011. Microscopic origin of the giant ferroelectric polarization in tetragonal-like BiFeO$_3$. *Physical Review Letters* 107:147602.

58. Kim, K.-E., B.-K. Jang, Y. Heo, J. H. Lee, M. Jeong, J. Y. Lee, J. Seidel, and C.-H. Yang. 2014. Electric control of straight stripe conductive mixed-phase nanostructures in La-doped BiFeO$_3$. *NPG Asia Materials* 6:e81.

59. He, Q., Y.-H. Chu, J. Heron, S. Yang, W. Liang, C. Kuo, H. Lin, P. Yu, C. Liang, and R. Zeches. 2011. Electrically controllable spontaneous magnetism in nanoscale mixed phase multiferroics. *Nature Communications* 2:225.

60. Liu, H.-J., C.-W. Liang, W.-I. Liang, H.-J. Chen, J.-C. Yang, C.-Y. Peng, G.-F. Wang, F.-N. Chu, Y.-C. Chen, and H.-Y. Lee. 2012. Strain-driven phase boundaries in $BiFeO_3$ thin films studied by atomic force microscopy and x-ray diffraction. *Physical Review B* 85:014104.

61. You, L., Z. Chen, X. Zou, H. Ding, W. Chen, L. Chen, G. Yuan, and J. Wang. 2012. Characterization and manipulation of mixed phase nanodomains in highly strained $BiFeO_3$ thin films. *ACS Nano* 6:5388–5394.

62. Beekman, C., W. Siemons, T. Z. Ward, M. Chi, J. Howe, M. D. Biegalski, N. Balke, P. Maksymovych, A. Farrar, and J. Romero. 2013. Phase transitions, phase coexistence, and piezoelectric switching behavior in highly strained $BiFeO_3$ films. *Advanced Materials* 25:5561–5567.

63. Rossell, M., R. Erni, M. P. Prange, J.-C. Idrobo, W. Luo, R. Zeches, S. T. Pantelides, and R. Ramesh. 2012. Atomic structure of highly strained $BiFeO_3$ thin films. *Physical Review Letters* 108:047601.

64. Streiffer, S., C. Parker, A. Romanov, M. Lefevre, L. Zhao, J. Speck, W. Pompe, C. Foster, and G. Bai. 1998. Domain patterns in epitaxial rhombohedral ferroelectric films. I. Geometry and experiments. *Journal of Applied Physics* 83:2742–2753.

65. Romanov, A., M. Lefevre, J. Speck, W. Pompe, S. Streiffer, and C. Foster. 1998. Domain pattern formation in epitaxial rhombohedral ferroelectric films. II. Interfacial defects and energetics. *Journal of Applied Physics* 83:2754–2765.

66. Yang, J.C., Y. L. Huang, Q. He, and Y. H. Chu. 2014. Multifunctionalities driven by ferroic domains. *Journal of Applied Physics* 116:066801.

67. Chen, Z., J. Liu, Y. Qi, D. Chen, S.-L. Hsu, A. R. Damodaran, X. He, A. T. N'Diaye, A. Rockett, and L. W. Martin. 2015. 180° Ferroelectric stripe nanodomains in $BiFeO_3$ thin films. *Nano Letters* 15:6506–6513.

68. Zavaliche, F., S. Yang, T. Zhao, Y. Chu, M. Cruz, C. Eom, and R. Ramesh. 2006. Multiferroic $BiFeO_3$ films: Domain structure and polarization dynamics. *Phase Transitions* 79:991–1017.

69. Cruz, M., Y. Chu, J. Zhang, P. Yang, F. Zavaliche, Q. He, P. Shafer, L. Chen, and R. Ramesh. 2007. Strain control of domain-wall stability in epitaxial $BiFeO_3$ (110) films. *Physical Review Letters* 99:217601.

70. Seidel, J., L. W. Martin, Q. He, Q. Zhan, Y.-H. Chu, A. Rother, M. Hawkridge, P. Maksymovych, P. Yu, and M. Gajek. 2009. Conduction at domain walls in oxide multiferroics. *Nature Materials* 8:229–234.

71. Aird, A., and E. K. Salje. 1998. Sheet superconductivity in twin walls: experimental evidence of WO_{3-x}. *Journal of Physics: Condensed Matter* 10:L377–L380.

72. Meier, D., J. Seidel, A. Cano, K. Delaney, Y. Kumagai, M. Mostovoy, N. A. Spaldin, R. Ramesh, and M. Fiebig. 2012. Anisotropic conductance at improper ferroelectric domain walls. *Nature Materials* 11:284–288.

73. Choi, T., Y. Horibe, H. Yi, Y. Choi, W. Wu, and S.-W. Cheong. 2010. Insulating interlocked ferroelectric and structural antiphase domain walls in multiferroic $YMnO_3$. *Nature Materials* 9:253–258.

74. Sluka, T., A. K. Tagantsev, P. Bednyakov, and N. Setter. 2013. Free-electron gas at charged domain walls in insulating $BaTiO_3$. *Nature Communications* 4:1808.

75. Martin, L. W., Y.-H. Chu, M. B. Holcomb, M. Huijben, P. Yu, S.-J. Han, D. Lee, S. X. Wang, and R. Ramesh. 2008. Nanoscale control of exchange bias with $BiFeO_3$ thin films. *Nano Letters* 8:2050–2055.

76. He, Q., C.-H. Yeh, J.-C. Yang, G. Singh-Bhalla, C.-W. Liang, P.-W. Chiu, G. Catalan, L. Martin, Y.-H. Chu, and J. Scott. 2012. Magnetotransport at domain walls in $BiFeO_3$. *Physical Review Letters* 108:067203.

77. Bibes, M. 2008. Multiferroics Towards a magnetoelectric memory. The room-temperature manipulation of magnetization by an electric field using the multiferroic $BiFeO_3$ represents an essential step towards the magnetoelectric control of spintronics devices. *Nature Materials* 7:425.

78. Yang, J. C., Y. H. Chu, and R. Ramesh. 2015. Multiferroic thin films. In *Wiley Encyclopedia of Electrical and Electronics Engineering*, 1–15. New Jersey: John Wiley & Sons.

79. Moreira Dos Santos, A. F., A. K. Cheetham, W. Tian, X. Pan, Y. Jia, N. J. Murphy, J. Lettieri, and D. G. Schlom. 2004. Epitaxial growth and properties of metastable $BiMnO_3$ thin films. *Applied Physics Letters* 84:91

80. Atou, T., H. Chiba, K. Ohoyama, Y. Yamaguchi, and Y. Syono. 1999. Structure determination of ferromagnetic perovskite $BiMnO_3$. *Journal of Solid State Chemistry* 145:639–642.

81. Montanari, E., G. Calestani, L. Righi, E. Gilioli, F. Bolzoni, K. Knight, and P. Radaelli. 2007. Structural anomalies at the magnetic transition in centrosymmetric $BiMnO_3$. *Physical Review B* 75:220101.

82. McLeod, J., Z. Pchelkina, L. Finkelstein, E. Kurmaev, R. Wilks, A. Moewes, I. Solovyev, A. Belik, and E. Takayama-Muromachi. 2010. Electronic structure of $BiMO_3$ multiferroics and related oxides. *Physical Review B* 81:144103.

83. Sugawara, F., S. Iiida, Y. Syono, and S.-i. Akimoto. 1968. Magnetic properties and crystal distortions of $BiMnO_3$ and $BiCrO_3$. *Journal of the Physical Society of Japan* 25:1553–1558.

84. Chiba, H., T. Atou, and Y. Syono. 1997. Magnetic and electrical properties of Bi 1– x Sr x MnO 3: Hole-doping effect on ferromagnetic perovskite $BiMnO_3$. *Journal of Solid State Chemistry* 132:139–143.

85. Solovyev, I., N. Hamada, and K. Terakura. 1996. Crucial role of the lattice distortion in the magnetism of $LaMnO_3$. *Physical Review Letters* 76:4825.

86. Belik, A. A., S. Iikubo, T. Yokosawa, K. Kodama, N. Igawa, S. Shamoto, M. Azuma, M. Takano, K. Kimoto, and Y. Matsui. 2007. Origin of the monoclinic-to-monoclinic phase transition and evidence for the centrosymmetric crystal structure of $BiMnO_3$. *Journal of the American Chemical Society* 129:971–977.

87. Yang, H., Z. Chi, J. Jiang, W. Feng, J. Dai, C. Jin, and R. Yu. 2008. Is ferroelectricity in $BiMnO_3$ induced by superlattice? *Journal of Materials Science* 43:3604–3607.

88. Yang, H., Z. Chi, J. Jiang, W. Feng, Z. Cao, T. Xian, C. Jin, and R. Yu. 2008. Centrosymmetric crystal structure of $BiMnO_3$ studied by transmission electron microscopy and theoretical simulations. *Journal of Alloys and Compounds* 461:1–5.

89. Toulemonde, P., P. Bordet, P. Bouvier, and J. Kreisel. 2014. Single-crystalline $BiMnO_3$ studied by temperature-dependent x-ray diffraction and Raman spectroscopy. *Physical Review B* 89:224107.

90. Baettig, P., R. Seshadri, and N. A. Spaldin. 2007. Anti-polarity in ideal $BiMnO_3$. *Journal of the American Chemical Society* 129:9854–9855.

91. Solovyev, I., and Z. Pchelkina. 2010. Magnetic-field control of the electric polarization in $BiMnO_3$. *Physical Review B* 82:094425.

92. Gajek, M., M. Bibes, S. Fusil, K. Bouzehouane, J. Fontcuberta, A. Barthelemy, and A. Fert. 2007. Tunnel junctions with multiferroic barriers. *Nature Materials* 6:296–302.

93. Troyanchuk, I., O. Mantytskaja, H. Szymczak, and M. Y. Shvedun. 2002. Magnetic phase transitions in the system $La_{1-x}Bi_xMnO_{3+\lambda}$. *Low Temperature Physics* 28:569–573.

94. Gajek, M., M. Bibes, F. Wyczisk, M. Varela, J. Fontcuberta, and A. Barthélémy. 2007. Growth and magnetic properties of multiferroic $La_xBi_{1-x}MnO_3$ thin films. *Physical Review B* 75:174417.

95. Choi, E.-M., A. Kursumovic, O. J. Lee, J. e. E. Kleibeuker, A. Chen, W. Zhang, H. Wang, and J. L. MacManus-Driscoll. 2014. Ferroelectric Sm-doped BiMnO3 thin films with ferromagnetic transition temperature enhanced to 140 K. *ACS Applied Materials and Interfaces* 6:14836–14843.

96 Kumar, A., G. Sharma, R. S. Katiyar, R. Pirc, R. Blinc, and J. Scott. 2009. Magnetic control of large room-temperature polarization. *Journal of Physics: Condensed Matter* 21:382204.

97. Smolenskii, G., A. Agranovskaia, S. Popov, and V. Isupov. 1958. New ferroelectrics of complex composition. *Soviet Physics-Technical Physics* 3:1981–1982.

98. Samara, G. A. 2003. The relaxational properties of compositionally disordered ABO_3 perovskites. *Journal of Physics: Condensed Matter* 15:R367.

99. Cross, L. E. 1987. Relaxor ferroelectrics. *Ferroelectrics* 76:241–267.

100. Sanchez, D. A., N. Ortega, A. Kumar, R. Roque-Malherbe, R. Polanco, J. Scott, and R. S. Katiyar. 2011. Symmetries and multiferroic properties of novel room-temperature magnetoelectrics: Lead iron tantalate–lead zirconate titanate (PFT/PZT). *AIP Advances* 1:042169.

101. Evans, D., A. Schilling, A. Kumar, D. Sanchez, N. Ortega, M. Arredondo, R. Katiyar, J. Gregg, and J. Scott. 2013. Magnetic switching of ferroelectric domains at room temperature in multiferroic PZTFT. *Nature Communications* 4:1534.

102. Kim, J.-H., K.-S. Ryu, J.-W. Jeong, and S.-C. Shin. 2010. Large converse magnetoelectric coupling effect at room temperature in CoPd/PMN-PT (001) heterostructure. *Applied Physics Letters* 97:252508.

103. Kumar, A., I. Rivera, R. Katiyar, and J. Scott. 2008. Multiferroic $Pb(Fe_{0.66}W_{0.33})_{0.80}Ti_{0.20}O_3$ thin films: A room-temperature relaxor ferroelectric and weak ferromagnetic. *Applied Physics Letters* 92:2913.

104. Fang, B.-J., C.-L. Ding, W. Liu, L.-Q. Li, and L. Tang. 2009. Preparation and electrical properties of high-Curie temperature ferroelectrics. *The European Physical Journal Applied Physics* 45:20302.

105. Yokosuka, M. 1999. Electrical, electromechanical and structural studies on solid solution ceramic $Pb(Fe_{1/2}Nb_{1/2})O_3$-$Pb(Zn_{1/3}Nb_{2/3})O_3$. *Japanese Journal of Applied Physics* 38:5488.

106. Rai, R., A. L. Kholkin, and S. Sharma. 2010. Multiferroic properties of $BiFeO_3$ doped $Bi(MgTi)O_3$–$PbTiO_3$ ceramic system. *Journal of Alloys and Compounds* 506:815–819.

107. Cheng, J., S. Yu, J. Chen, Z. Meng, and L. E. Cross. 2006. Dielectric and magnetic enhancements in $BiFeO_3$-$PbTiO_3$ solid solutions with La doping. *Applied Physics Letters* 89:2911.

108. Scott, J. F. 2013. Room-temperature multiferroic magnetoelectrics. *NPG Asia Materials* 5:e72.

109. Roy, S., and S. Majumder. 2012. Recent advances in multiferroic thin films and composites. *Journal of Alloys and Compounds* 538:153–159.

110. Hill, N. A., P. Bättig, and C. Daul. 2002. First principles search for multiferroism in $BiCrO_3$. *The Journal of Physical Chemistry B* 106:3383–3388.

111. Murakami, M., S. Fujino, S. Lim, C. Long, L. Salamanca-Riba, M. Wuttig, I. Takeuchi, V. Nagarajan, and A. Varatharajan. 2006. Fabrication of multiferroic epitaxial $BiCrO_3$ thin films. *Applied Physics Letters* 88:152902.

112. Oka, K., M. Azuma, W.-t. Chen, H. Yusa, A. A. Belik, E. Takayama-Muromachi, M. Mizumaki, N. Ishimatsu, N. Hiraoka, and M. Tsujimoto. 2010. Pressure-induced spin-state transition in $BiCoO_3$. *Journal of the American Chemical Society* 132:9438–9443.

113. Uratani, Y., T. Shishidou, F. Ishii, and T. Oguchi. 2005. First-principles predictions of giant electric polarization. *Japanese Journal of Applied Physics* 44:7130.

114. Diéguez, O., and J. Íñiguez. 2011. First-principles investigation of morphotropic transitions and phase-change functional responses in $BiFeO_3$–$BiCoO_3$ multiferroic solid solutions. *Physical Review Letters* 107:057601.

115. Martin, L. W., Q. Zhan, Y. Suzuki, R. Ramesh, M. Chi, N. Browning, T. Mizoguchi, and J. Kreisel. 2007. Growth and structure of $PbVO_3$ thin films. *Applied Physics Letters* 90:062903.

116. Kumar, A., L. W. Martin, S. Denev, J. B. Kortright, Y. Suzuki, R. Ramesh, and V. Gopalan. 2007. Polar and magnetic properties of $PbVO_3$ thin films. *Physical Review B* 75:060101.

117. Yakel, H., W. Koehler, E. Bertaut, and E. Forrat. 1963. On the crystal structure of the manganese (III) trioxides of the heavy lanthanides and yttrium. *Acta Crystallographica* 16:957–962.

118. Lottermoser, T., T. Lonkai, U. Amann, D. Hohlwein, J. Ihringer, and M. Fiebig. 2004. Magnetic phase control by an electric field. *Nature* 430:541–544.

119. Kimura, T., G. Lawes, T. Goto, Y. Tokura, and A. Ramirez. 2005. Magnetoelectric phase diagrams of orthorhombic R MnO_3 (R = Gd, Tb, and Dy). *Physical Review B* 71:224425.

120. Bertaut, E., and M. Mercier. 1963. Structure magnetique de $MnYO_3$. *Physics Letters* 5:27–29.

121. Van Aken, B. B., T. T. Palstra, A. Filippetti, and N. A. Spaldin. 2004. The origin of ferroelectricity in magnetoelectric $YMnO_3$. *Nature Materials* 3:164–170.

122. Lorenz, B. 2013. Hexagonal manganites—($RMnO_3$): Class (I) multiferroics with strong coupling of magnetism and ferroelectricity, *ISRN Condensed Matter Physics*, 1–43, Article ID 497073, http://dx.doi.org/10.1155/2013/497073.

123. Muñoz, A, J. A. Alonso, M. J. Martínez-Lope, M. T. Casáis, J. L. Martínez, and M. T. Fernández-Díaz. 2000. Magnetic structure of hexagonal $RMnO_3$ (R = Y, Sc): Thermal evolution from neutron powder diffraction data. *Physical Review B* 62:9498.

124. Kreisel, J. and N. Kenselmann. 2009. Multiferroics—The challenge of coupling magnetism and ferroelectricity. *Europhysics News* 40:17–20.

125. Ismailzade, I., and S. Kizhaev. 1965. Determination of the Curie point of ferroelectrics $YMnO_3$ and $YbMnO_3$. *Soviet Phys.-Solid State (English Transl.)* 7:298–301.

126. Cho, D.-Y., J.-Y. Kim, B.-G. Park, K.-J. Rho, J.-H. Park, H.-J. Noh, B. Kim, S.-J. Oh, H.-M. Park, and J.-S. Ahn. 2007. Ferroelectricity Driven by Y d^0-ness with Rehybridization in $YMnO_3$. *Physical Review Letters* 98:217601.

127. Fiebig, M., T. Lottermoser, D. Fröhlich, A. Goltsev, and R. Pisarev. 2002. Observation of coupled magnetic and electric domains. *Nature* 419:818–820.

128. Fusil, S., V. Garcia, A. Barthélémy, and M. Bibes. 2014. Magnetoelectric devices for spintronics. *Annual Review of Materials Research* 44:91–116.

129. Salvador, P. A., T.-D. Doan, B. Mercey, and B. Raveau. 1998. Stabilization of $YMnO_3$ in a perovskite structure as a thin film. *Chemistry of Materials* 10:2592–2595.

130. Yoo, D. C., J. Y. Lee, I. S. Kim, and Y. T. Kim. 2002. Microstructure control of $YMnO_3$ thin films on Si (100) substrates. *Thin Solid Films* 416:62–65.

131. Bosak, A. A., A. A. Kamenev, I. E. Graboy, S. V. Antonov, O. Y. Gorbenko, A. R. Kaul, C. Dubourdieu, J. P. Senateur, V. L. Svechnikov, H. W. Zandbergen, and B. Holländer. 2001. Epitaxial phase stabilisation phenomena in rare earth manganites. *Thin Solid Films* 400:149–153.

132. Ito, D., N. Fujimura, T. Yoshimura, and T. Ito. 2003. Ferroelectric properties of $YMnO_3$ epitaxial films for ferroelectric-gate field-effect transistors. *Journal of Applied Physics* 93:5563–5567.

133. Fujimura, N., H. Sakata, D. Ito, T. Yoshimura, T. Yokota, and T. Ito. 2003. Ferromagnetic and ferroelectric behaviors of A-site substituted $YMnO_3$-based epitaxial thin films. *Journal of Applied Physics* 93:6990–6992.

134. Choi, T., and J. Lee. 2004. Bi modification for low-temperature processing of $YMnO_3$ thin films. *Applied Physics Letters* 84:5043–5045.

135. Namdeo, S., A. K. Sinha, M. N. Singh, and A. M. Awasthi. 2013. Investigation of charge states and multiferroicity in Fe-doped h-$YMnO_3$. *Journal of Applied Physics* 113:104101.

136. Gutierrez, D., O. Peña, P. Duran, and C. Moure. 2002. Crystalline structure and electrical properties of $YCo_xMn_{1-x}O_3$ solid solutions. *Journal of the European Ceramic Society* 22:1257–1262.

137. Aikawa, Y., T. Katsufuji, T. Arima, and K. Kato. 2005. Effect of Mn trimerization on the magnetic and dielectric properties of hexagonal $YMnO_3$. *Physical Review B* 71:184418.

138. Nugroho, A. A., N. Bellido, U. Adem, G. Nénert, C. Simon, M. O. Tjia, M. Mostovoy, and T. T. M. Palstra. 2007. Enhancing the magnetoelectric coupling in $YMnO_3$ by Ga doping. *Physical Review B* 75:174435.

139. Mori, S., J. Tokunaga, Y. Horibe, Y. Aikawa, and T. Katsufuji. 2005. Magnetocapacitance effect and related microstrucuture in Ti-doped $YMnO_3$. *Physical Review B* 72:224434.

Single-Phase Type-I Multiferroics

Charge Order–Driven Multiferroicity

Manuel Angst

Contents

3.1 Introduction

In this chapter, multiferroics with ferroelectricity originating from charge ordering are explored in depth. After an examination of the basic concepts, the main focus is on the discussion of materials for which this mechanism has been proposed. I close this chapter with an outlook on considering application prospects of this class of multiferroics.

Charge order (CO) was first conceptualized by Wigner [1] as the crystallization of carriers in a free electron gas into an "electron crystal," driven by the minimization of their mutual Coulomb repulsion. In real materials, the nuclei cannot be neglected, and correspondingly, the concept was amended to denote the localization of electrons on some of the ions in an ordered fashion, still Coulomb-driven. As this leads to an ordered arrangement of the valence states (or oxidation states) of the ions, "ion valence order" is an often-used and essentially an equivalent term.* This is true for oxides and similar materials. In organic compounds, such as charge transfer salts, the ions/atoms are replaced by whole molecules, the bonding of which mainly is of the van der Waals (or H-bridge) type rather than the covalent type.

Because the electric dipoles making up an electric polarization consist of imbalanced charges, any CO breaking inversion symmetry[†] should automatically lead to an electric polarization of the material. In contrast to the polarization induced by spin spirals (see Chapter 4), this polarization can be potentially large, of the same order as the polarization of traditional ferroelectrics like $BaTiO_3$. For transition metal oxides, the possibility of CO, that is, ion valence order, implies partially filled d-shells: these are the orbitals involved in the CO. Partially filled d-shells imply that there is an active spin degree of freedom, and thus, the "magnetism is for free." Furthermore, because the d-orbitals on the same sites are involved in both charge and spin orders, we can expect them to be strongly coupled, which may translate into a sizeable magnetoelectric coupling. The situation is in principle similar for molecular compounds, just with the transition metal ions replaced by whole molecules. CO then also implies unpaired electrons on some of the molecules and thus an active spin degree of freedom, which can lead to magnetic order [2].

If CO is driven by Coulomb repulsion, a centrosymmetric arrangement would normally be expected. So how can noncentrosymmetric CO occur? One possibility was first proposed in 2004 [3] and is illustrated in Figure 3.1, based on Reference [4]. Instead of localizing on atomic sites, they could alternatively localize on bonds between sites ("bond-centered" CO), which leads to dimerization. This corresponds to the classical Peierls transition often observed in low-dimensional systems [5]. Neither the normal "site-centered" nor the "bond-centered" CO individually breaks inversion symmetry, but their combination does, because the location of the centers of inversion is different for these two orders. Depending on the topology of the ion lattice,

---

* I will examine the meaning of terms like "oxidation state": this naïve picture of CO has to be modified.

[†] More accurately, any CO breaking enough symmetry elements that upon application of all remaining symmetry elements at least two points are unmoved.

Chain of mixed-valentions (e.g., $Fe^{2.5+}$)

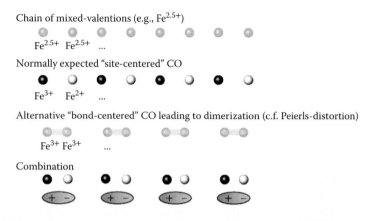

$Fe^{2.5+}$ $Fe^{2.5+}$ ...

Normally expected "site-centered" CO

Fe^{3+} Fe^{2+} ...

Alternative "bond-centered" CO leading to dimerization (c.f. Peierls-distortion)

Fe^{3+} Fe^{3+} ...

Combination

FIGURE 3.1 Possible ways of charge order to break inversion symmetry.

purely site-centered CO may break inversion symmetry as well. Simple centrosymmetric CO can be prevented by geometrical frustration, which is the inability to simultaneously fulfill all interactions, for example, for ion sites on an equilateral triangle. Geometrical frustration is particularly well studied for spin order [6], but equivalently applicable to other ordering phenomena like CO [7]. It leads to disordered or complex-ordered ground states, which can be polar. An example of ferroelectricity arising from charge frustration was proposed in 2005 [8] and is further discussed later in this chapter.

The scenario outlined earlier is nice and straightforward. However, as often, the devil is in the detail. One problem is conceptual, and lies in the answer to the question of "what does CO actually mean in real materials?" I briefly discuss this and the implications for ferroelectricity in the next section. The other problem is related to experimental issues and may be encompassed by the question "what experimentally verified examples of this mechanism exist?" This will occupy the bulk of this chapter, in which oxides and organic charge-transfer salts will be covered. At the end of the chapter, the implications for the further development of this field and for application prospects are summarized.

3.2 What Does Charge Order Mean?

3.2.1 Transition-Metal Oxides

The CO scenario outlined earlier implies an essentially ionic bonding picture. However, 100% ionic bonds do not exist in nature. The character of the bond between two ions is always at least partially covalent. For example, the Fe–O bond is about 50% ionic and 50% covalent. The oxidation states or valences that are ordered in CO are defined as the charge the atom would have if all bonding was 100% ionic [9]. However, because the bonding is far from purely ionic, the "charge on the atom" (defined for example via the charge density integrated in a sphere around the nucleus) is not equal to its oxidation state.

Indeed, there is a negative feedback mechanism at play that leads the total charge near the nucleus to be almost independent of the oxidation state [10]. Given this fact and based on the dependence of the calculation results on the (arbitrary) radius of the spheres around the nuclei, various alternative definitions of valence have been considered. These include the total d-orbital occupation [11], counting only fully occupied orbitals [12], and Wannier function–based approaches [13,14], but there is no general consensus.

The experimental situation concerning the assessment of oxidation state and CO is somewhat clearer. The oxidation state influences the screening of the core electrons, leading to a chemical shift of their energies, observable by x-ray absorption and other spectroscopies. Also influenced are the magnetic moment (accessible by neutron diffraction) and in particular the bond lengths to the neighboring ions, accessible via the structure refined from diffraction data. The latter gave rise to the so-called bond-valence-sum (BVS) method [15]: for each bond of the ion i with neighboring atom j, a partial "bond-valence" s_{ij} is assigned, which is a function of the difference between the actual bond length d_{ij} and an empirical tabulated [16] parameter $d_{0,ij}$ depending on the identity of the two atoms. The oxidation state or valence V_i is then the sum of these bond-valences:

$$V_i = \sum_{ij} s_{ij} = \sum_{ij} \exp\left(\frac{d_{0,ij} - d_{ij}}{0.37}\right) \tag{3.1}$$

Valence obtained from BVS correlates reasonably well with that obtained by spectroscopic means (see, e.g., [17]), but has the advantages of (i) yielding not only the valences of the ions, but also their arrangement in space, and (ii) being obtainable in many cases from laboratory x-ray diffraction. Pending further theoretical development, the oxidation state may therefore simply be identified with the bond-valence-sum (though in most cases, this valence is a noninteger with a separation of the BVS for the two sites inequivalent due to CO being smaller than one).

3.2.2 What Are the Implications of This CO Picture for Ferroelectric Polarization?

With regard to ferroelectricity originating from CO, the nonequivalence of the "ion valence" and the "charge at the ion" have considerable consequences. The picture outlined at the beginning of the chapter cannot be expected to hold if the actual charge localized at the ion is almost independent of the oxidation state. Instead, a ferroelectric polarization can be expected to arise only from (i) positional shifts of the ions in response to the CO, or (ii) from more complex shifts of electron density further away from the charge-ordered ions. A charge concept directly related to polarization is the "Born effective charge" of ions, defined as the derivative of the polarization change upon shifting the ions [18]. This is a well-defined and measureable (e.g., by vibrational spectroscopy) quantity, but it is a tensor and particularly for ferroelectrics, it often deviates enormously from the formal oxidation state [18]. The charge definition of Jiang et al. [13] as a directional average of the effective charge tensor is close to the use of CO in polarization, but the relation with the valence separation as obtained, for example, from BVS needs to be further elaborated. Although the size of the polarization upon CO is not simple to estimate, some polarization is expected due to the broken inversion symmetry. Density functional theory (DFT)-based calculations with the polarization obtained by the Berry-phase approach [19] found large values for several proposed model systems [20,21].

Ferroelectricity implies not only a spontaneous polarization but also the fact that it can be switched by an applied electric field. This is possible only if the material is insulating enough—residual conductivity or "leakage" leads to artifacts in standard methods of polarization and dielectric measurements. By surveying the proposed example materials, it will become clear that residual conductivity is often a concern in CO materials, casting doubt on macroscopic measurement results.

3.2.3 Molecular Compounds

In compounds such as organic charge-transfer salts, the transition metal ions are replaced by whole molecules as the entities exhibiting CO. Given the larger separation between molecules, held together mainly by van der Waals bonding, the negative feedback mechanism proposed in [10] for transition-metal oxides can be expected to be less effective. Indeed, substantially different electron densities have been found, for example, by NMR, between molecules becoming inequivalent due to CO in such compounds [22,23]. In addition to nuclear magnetic resonance (NMR), vibrational spectroscopies are often applied [24,25]. As seen earlier, these are sensitive to the dynamical or Born effective charges that are directly relevant for yielding an electric polarization. In contrast oxides, diffraction methods are less often applied, mainly due to low crystal symmetries already above the CO transition and the absence of superstructure reflections (see the examples later in this chapter), but also due to the inapplicability of the BVS method to assign valences to whole molecules. A conceptual contrast to CO in oxides is that, for molecular compounds, the charge pattern is typically what is expected from electron repulsion [25]. With regard to the magnetism also necessary for multiferroicity, molecular compounds are at a disadvantage because the intermolecular magnetic interactions are weak, leading to low magnetic ordering temperatures [2]. I will further compare oxides and molecular compounds with ferroelectricity from CO after reviewing the examples that have been proposed.

3.3 Proposed Examples of Charge Order– Driven Multiferroicity

3.3.1 Charge Order in Oxides and Fluorides

In transition metal oxides, CO is ubiquitous [26]. When the compound is both charge ordered and ferroelectric, it can be that the CO is driving ferroelectricity (implying that CO and ferroelectric ordering temperatures coincide), or that something else is driving ferroelectricity at a typically lower temperature. In the latter case, CO still might be necessary for ferroelectricity. For example, following Figure 3.1, site-centered CO could occur at a high temperature, and the dimerization necessary to break inversion symmetry could be established at a lower temperature, for example, by magnetostriction when the spin-ordering along the chain is of the type ↑↑↓↓↑↑. This situation seems to be realized for manganites RMn_2O_5, one of the early examples of multiferroics with strong magnetoelectric coupling [27], and was also proposed for nickelates $RNiO_3$ (R is a rare earth). Given that the ferroelectric mechanism is exchange striction rather than charge ordering, I will not discuss these examples in detail (see [4] for an in-depth discussion). Another situation is exemplified by the bilayer manganite $Pr(Sr_{0.1}Ca_{0.9})_2Mn_2O_7$ [28], where Mn^{3+}/Mn^{4+} CO occurs on cooling through ~350 K, concomitant with an orbital ordering of the Mn^{3+}. Upon further cooling, there is a second phase transition close to 300 K in which both CO and orbital orders are modified—and the compound becomes polar as shown by second-harmonic generation. Although the CO is also modified at 300 K, it is the modification of the orbital order, the reorientation of orbital stripes and the associated oxygen displacements that lead to the emergence of a polarization. Thus, ferroelectricity is driven by orbital order rather than CO [29]. Orbital order–driven ferroelectricity was also found for GeV_4S_8 [30], and it can be obtained by strain-engineering in thin films [31,32]. Although certainly very interesting, I omit a detailed discussion of orbital order–driven ferroelectricity (see [21] for a theoretical perspective), and in the following sections focus on examples where ferroelectricity has been proposed to be driven by CO.

3.3.1.1 Mixed-Valence Manganites

Mixed-valence perovskite manganites are among the oldest proposed examples of ferroelectricity driven by CO. For $Pr_{1-x}Ca_xMnO_3$ with $x\sim0.4$, either a checkerboardlike site-centered [33] or a "Zener-polaron" bond-centered [34] CO has been advocated. Efremov et al. [3] argued that either of them could be the ground state depending on the doping level, and that there was an intermediate phase combining both features. This corresponds to the 3D analogue of the situation shown in Figure 3.1, and consequently ferroelectricity may be expected. Experimental verification of this scenario is complicated by a low resistivity (well below 100 Ωcm at room temperature [35]), which prevents polarization measurements. Similarly, earlier dielectric spectroscopy measurements, for example, [36], were shown to be dominated by sample-contact interface contributions obscuring the intrinsic response [35]. Indirect evidence for polar regions has been proposed based on electron holography [37], and later by studies of the electric field gradient at the Ca/Pr site [38]. An anomaly indicating a first-order transition had been found below the charge-ordering transition for $Pr_{1-x}Ca_xMnO_3$, not only for $x\sim0.4$, but in contrast to expectation [3] also for x up to 0.85. The electric field gradient had been correlated to the ferroelectric polarization, for example, in Rochelle salt [39], and the transition indicated by the anomaly had been interpreted as a ferroelectric transition. The perhaps most convincing evidence of

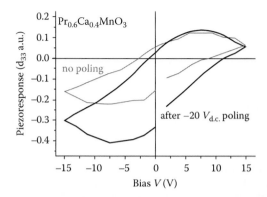

FIGURE 3.2 Local piezoresponse versus voltage loop in a region of $Pr_{0.6}Ca_{0.4}MnO_3$ at room temperature. (Reproduced from Figueiras, F. G. N. et al., *Physical Chemistry Chemical Physics*, 16, 4977–4981, 2014. With permission of the PCCP owner societies.)

at least nanoscale ferroelectric domains has been obtained recently by piezoresponse force microscopy [40]. Convincing hysteresis loops of the piezoresponse (Figure 3.2) were obtained, though only in small regions of the sample. In contrast to Reference [38], these indications of ferroelectricity were obtained above the charge-ordering temperature. This is not necessarily a contradiction, as the piezoresponse in small regions could be a precursor effect. Interestingly, a similar piezoresponse was observed for $(La,Ca)MnO_3$—this system had also been predicted to be ferroelectric, possibly with a mechanism that requires CO, but is ultimately magnetically driven [41]. Obtaining piezoresponse loops at lower temperature in a perhaps larger sample fraction would go some way to establishing ferroelectricitylike behavior, as would a definite structural study settling the "checkerboard versus Zener polaron" controversy. However, even in case more evidence can be accumulated, ferroelectricity would remain doubtful given the high conductivity of these systems. In any case, the high conductivity is a barrier to most potential applications.

3.3.1.2 LuFe₂O₄

Ferroelectricity from iron valence order in $LuFe_2O_4$ (and presumably other rare earth ferrites) had been proposed in 2005 [8], based on a combination of macroscopic and microscopic evidences. The former included dielectric spectroscopy featuring extremely large dielectric permittivity with a frequency-dependent drop as a function of temperature that was indicated as signature of a relaxorlike ferroelectric transition. The primary evidence came from pyroelectric current measurements obtained by cooling in a positive or negative 10 kV/cm electric field and then monitoring the current flowing upon warming in zero field. For a ferroelectric, integration then yields the thermo-remnant polarization. A large (~28 µC/cm²) polarization that disappears at about 350 K was deduced from these measurements. Previous x-ray diffraction experiments [42] had indicated the onset of a superstructure, with propagation near (1/3,1/3,3/2) at about 320 K, that is, quite close to the disappearance of the polarization. Resonant x-ray diffraction at the Fe K edge indicated that this superstructure involves iron valence order [8,43]. Assuming a bimodal Fe valence distribution, as indicated by absorption and Mössbauer spectroscopies (e.g., [44–46]), this superstructure propagation vector is consistent with the charge configurations in the Fe–O bilayers of $LuFe_2O_4$ shown in Figure 3.3 right. If we ignore the L-component, the configurations on the left are also possible. For a single bilayer, the configurations as shown in Figure 3.3 bottom do not minimize the Coulomb-repulsion between electrons (the extra-electrons of the Fe^{2+} are quite close to each other in the bilayers with Fe^{2+}-majority); thus, it is natural to assume that the CO in each bilayer is as shown in Figure 3.3 top. This was also assumed to be the charge configuration in [8]: one of the sheets in each bilayer is rich in Fe^{2+} and the other is rich in Fe^{3+}. In the simple approximation of a point-charge model, this immediately implies an electric dipole moment, that is, the bilayer is polar (indicated by semitransparent arrows in Figure 3.3). Therefore, the unusual CO, which can arise due to geometrical frustration [47,48] that also leads to very complex

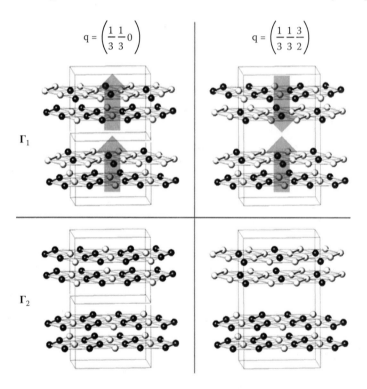

FIGURE 3.3 Possible "(1/3,1/3)-type" charge order configurations in rare earth ferrites, assuming a bi-modal valence distribution of Fe^{2+} (black) and Fe^{3+} (white). (Angst, M.: Ferroelectricity from iron valence ordering in rare earth ferrites? *Phys. Stat. Sol.-Rapid Res. Lett.*, 2013, 7, 383–400. 2013. Copyright Wiley-VCH Verlag GmbH & Co. KGaA. Reproduced with permission.)

magnetic order [49], is indeed polar (if this configuration is realized), strongly supporting the notion of CO-based ferroelectricity.

Subsequently, this example has long been considered a prototype and the "proof of principle" that the mechanism of CO-driven ferroelectricity really works. As a consequence, $LuFe_2O_4$ and related rare earth ferrites have attracted a tremendous amount of interest (see [48] for a recent review). Unfortunately, later work cast doubt on the attractive picture originally proposed.

On a microscopic level, it was noted first in 2008 [50] that the observed (1/3, 1/3, 3/2) propagation is not consistent with the configuration shown in Figure 3.3 top left, the only one with a net ferroelectric polarization. Instead, still assuming polar bilayers, the antipolar configuration shown in Figure 3.3 top right was proposed, though it was noted that the ferroelectric configuration might be obtainable by cooling in a sufficiently large electric field. However, if this was the case, electric-field cooling should lead to a drastic change in the diffraction pattern with intensity shifts to (1/3, 1/3, integer) positions when cooled in electric fields. This was found not to be the case in diffraction experiments performed by several groups [51–53]. Later, the charge-ordered crystal structure could be refined* from x-ray diffraction data on a small single crystal in which essentially only one of three CO domains existed [54]. A structural symmetry consistent with the CO shown in Figure 3.3 bottom right provided a good refinement, whereas the alternative (top right) did not (the other two configurations in the figure are inconsistent with the reflection pattern even ignoring intensities). Assessment of the valence state of iron in the four Fe sites confirmed this charge configuration, although the deviation of the valence of the Fe majority sites was found to be smaller than the one of the Fe minority sites. This is likely due to the intensities of the superstructure reflections with propagation $(\tau,\tau,3/2)$ [42,50] having been underestimated because they are farther from the commensurate positions. The addition of this propagation [in commensurate approximation (0, 0, 3/2)] is necessary to obtain the bimodal valence distribution indicated by spectroscopy. Recent work by resonant and powder x-ray diffraction [55] obtain similar results below the CO transition temperature, but does not take into account the reflections with $(\tau, \tau, 3/2)$ propagation at all. Regardless of the deviation of the valence from ideal 2+ and 3+, the charge configuration has the symmetry properties of Figure 3.3 bottom right: the bilayers can be alternatively charged, but they cannot possibly be polar.

Additional support for this charge configuration comes from combining the magnetic order [49] with spin-charge coupling. Spin-charge coupling can be deduced from Fe Mössbauer spectroscopy, particularly with applied magnetic fields [56], and from x-ray magnetic circular dichroism at the Fe L edges [46,53,57]. All these studies conclude that the relation between valence states and spin direction[†] is such that all Fe^{2+} spins point in the direction of the applied magnetic field, as do 1/3 of the Fe^{3+} spins, whereas the other 2/3 of the Fe^{3+} spins point opposite to the magnetic field direction (field applied

---

* In a commensurate approximation, that is, ignoring a small incommensurability of the super-structure, which is likely due to antiphase boundaries [48].

† Spins are very strongly Ising pointing perpendicular to the Fe/O layers.

perpendicular to the layers). The combination of this spin-charge coupling with the spin order [49] implies the nonpolar CO as shown in Figure 3.3 bottom right, in agreement with the structure refinement.

The macroscopic evidences of ferroelectricity have also been strongly questioned. Already in 2008 it was reported in Reference [58] that the localization of free carriers at interfaces upon electric-field cooling of a leaky dielectric can yield pyroelectric current results strongly resembling those measured on $LuFe_2O_4$. Despite the high attention this compound commanded, no group succeeded in observing a convincing ferroelectric hysteresis loop. Out of the few published P versus E loops [59,60], some showed hysteresis, but not the saturation necessary [61] to demonstrate that the hysteresis is not due to an artifact. The P versus E measurements were affected by being performed (i) at temperatures where the leakage (conductivity) leading to artifacts is relatively high, or (ii) up to only moderate electric field values (potentially much lower than the coercive field). However, a recent study by Lafuerza et al. [62] presents P versus E measurements at 10 K to 30 kV/cm, which certainly is high enough that an effect should be seen if there was ferroelectricity. In contrast, P is simply proportional to E, strongly suggesting the absence of ferroelectricity. They also present P versus E loops measured at higher temperatures, where the sample is more conductive. In the opening of hysteresis there is confirmation that previously observed hysteresis is an artifact. In the whole literature, there is only one P versus E loop in a related material, intercalated $Lu_2Fe_{2.14}Mn_{0.86}O_7$, which has a shape expected for an intrinsic hysteresis loop, however, with an associated polarization that is extremely small, about 0.01 $\mu C/cm^2$ [63]. Dielectric permittivity as the response function associated with the polarization is a slightly more indirect way of assessing ferroelectricity and ferroelectric or antiferroelectric transitions. In-depth analysis of dielectric spectroscopy on $LuFe_2O_4$ including variation of contact materials and equivalent circuit modeling [60,64,65] show that the initially proposed giant dielectric constants [8] and soon after observed giant (>20%) magneto-capacitance [66] arise from an interplay of sample conductivity and interface capacitances due to Shottky-type barriers at grain boundaries and sample-contact interfaces. The extracted intrinsic dielectric constant does not show any anomalies. Recently, it was proposed that the extrinsic effects are largely suppressed when using gold as the electrode material. With such a setup, an approximately 1% magneto-capacitance effect was deduced at 220 K across a meta-magnetic transition [67]. However, given a simultaneous change of the resistivity by about 18% and the correlation of resistivity and extrinsic capacitance contributions, the conclusion of an intrinsic magneto-dielectric effect remains tentative. The principle result of dielectric spectroscopy is essentially the same as for $Pr_{1-x}Ca_xMnO_3$: features are dominated by extrinsic interface effects mainly due to too high sample conductivity (although the resistivity of $LuFe_2O_4$ is somewhat higher, about 550 Ωcm at room temperature [68]).

3.3.1.3 Magnetite Fe_3O_4
Magnetite is not only the oldest magnetic material known to mankind, but, with the Verwey transition of around 120 K, also provides the classical example of a metal–insulator transition attributed to charge ordering (of Fe^{2+} and Fe^{3+} on

the B-site of its spinel structure) [69]. As such (and because of tremendous practical problems related, for instance, to microtwinning of the CO domains), it has drawn the attention of generations of condensed matter physicists and materials scientists. CO in magnetite has been highly controversial, not only concerning the CO pattern but also concerning its existence in terms of Fe^{2+} and Fe^{3+} (see for example the critical review [70]). It was only very recently that the extremely complex CO was resolved, in a large Cc cell with 16 Wyckoff sites for the B-site Fe, based on high-energy x-ray diffraction on a tiny monodomain crystal [71]. A BVS analysis of the 16 sites did not yield a bimodal distribution, which can be explained by replacing the initially expected strict $Fe^{2+/3+}$ order by a complex order of "*trimerons*" plus some nonparticipating Fe^{3+}. These trimerons are three Fe sites in a chain, to first approximation Fe^{3+}–Fe^{2+}–Fe^{3+}, but with the extra electron of the central site partially redistributed over the other two sites. The order is, even in first approximation, inconsistent with being (solely) driven by Coulomb repulsion. Rather, lattice effects have to be of primary importance.

The first indications of ferroelectricity and magnetoelectric coupling in magnetite (far) below the Verwey transition were reported almost 40 years ago [72,73], although, for example, the P versus E loops reported in [73] are not very convincing given the lack of saturation. These early reports did not generate a lot of interest until the revival of multiferroicity and magnetoelectric effects during the last decade. The experimental case for ferroelectricity in magnetite was strengthened in 2009 by measurements [74] on a magnetite thin film (150 nm) indicating a switchable (see Figure 3.4) polarization below 40 K of up to about 12 $\mu C/cm^2$. While the form of the electric hysteresis loops measured still do

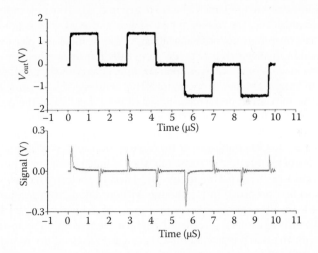

FIGURE 3.4 Positive-up-negative-down (PUND) measurements on a magnetite thin film at 15 K: Currents flowing in response to an input-voltage across the film (top) lead to a signal-voltage across a 50 Ω load resistance (bottom). Polarization switching is indicated by the larger signal at the start of the first pulse, see text. (Alexe, M. et al., *Adv. Mater.*, 21, 4452–4455, 2009. 2009. Copyright Wiley-VCH Verlag GmbH & Co. KGaA. Reproduced with permission.)

not show proper saturation, the "PUND" measurements with 1 kHz frequency (Figure 3.4) eliminate some of the main artifacts often present. Apart from the current associated with ferroelectric switching, Ohmic (current proportional to the voltage) and capacitive (current proportional to the time-derivative of the voltage) current contributions are present. The latter could be mistaken for polarization switching, but it should be the same at each rising flank of the voltage pulse. In contrast, switching occurs only when the first of two pulses in the same direction starts, that is, the signal surplus at the two switching events (at slightly over 0 and 5.5 μs) could be uniquely tied to the polarization (of the system as a whole, including any interfaces, however). It could be questioned how characteristic thin films are of the bulk material [75]. Bulk magnetite was studied in depth by Schrettle et al. [76]. Dielectric spectroscopy revealed several relaxation phenomena including interface contributions (similar as in $LuFe_2O_4$), but also an apparently intrinsic contribution behaving similar as ac-susceptibility in spin-glass systems. Such behavior is expected for relaxor ferroelectrics and ascribed to the freezing of polar nanodomains [77]. The indicated freezing temperature of roughly 40 K matches the temperature below which polarization switching was observed on thin films [74]. Indeed, even in relaxor ferroelectrics P vs. E loops with large hysteresis are typically obtained at temperatures well below the freezing temperature [77]. On bulk magnetite, convincing ferroelectric hysteresis loops with indications of saturation could be obtained [76] at 5.6 K with a frequency of 513 Hz, suggesting a switchable polarization of about 0.5 μC/cm². This is more than an order of magnitude smaller than the value for thin films reported in [74], which may be partly due to the strain on the film favoring some of the possible domains. At higher temperatures or lower frequencies, leakage current and interface effects were too large to be corrected and thus, the polarization could not be obtained [76]. The relaxor-type freezing at around 40 K is not an actual ferroelectric transition: no thermal or structural indications of a phase transition were found in this temperature range. Above the freezing temperature, there is no longer a macroscopic switchable polarization, but locally the polarization may survive up to much higher temperatures. This was tested by dynamic and static pyroelectric measurements, which indicated no anomalies of the pyroelectric coefficient at low temperatures [78]. Instead, a finite pyroelectric coefficient up to the Verwey transition was found, with a broad maximum around 70 K. This suggests that the CO established at the Verwey transition is what underlies the (relaxor) ferroelectricity eventually established at much lower temperatures.

Theoretically, the experimentally suggested ferroelectricity, which is due to CO, was first proposed in [4], though without taking into account the full structure from either experiment or first-principles calculations. DFT calculations with fully optimized Cc cell and establishment of the ferroelectric polarization with the Berry-phase method, taking a $P2/c$ structure as paraelectric reference, were presented in [74,79]. The DFT results were essentially the same as published earlier by Jeng et al. [80], who had not, however, considered the electric polarization emerging from their solution.

In ionic approximation, the CO on the tetrahedral network of Fe B sites as obtained by DFT is similar to the experimental one shown in Figure 3.5 [71]. On a majority of the tetrahedra (colored cyan), there are either three Fe^{2+} and one

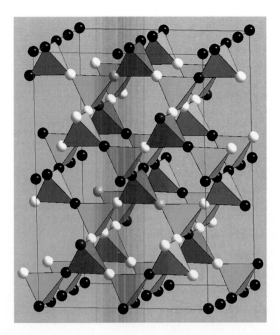

FIGURE 3.5 Charge order of the B-site iron in magnetite below the Verwey transition, in a *Cc* cell. The Fe valences in ionic approximation are indicated by dark (Fe²⁺) and bright (Fe³⁺) coloring. Fe sites that have opposite valence in the centrosymmetric *P2/c* model are hued in red. Tetrahedra with "3:1" configuration are cyan, those with "2:2" configuration orange. Atomic coordinates from [71].

Fe³⁺ ions ("3:1") or vice versa. On a minority of the tetrahedra (orange), there are two each of Fe²⁺ and Fe³⁺ ("2:2"), as for the original (centrosymmetric) CO model proposed by Verwey [81] (which is also the only model consistent with CO driven exclusively by Coulomb repulsion between electrons). In contrast, even for the centrosymmetric *P2/c* model, which is only very slightly less stable than the calculated *Cc* groundstate [80], all tetrahedra are "3:1." It is the mixing of 3:1 and 2:2 configurations that breaks inversion symmetry. The *Cc*-structure can be obtained by exchanging the valences of just a few (four pairs in the *Cc* cell) neighboring Fe atoms. In Figure 3.5, those sites, which would have different valence in the *P2/c* structure, are hued red. In a simple ionic point-charge approximation, these valence exchanges create four electric dipole moments, which are uncompensated and thus correspond to an electric polarization. Yamauchi et al. [79] compared the so obtained polarization with the polarization calculated in a point-charge model with atomic coordinates optimized in *Cc*, as well as with the polarization calculated using the Berry phase approach. They found quite similar values ($\mathbf{P}_{Berry} = (-4.41, 0, 4.12)$ µC/cm² along the three axes) and concluded that the polarization is mainly due to the charge shifts captured by the "dipole from valence exchange" model, rather than being due to ionic displacements.

However, while the DFT [80] and experimental [71] structures are very similar, there *are* some subtle differences as discussed in [82]. The CO is not superimposable, unless the optimized atomic coordinates were shifted by (0,1/4,0) with respect to the c-glide operator.* Refinements of the model obtained by DFT (without origin shift) give much worse residuals than the experimental structure. It would be very interesting to perform a DFT calculation including atomic position optimization using the experimental structure as a starting point, and check the relative stability of this structure with the previously calculated [80] structure. Another difference between calculated and experimental structures is that the latter have much less valence separation, as ascertained from BVS and Jahn-Teller distortions, than the former. The polarization for the experimental structure, calculated in a point-charge model considering the formal valences, is mainly along the c-direction and has a magnitude of 42 μC/cm^2, which is much larger than that found in DFT calculations or experiments. Interestingly, a repetition of the calculation with the Fe valence set to its average, 2.5+, yields a rather similar value of 36 μC/cm^2. Thus, for the experimental structure, it seems that the largest part of the polarization comes from atomic displacements, rather than from any charge transfer associated with the CO [71], opposite to the finding from DFT [79]. Once an optimized DFT calculation based on the experimental structure is available, it would be of interest to also calculate its polarization with the Berry phase, because in addition to ionic displacements, electron density displacements also contribute to the polarization.

With the polar crystal structure established [71], only switchability is necessary to establish ferroelectricity. Switchability is not present immediately below the Verwey transition, mainly due to persisting very high conductivity.† It is very likely that the low-temperature results [74,76] are due to the ferroelectric nature of the magnetite ground state, although a direct confirmation would involve the demonstration that the Cc-structure can be switched to its inversion twin with an electric field. Therefore, it does seem that magnetite is not only the first example of a magnetic material but also the first example of ferroelectricity originating from CO—at least in the sense that the polar structure is established due to the CO. As far as the polarization considerations based on the experimental structure [71] are concerned, the connection between CO and electric polarization is not as direct and intuitive as in the picture painted at the beginning of this chapter.

For multiferroics to be useful, the magnetoelectric coupling needs to be present and strong as well. Though the CO established below the Verwey transition does not change the basic magnetic structure (spin on all B sites parallel to the field, spin on all A sites opposite), evidence for magnetoelectric coupling in magnetite has accumulated for an even longer time than indications of ferroelectricity, see for example, [72]. According to a recent DFT study [84], the polarization associated with the magnetoelectric effect is much smaller (below 0.002 μC/cm^2) than the ground-state polarization associated with CO, and it is due to spin-dependent p–d hybridization [85].‡ See the recent review [21] for

---

* This is not a Cc-symmetry-allowed origin shift.
† For example, at 100 K the resistivity is still below 100 Ωcm [83].
‡ It should be noted that the DFT study [84] is based on the previous DFT structure [80], which as discussed earlier is somewhat different from the experimental one [71].

more discussion on the charge and orbital orderings, polarization, and magnetoelectricity in magnetite from a mainly theoretical perspective.

3.3.1.4 Fluorides

While there are quite a lot of multiferroic fluorides [86] and the ionic approximation should work a little bit better than in oxides due to the higher electronegativity of F, a CO-based mechanism for the ferroelectricity has so far been proposed only for one compound, $K_3Fe^{2+}{}_3Fe^{3+}{}_2F_{15}$ [87]. The starting point was the experimentally suggested [88] *Pba2* structure at room temperature, which according to BVS has half of the 20 Fe sites in the cell 2+, 30% 3+, and the remaining 20% (a single Wyckoff site) mixed valent. The DFT calculations [87] could stabilize insulating states only if the mixed-valent Fe is forced to become CO as well with three possible types of charge configurations, the most stable of which is polar with an associated polarization of about 0.5 $\mu C/cm^2$. The ferroelectric transition far above room temperature is problematic, and still the mixed valence sites are experimentally indicated at room temperature. Furthermore, the partial Fe substitution by Cr or Cu, which leave the remaining Fe in a single valence state, leaves ferroelectricity intact [86].

3.3.2 Charge Order in Organics

Ferroelectricity was first discovered at room temperature in Rochelle salt, which is an organic (metal-organic) material [89]. It appears therefore likely that multiferroicity and magnetoelectric effects may occur in organics as well. The particular mechanism of CO-driven ferroelectricity has been suggested to be active in several organic materials, typically charge transfer salts. In these compounds, the units exhibiting CO are not transition metal ions but rather whole molecules, which are held together mainly by van der Waals bonding, rather than covalent bonding.

3.3.2.1 (TMTTF)$_2$X

The suggestion of ferroelectricity from CO in an organic material was first made for the Fabre salt $(TMTTF)_xX$ [90], a few years before corresponding scenarios in oxides (outlined earlier) have been widely discussed. In Fabre/Bechgaard salts, TMTTF/TMTSF molecules donate electrons to a counterion X, for example, Br, PF_6, AsF_6, SbF_6, ClO_4, SCN, etc., resulting in a structure as shown in Figure 3.6. Upon lowering the temperature, in many of these salts, a phase transition had been observed, for example, in resistivity or thermo-power measurements, but not in x-ray diffraction or magnetic susceptibility [91]. ^{13}C nuclear magnetic resonance with the carbon sites at the center of the TMTTF cations 100% enriched has shown that this transition involves a charge disproportionation of the cations, with a higher and a lower electron density at the enriched C positions [22]. This disproportionation was later confirmed by other methods such as vibrational spectroscopy, with a deduced difference of about 0.1–0.3 electrons [25]. It is likely that this disproportionation is ordered, that is, it is CO. A possible CO configuration consistent with the aforementioned measurements is indicated in Figure 3.6.

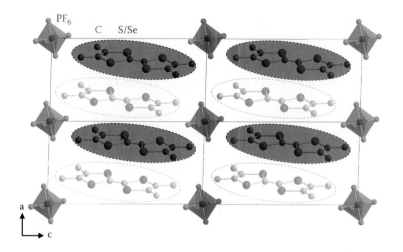

FIGURE 3.6 $P\bar{1}$ structure of the Bechgaard salt $[(TMTSF)_2]^+[PF_6]^-$ [92]. H atoms are omitted. The Fabre salt $(TMTTF)_2PF_6$ is isostructural, with Se replaced by S. The Anion PF_6 can be replaced by various other units. The hole is initially delocalized over the two cations of the unit cell. A possible selective localization on one of the cations, that is, CO, is indicated by bright and dark shading.

The fact that this CO may be ferroelectric is indicated by dielectric measurements (for $X = PF_6$, AsF_6, and SbF_6) showing a near-divergence of the real part of the dielectric permittivity at the CO transition temperature T_{CO} (regardless of frequency) well described by the Curie–Weiss law [25,90]. Although, as seen for the oxide examples discussed earlier, dielectric measurements have to be analyzed carefully to exclude interface effects, the peak of the dielectric permittivity observed at T_{CO} for $(TMTTF)_2X$ is rather strong evidence because interface effects would lead to high (apparent) dielectric constants at high temperatures rather masking a peak at T_{CO} (see, for example, the discussion on $LuFe_2O_4$ in the previous section). Unfortunately, dielectric spectroscopy is the only macroscopic evidence of ferroelectricity. Neither polarization loops nor pyroelectric currents have been measured.* This is no doubt due to the quite high conductivity [93], which is highly anisotropic, but of the same order of magnitude as for $Pr_{0.6}Ca_{0.4}MnO_3$. As in that case, the assignation "ferroelectric" may be considered doubtful given the high conductivity.

Microscopically, support for the ferroelectricity, or at least a polar state, would require the determination of the CO pattern. That the CO transition does involve structural changes is most clearly indicated by sensitive thermal expansion measurements. de Souza et al. [94] observed clear anomalies at T_{CO} for $(TMTTF)_2PF_6$, and from the crystallographic anisotropy of the effect suggested the CO pattern shown in Figure 3.6 by bright and dark shading

---

* Within a simple point-charge model assuming the CO shown in Figure 3.6 and the charge difference of 0.1–0.3 electrons [25], a substantial polarization of several μC/cm² is obtained.

of the TMTTF cations. This charge pattern is consistent with the theoretical expectation of a "$4k_F$ CDW" driven by Coulomb repulsion [95]. Given the offset of TMTTF along c, it is also clearly polar. Because the cell size is not increased by the CO, no superstructure reflections will appear. This makes the CO difficult to tackle with diffraction methods because only intensity changes of Bragg peaks can be expected, which may be further obscured by twinning. Nevertheless, intensity changes have been observed by powder neutron diffraction [96]. Although a full structural refinement was not possible, the intensity changes could be interpreted as a shift of the anions, which demonstrates the involvement of the lattice degrees of freedom in CO. In any case, the absence of any superstructure reflections, coupled with, for instance, the NMR results, leaves the CO shown in Figure 3.6 as the only possibility. It corresponds to the situation sketched in Figure 3.1, except that the "dimerization" is present already above T_{CO}. In case of the noncentrosymmetric anion SCN, the CO coincides with an anion ordering, and superstructure reflections do emerge [97]—in contrast to, for instance, PF_6, the dielectric permittivity has a step down rather than a sharp peak at T_{CO}, which can be interpreted as antipolar ordering [25]. I close this subsection by mentioning that some indications of coupling of the polar CO to magnetism exist, both from theoretical [98] and experimental [99] studies.

3.3.2.2 (BEDT-TTF)$_2$X

Another large family of charge transfer salts is built on the BEDT-TTF (or ET) molecule, which is similar to the TMTTF molecule shown in Figure 3.6, but has four full loops of C/S rather than two. There is a vast amount of possible anions X. The structure is invariably layered with ET-layers separated by anion layers, and there are furthermore a considerable variety of possible arrangements of ET molecules within their layers, two of which are shown in Figure 3.7 [101,102].

The case for CO-based ferroelectricity in these materials was first made for α-ET$_2$I$_3$ [103]. From the first synthesis of this salt, a prominent metal-insulator transition at $T_{CO} = 135$ K was noticed [104]; this was later shown to be associated with a charge disproportionation by NMR and vibrational spectroscopies (e.g., [23,24]). In this case, it was even possible to solve the low-temperature structure based on x-ray diffraction data [101]. The CO pattern deduced from

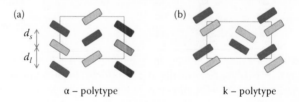

(a) d_s d_l (b)

α – polytype   k – polytype

FIGURE 3.7 (a) and (b) Sketch of two of the possible arrangements of the BEDT-TTF molecules in the cation layer (paper-plane). The cation layers are separated by anion layers (after [100]). Bright versus dark shading indicates proposed [101,102] CO.

the diffraction data (Figure 3.7a) agrees with theoretical predictions based on a model Hamiltonian including intersite Coulomb interaction [105].* Because of weak dimerization existing already above T_{CO} ($d_s < d_l$ in Figure 3.7a), an electric polarization is expected to emerge, again quite closely corresponding to the sketch in Figure 3.1. Whereas dielectric spectroscopy has been inconclusive, a ferroelectric polarization is indicated by a large second-harmonic generation signal emerging below T_{CO} [103,106]. Fast response in pump-probe experiments was furthermore interpreted as evidence of a polarization emerging mainly due to electron density shifts rather than ion shifts [103]. The optical nonlinearity associated with the polar state may find application, for example, in THz wave generation [107]. However, as in the examples outlined earlier, the switchability requirement of ferroelectricity is hampered by too high conductivity. In fact, with about 100 Ωcm resistivity at 100 K (well below T_{CO}) [104], the conductivity is even much higher than that of $(\text{TMTTF})_2\text{X}$.

More recently, CO-driven ferroelectricity has been proposed for κ-ET$_2X$ with $X = \text{Cu}_2(\text{CN})_3$ [108] and $X = \text{Cu}[\text{N}(\text{CN})_2]\text{Cl}$ [102]. For the former compound, which has a spin-liquid ground-state, dielectric relaxation versus temperature was found to exhibit a frequency-dependent maximum that was interpreted in terms of relaxor ferroelectricity [108]. For the latter compound [102], $\varepsilon'(\text{T})$ measurements exhibit a peak at a nearly frequency-independent temperature that coincides with the magnetic ordering temperature $T_N \sim 27$ K. In contrast to the earlier discussed organic compounds, these indications of ferroelectricity are supported by polarization measurements. In particular PUND measurements (Figure 3.8 left; compare with the corresponding measurements on magnetite, shown in Figure 3.4) demonstrate a substantial switchable polarization. Polarization versus electric field measurements (Figure 3.8

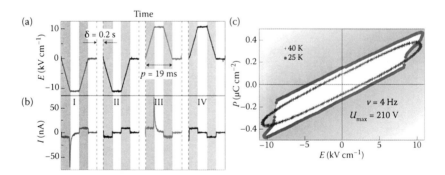

FIGURE 3.8 Polarization switching in κ-ET$_2$Cu[N(CN)$_2$]Cl. (Left) Positive-up-negative-down (PUND) measurements at 25 K. (Right) Polarization versus electric field loops at 25 and 40 K. The direction of E corresponds to horizontal in Figure 3.7b. (Reprinted by permission from Macmillan Publishers Ltd. Lunkenheimer, P. et al., Nat. Mater., 11:755–758, copyright 2012.)

---

* Reference [105] also predicts magnetic order in the ground state for likely parameter ranges. This is counterindicated by experiments, however.

right) above T_N show hysteresis but no indications of saturation, as is typical for leaky dielectrics [61]. Those below T_N *do* show indications of saturation, although their form is somewhat different from usual. Overall, these indications of ferroelectricity are quite strong, which is just about possible because the resistivity at 25 K, ~10 MΩcm, is sufficiently increased from the moderate room temperature value of below 1 kΩcm. A ferroelectric transition coinciding with a magnetic transition might be taken as an indication of a magnetic mechanism. However, Lunkenheimer et al. [102] found no magnetic-field dependence of the dielectric permittivity, from which they concluded that ferroelectricity is rather driven by something else, and it is the magnetic order that follows the ferroelectric order. Given the dimerized structure of κ-ET$_2$X (Figure 3.7b, the pairs adjacent parallel ET molecules build the dimers with one hole each), the selective localization of the holes on one of the two ET molecules of the dimer in a CO pattern is a natural assumption for a source of the polarization. Correspondingly, this was proposed in both [102] and [108], and it also received support from calculations with model Hamiltonians [109]. The problem with the proposed CO-origin of ferroelectricity in κ-ET$_2$X is the lack of evidence of any charge disproportionation (in contrast to the previously discussed organic examples). In particular, a comprehensive study by infrared spectroscopy for both X = Cu[N(CN)$_2$]Cl and X = Cu$_2$(CN)$_3$, and additionally for X = Cu[N(CN)$_2$]Br (a superconductor) found similar behavior in all these salts and an absence of any indication of charge disproportionation within the error bars of 0.5% (and static or fluctuating on time scales lower than 10^{11} Hz) [110,111]. In [111], an alternative scenario involving charges trapped at magnetic domain walls and discommensurations is proposed to explain the dielectric response. It will be necessary to elucidate the magnetic structure in more detail to test this scenario. However, the polarization response shown in Figure 3.8 is not simple to reconcile with this alternative, and other recent experiments (such as THz spectroscopy [112]) can be well accounted for in terms of the CO model. These experiments include dielectric spectroscopy and pyroelectric current measurements on an additional polytype, β'-ET$_2$ICl$_2$ [113].

3.3.2.3 Acid-Base Neutral-Ionic Materials

Another class of charge-transfer salts containing ferroelectricity at least related to charge transfer, if not exactly CO, is exemplified by TTF-CA. Here, the TTF-molecule, which corresponds to the TMTTF molecule (Figure 3.6) without the four arms at the ends, acts as an electron donor (D) and CA (chloranil, a ring-structure with formula C$_6$Cl$_4$O$_2$) as an electron acceptor (A). In a simplified picture, the high-temperature structure can be described as an alternating chain of neutral molecules DADADA..., but below a "neutral-ionic" transition at T_{NI} ~ 81 K the molecules become ionized D$^+$ and A$^-$ [114].*
The charge transfer involved is somewhat similar to CO, although there is a clear distinction: in contrast to CO, the charge transfer occurring here does

---

* In reality, the ionization is partial, and increases from about 0.3 to 0.6 on cooling through T_{NI}, see, for example, [116].

not make sites inequivalent (rather, the sites involved are inequivalent already before the charge transfer). Inversion symmetry is broken because the neutral-ionic transition is accompanied by a dimerization [115], c.f. Figure 3.1. This leads to the two possibilities of $(D^+A^-)(D^+A^-)(D^+A^-)...$ and $(A^-D^+)(A^-D^+)$ $(A^-D^+)...$ with opposite polarization.

As in all the examples discussed earlier, there are some problems with current leakage, requiring polarization measurements to be performed at low enough temperatures and high enough frequencies (c.f. the corresponding discussion on magnetite). On the other hand, rate-limiting domain wall dynamics impose a limit on the opposite, which overall leaves a relatively small temperature-frequency window in which polarization switching may be observed. Kobayashi et al. [116] have succeeded in measuring very convincing polarization hysteresis loops, which demonstrate a saturated polarization of about 6 μC/cm² (see Figure 3.9). Without electric field, this polarization decays toward 0, on a time-scale of the order of hours. Intrinsic ferroelectricity was furthermore proven by directly observing, with x-ray diffraction, the structural switching upon applying electric fields. Intriguingly, the direction of the "ionic polarization" resulting from ion-shifts is opposite to the much larger total polarization [116,117]. This implies that the by far dominating part of the polarization originates from shifts of electron density relative to the ions or "electronic polarization," in qualitative agreement with DFT calculations [118].* Although the ground state is nonmagnetic due to spin-singlet formation on the dimers, magnetic fields were shown to suppress the polarization by breaking the spin-singlets in the closely related TTF-BA [119].

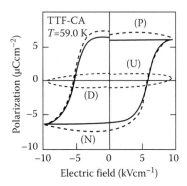

FIGURE 3.9 *P* versus *E* loop on TTF-CA at 59 K and 500 Hz, measured with the PUND technique (c.f. Figure 3.4) using triangular pulses. The dashed lines are the measured response for switching and nonswitching pulses; the full line is the subtraction of the latter from the former. (Reprinted with permission from Kobayashi, K. et al., *Phys. Rev. Lett.*, 108, 237601. Copyright 2012 by the American Physical Society.)

---

* However, the calculations either overestimate the magnitude of the ionic polarization or underestimate the magnitude of the electronic polarization.

3.4 Conclusions and Outlook

Basic physical properties of the compounds discussed are listed in Table 3.1. The charge-ordering temperature gives an upper limit for the CO-induced ferroelectricity. Generally, it is higher for transition metal oxides than for organics. One reason for this is likely that lattice contributions are strong in the former, which can be seen in the deviations of the CO pattern from what would be expected from a Coulomb-driven "Wigner crystallization" scenario.* For the organic charge-transfer salts, the CO is consistent with "Wigner crystallization" [127], although lattice effects are certainly also relevant.† The highest T_{CO} is realized for $K_3Fe_5F_{15}$, where, however, ferroelectricity driven by CO is cast in doubt by isostructural compounds with no mixed-valence ions being ferroelectric as well [86]. $LuFe_2O_4$, the only other compound charge ordered at room temperature, has a nonpolar CO [54] and is not ferroelectric [60,62]. The strongest case for ferroelectricity originating from CO in an oxide is presented for classical magnetite. Unfortunately, this is the oxide with the lowest T_{CO}, and switchability of the polarization has been demonstrated only at even much lower temperatures [74,76]. For many of the listed charge-transfer salts, the occurrence of a polar CO is clear, the most direct evidence being presented for α-BEDT-TTF$_2$I$_3$ by structural refinement [101] and second harmonic generation [103]. However, switchability has been demonstrated only for two compounds. The first of these, TTF-CA, exhibits charge transfer, but not CO, in a usual definition [116]. For the second, κ-BEDT-TTF$_2$Cu[N(CN)$_2$]Cl, CO has been proposed only to explain the observed polarization [102]: in contrast to the other listed salts, there is no spectroscopic indication of a charge disproportionation [110]. Therefore, magnetite remains the only demonstrated example of ferroelectricity (polar structure *and* switchable polarization) originating from CO.

Multiferroicity driven by CO requires not only the corresponding ferroelectricity but also the presence of magnetism. As discussed at the beginning of this chapter, spins are inherently present whenever CO occurs. However, for the charge-transfer salts, this spin degree of freedom is often removed by spin singlet formation (Spin-Peierls transition). Only two of the listed salts clearly order magnetically, and because of the weak intermolecular interactions they do so only at very low temperatures. Evidence for magnetoelectric coupling‡ is scarce. Indeed, the absence of such coupling for κ-BEDT-TTF$_2$Cu[N(CN)$_2$]Cl was taken as evidence of the absence of a magnetic mechanism of ferroelectricity in [102]. The oxides listed are all magnetically ordered. However, clear evidence of magnetoelectric coupling had been obtained only for magnetite [72]. Even for magnetite, the polarization associated with the magnetoelectric effect is much smaller than the CO-induced polarization, and arises from a different mechanism [84]. The reason for the decoupling of the principal polarization component from the magnetism is the robustness of the ferrimagnetic order,

---

* See in particular $LuFe_2O_4$ and Fe_3O_4.
† This can be seen for example from the strong X-dependence of T_{CO} for TMTTF$_2X$ in Table 3.1.
‡ Magnetoelectric coupling may also occur for Spin-Peierls systems in the sense that breaking the spin singlets may affect the ferroelectric polarization, demonstrated for TTF-BA [119].

TABLE 3.1   Overview of Basic Properties of the Discussed Oxide (/Fluoride) and Organic Compounds, Compared to Two Other Multiferroics and a Classical Ferroelectric

Compound	Resistivity (Ωcm)		T_{CO} (K) (or T_{FE})	T_N (K) (or T_{SP})
	295 K	Low Temperature		
$Pr_{0.6}Ca_{0.4}MnO_3$	10–60 [35]	10^7 @ 80 K [35]	232 [33]	170 [33]
$LuFe_2O_4$	550 [68]	10^8 @ 160 K [68]	320 [42]	240 [49]
Fe_3O_4	0.01 [83]	10^7 @ 40 K [83]	120 [69]	858 [83]
$K_3Fe_5F_{15}$	10^9 [120]	10^7 @ 450 K [120]	490 [88]	120 [88]
$TMTTF_2PF_6$	0.08$\|\|a$, 160$\perp$ ab [93]	$10^6 \perp ab$ @ 50 K	60 [90]	19 (SP) [25]
$TMTTF_2AsF_6$	0.2$\|\|a$, 200$\perp$ ab [93]	$10^6 \perp ab$ @ 50 K	100 [90]	13 (SP) [99]
$TMTTF_2SbF_6$	0.4$\|\|a$, 240$\perp$ ab [93]	$10^7 \perp ab$ @ 50 K	150 [90]	8 [25]
α-BEDT-TTF$_2$I$_3$	0.01$\|\|ab$ [104]	$10^5 \|\| ab$ @ 80 K	135 [104]	?
κ-BEDT-TTF$_2$ Cu[N(CN)$_2$]Cl	100 [102]	10^9 @ 5 K [102]	27 [102]	27 [102]
TTF-CA	10^6 [121]	10^8 @ 65 K [121]	81 [116]	81 (SP) [116]
$TbMnO_3$	2500 [122]	10^9 @ 100 K [122]	27 (FE) [123]	42 [123]
$BiFeO_3$	10^{11} [125]		1143 (FE) [124]	643 [124]
$BaTiO_3$	10^{12} [126]		400 (FE) [126]	-

Note: Listed are the resistivities at room and low temperatures, the charge ordering temperatures (or alternatively ferroelectric FE Curie temperatures) and the antiferromagnetic ordering temperatures (or alternatively Spin-Peierls SP transition temperatures).

as seen by the magnetic ordering temperature being (much) higher than T_{CO}, a feature that is unique among all compounds discussed.*

The values of the electric polarization are not included in Table 3.1, because for most compounds these values have not been ascertained experimentally. The only exceptions are TTF-CA with about 6 $\mu C/cm^2$ (Figure 3.9) [116] and magnetite with reported values of 12 $\mu C/cm^2$ for thin films [74] and ~0.5 $\mu C/cm^2$ for bulk samples [76]. The difficulty in such measurements arises mainly from a too high residual conductivity. As can be seen in Table 3.1, the room-temperature resistivity of most examples is rather low and even at low temperatures it does not reach the ideal value of the classical ferroelectric $BaTiO_3$ or even the slightly lower value of multiferroic $BiFeO_3$. This leads to conductivity-related artifacts, hampering polarization and dielectric measurements, and it is of course also an obstacle for prospective applications, except the high-frequency ones.

The relatively high value of residual conductivity is not a coincidence. By definition, a charge-ordering transition implies that the carriers involved are mobile at higher temperatures. Cooling through T_{CO}, the resistivity increases of course, but it does so often only marginally. The situation for $LuFe_2O_4$, where T_{CO} is just barely discernible from resistivity versus temperature curves [62] is rather typical. Even for magnetite, the resistivity jumps by only three orders of magnitude at T_{CO} [83].† For the eventual observation of ferroelectricity from CO at low frequencies, the conductivity issue is the main obstacle to be overcome. For this, the focus should be on materials with high T_{CO}, considerably above room temperature, and large charge gap opening. Materials like the charge transfer salts seem to be less suited than oxides or fluorides and this is a conclusion that is reinforced when also considering the magnetic ordering as part of the multiferroicity. Of the materials listed, $K_3Fe_5F_{15}$ and related compounds probably deserve more attention, despite the questions about the mechanism. However, in terms of eventual applications, new examples will have to be found.

Acknowledgment

I thank T. Mueller and H. Williamson for proofreading the chapter.

References

1. Wigner, E. 1934. On the interaction of electrons in metals. *Physical Review* 46:1002–1011.
2. Gatteschi, D. 1994. Molecular magnetism—A basis for new materials. *Advanced Materials* 6:635–645.

---

* Consequently, it rather is the CO that can be influenced by the magnetic ordering. For example, the possible CO domains are selected by the magnetization direction, which is used to measure samples with fewer domains by cooling through the Verwey temperature in a magnetic field, for example, in [71]. There are, however, no reports, of switching the crystal structure to other domains upon changing the magnetic field direction once CO occurred.

† For comparison, the resistivity changes by more than five orders of magnitude at the Mott-transition of V_2O_3 [128].

3. Efremov, D. V., J. Van den Brink, and D. I. Khomskii. 2004. Bond-versus site-centred ordering and possible ferroelectricity in manganites. *Nature Materials* 3:853–856.

4. van den Brink, J., and D. I. Khomskii. 2008. Multiferroicity due to charge ordering. *Journal of Physics–Condensed Matter* 20:434217.

5. Peierls, R. 1930. Regarding the theory of electric and thermal conductibility of metals. *Annalen Der Physik* 4:121–148.

6. Moessner, R., and A. R. Ramirez. 2006. Geometrical frustration. *Physics Today* 59:24–29.

7. Angst, M., R. P. Hermann, W. Schweika, J. W. Kim, P. Khalifah, H. J. Xiang, M. H. Whangbo, D. H. Kim, B. C. Sales, and D. Mandrus. 2007. Incommensurate charge order phase in Fe_2OBO_3 due to geometrical frustration. *Physical Review Letters* 99:256402.

8. Ikeda, N., H. Ohsumi, K. Ohwada, K. Ishii, T. Inami, K. Kakurai, Y. Murakami, K. Yoshii, S. Mori, Y. Horibe, and H. Kito. 2005. Ferroelectricity from iron valence ordering in the charge-frustrated system $LuFe_2O_4$. *Nature* 436:1136–1138.

9. Karen, P., P. McArdle, and J. Takats. 2014. Toward a comprehensive definition of oxidation state (IUPAC Technical Report). *Pure and Applied Chemistry* 86:1017–1081.

10. Raebiger, H., S. Lany, and A. Zunger. 2008. Charge self-regulation upon changing the oxidation state of transition metals in insulators. *Nature* 453:763–766.

11. Quan, Y., V. Pardo, and W. E. Pickett. 2012. Formal valence, 3d-electron occupation, and charge-order transitions. *Physical Review Letters* 109:216401.

12. Sit, P. H. L., R. Car, M. H. Cohen, and A. Selloni. 2011. Simple, unambiguous theoretical approach to oxidation state determination via first-principles calculations. *Inorganic Chemistry* 50:10259–10267.

13. Jiang, L., S. V. Levchenko, and A. M. Rappe. 2012. Rigorous definition of oxidation states of ions in solids. *Physical Review Letters* 108:166403.

14. Pickett, W. E., Y. D. Quan, and V. Pardo. 2014. Charge states of ions, and mechanisms of charge ordering transitions. *Journal of Physics-Condensed Matter* 26:274203.

15. Brown, I. D., and D. Altermatt. 1985. Bond-valence parameters obtained from a systematic analysis of the inorganic crystal-structure database. *Acta Crystallographica Section B-Structural Science* 41:244–247.

16. Brese, N. E., and M. Okeeffe. 1991. Bond-valence parameters for solids. *Acta Crystallographica Section B-Structural Science* 47:192–197.

17. Angst, M., P. Khalifah, R. P. Hermann, H. J. Xiang, M. H. Whangbo, V. Varadarajan, J. W. Brill, B. C. Sales, and D. Mandrus. 2007. Charge order superstructure with integer iron valence in Fe_2OBO_3. *Physical Review Letters* 99:086403.

18. Ghosez, P., J. P. Michenaud, and X. Gonze. 1998. Dynamical atomic charges: The case of ABO_3 compounds. *Physical Review B* 58:6224–6240; Spaldin, N. A. 2012. A beginner's guide to the modern theory of polarization. *Journal of Solid State Chemistry* 195:2–10.

19. Resta, R. 1994. Macroscopic polarization in crystalline dielectrics—the geometric phase approach. *Reviews of Modern Physics* 66:899–915.

20. Xiang, H. J., and M. H. Whangbo. 2007. Charge order and the origin of giant magnetocapacitance in $LuFe_2O_4$. *Physical Review Letters* 98:246403.

21. Yamauchi, K., and P. Barone. 2014. Electronic ferroelectricity induced by charge and orbital orderings. *Journal of Physics–Condensed Matter* 26:103201.

22. Chow, D. S., E. Zamborszky, B. Alavi, D. J. Tantillo, A. Baur, C. A. Merlic, and S. E. Brown. 2000. Charge ordering in the TMTTF family of molecular conductors. *Physical Review Letters* 85:1698–1701.

23. Hirose, S., and A. Kawamoto. 2010. Local spin susceptibility in the zero-gap-semiconductor state of alpha-$(BEDT-TTF)_2I_3$ probed by ^{13}C NMR under pressure. *Physical Review B* 82:115114.

24. Wojciechowski, R., K. Yamamoto, K. Yakushi, M. Inokuchi, and A. Kawamoto. 2003. High-pressure Raman study of the charge ordering in alpha-$(BEDT-TTF)_2I_3$. *Physical Review B* 67:224105.

25. Monceau, P. 2012. Electronic crystals: An experimental overview. *Advances in Physics* 61:325–581.

26. Attfield, J. P. 2006. Charge ordering in transition metal oxides. *Solid State Sciences* 8:861–867.

27. Hur, N., S. Park, P. A. Sharma, J. S. Ahn, S. Guha, and S. W. Cheong. 2004. Electric polarization reversal and memory in a multiferroic material induced by magnetic fields. *Nature* 429:392–395.

28. Tokunaga, Y., T. Lottermoser, Y. Lee, R. Kumai, M. Uchida, T. Arima, and Y. Tokura. 2006. Rotation of orbital stripes and the consequent charge-polarized state in bilayer manganites. *Nature Materials* 5:937–941.

29. Keimer, B. 2006. Transition metal oxides—Ferroelectricity driven by orbital order. *Nature Materials* 5:933–934.

30. Singh, K., C. Simon, E. Cannuccia, M. B. Lepetit, B. Corraze, E. Janod, and L. Cario. 2014. Orbital-ordering-driven multiferroicity and magnetoelectric coupling in GeV_4S_8. *Physical Review Letters* 113:137602.

31. Ogawa, N., Y. Ogimoto, Y. Ida, Y. Nomura, R. Arita, and K. Miyano. 2012. Polar antiferromagnets produced with orbital order. *Physical Review Letters* 108:157603.

32. Gupta, K., P. Mahadevan, P. Mavropoulos, and M. Lezaic. 2013. Orbital-ordering-induced ferroelectricity in $SrCrO_3$. *Physical Review Letters* 111:077601.

33. Grenier, S., J. P. Hill, D. Gibbs, K. J. Thomas, M. Von Zimmermann, C. S. Nelson, V. Kiryukhin, Y. Tokura, Y. Tomioka, D. Casa, T. Gog, and C. Venkataraman. 2004. Resonant x-ray diffraction of the magnetoresistant perovskite $Pr_{0.6}Ca_{0.4}MnO_3$. *Physical Review B* 69:134419.

34. Daoud-Aladine, A., J. Rodriguez-Carvajal, L. Pinsard-Gaudart, M. T. Fernandez-Diaz, and A. Revcolevschi. 2002. Zener polaron ordering in half-doped manganites. *Physical Review Letters* 89:097205; Wu, L., R. F. Klie, Y. Zhu, and C. Jooss. 2007. Experimental confirmation of Zener-polaron-type charge and orbital ordering in $Pr_{1-x}Ca_xMnO_3$. *Physical Review B* 76:174210.

35. Biskup, N., A. de Andres, J. L. Martinez, and C. Perca. 2005. Origin of the colossal dielectric response of $Pr_{0.6}Ca_{0.4}MnO_3$. *Physical Review B* 72:024115.

36. Mercone, S., A. Wahl, A. Pautrat, M. Pollet, and C. Simon. 2004. Anomaly in the dielectric response at the charge-orbital-ordering transition of $Pr_{0.67}Ca_{0.33}MnO_3$. *Physical Review B* 69:174433.

37. Jooss, C., L. Wu, T. Beetz, R. F. Klie, M. Beleggia, M. A. Schofield, S. Schramm, J. Hoffmann, and Y. Zhu. 2007. Polaron melting and ordering as key mechanisms for colossal resistance effects in manganites. *Proceedings of the National Academy of Sciences of the United States of America* 104:13597–13602.

38. Lopes, A. M. L., J. P. Araujo, V. S. Amaral, J. G. Correia, Y. Tomioka, and Y. Tokura. 2008. New phase transition in the $Pr_{1-x}Ca_xMnO_3$ system: Evidence

for electrical polarization in charge ordered manganites. *Physical Review Letters* 100:155702.

39. Blinc, R., I. Zupancic, and J. Petkovse 1964. Na^{23} Magnetic-resonance study of ferroelectric transition in rochelle salt. *Physical Review A—General Physics* 136:1684–1692.

40. Figueiras, F. G. N., I. K. Bdikin, V. B. S. Amaral, and A. L. Kholkin. 2014. Local bias induced ferroelectricity in manganites with competing charge and orbital order states. *Physical Chemistry Chemical Physics* 16:4977–4981.

41. Giovannetti, G., S. Kumar, J. van den Brink, and S. Picozzi. 2009. Magnetically induced electronic ferroelectricity in half-doped manganites. *Physical Review Letters* 103:037601.

42. Yamada, Y., K. Kitsuda, S. Nohdo, and N. Ikeda. 2000. Charge and spin ordering process in the mixed-valence system $LuFe_2O_4$: Charge ordering. *Physical Review B* 62:12167–12174.

43. Mulders, A. M., S. M. Lawrence, U. Staub, M. Garcia-Fernandez, V. Scagnoli, C. Mazzoli, E. Pomjakushina, K. Conder, and Y. Wang. 2009. Direct observation of charge order and an orbital glass state in multiferroic $LuFe_2O_4$. *Physical Review Letters* 103:077602.

44. Tanaka, M., K. Siratori, and N. Kimizuka. 1984. Mossbauer study of RFe_2O_4. *Journal of the Physical Society of Japan* 53:760–772.

45. Xu, X. S., M. Angst, T. V. Brinzari, R. P. Hermann, J. L. Musfeldt, A. D. Christianson, D. Mandrus, B. C. Sales, S. McGill, J. W. Kim, and Z. Islam. 2008. Charge order, dynamics, and magnetostructural transition in multiferroic $LuFe_2O_4$. *Physical Review Letters* 101:227602.

46. Ko, K. T., H. J. Noh, J. Y. Kim, B. G. Park, J. H. Park, A. Tanaka, S. B. Kim, C. L. Zhang, and S. W. Cheong. 2009. Electronic origin of giant magnetic anisotropy in multiferroic $LuFe_2O_4$. *Physical Review Letters* 103:207202.

47. Harris, A. B., and T. Yildirim. 2010. Charge and spin ordering in the mixed-valence compound $LuFe_2O_4$. *Physical Review B* 81:134417.

48. Angst, M. 2013. Ferroelectricity from iron valence ordering in rare earth ferrites? *Physica Status Solidi-Rapid Research Letters* 7:383–400.

49. de Groot, J., K. Marty, M. D. Lumsden, A. D. Christianson, S. E. Nagler, S. Adiga, W. J. H. Borghols, K. Schmalzl, Z. Yamani, S. R. Bland, R. de Souza, U. Staub, W. Schweika, Y. Su, and M. Angst. 2012. Competing ferri- and antiferromagnetic phases in geometrically frustrated $LuFe_2O_4$. *Physical Review Letters* 108:037206.

50. Angst, M., R. P. Hermann, A. D. Christianson, M. D. Lumsden, C. Lee, M. H. Whangbo, J. W. Kim, P. J. Ryan, S. E. Nagler, W. Tian, R. Jin, B. C. Sales, and D. Mandrus. 2008. Charge order in $LuFe_2O_4$: Antiferroelectric ground state and coupling to magnetism. *Physical Review Letters* 101:227601.

51. Wen, J. S., G. Y. Xu, G. D. Gu, and S. M. Shapiro. 2010. Robust charge and magnetic orders under electric field and current in multiferroic $LuFe_2O_4$. *Physical Review B* 81:144121.

52. de Groot, J. 2012. Charge, spin and orbital order in the candidate multiferroic material $LuFe_2O_4$. PhD diss., RWTH Aachen University, Germany.

53. Bartkowiak, M., A. M. Mulders, V. Scagnoli, U. Staub, E. Pomjakushina, and K. Conder. 2012. Evolution of charge order through the magnetic phase transition of $LuFe_2O_4$. *Physical Review B* 86:035121.

54. de Groot, J., T. Mueller, R. A. Rosenberg, D. J. Keavney, Z. Islam, J. W. Kim, and M. Angst. 2012. Charge order in $LuFe_2O_4$: An unlikely route to ferroelectricity. *Physical Review Letters* 108:187601.

55. Lafuerza, S., G. Subias, J. Blasco, J. Garcia, G. Nisbet, K. Conder, and E. Pomjakushina. 2014. Determination of the charge-ordered phases in $LuFe_2O_4$. *EPL - Europhysics Letters* 107:47002; Lafuerza, S., J. Garcia, G. Subias, J. Blasco, and V. Cuartero. 2014. Strong local lattice instability in hexagonal ferrites RFe_2O_4 (R = Lu, Y, Yb) revealed by x-ray absorption spectroscopy. *Physical Review B* 89:045129; Blasco, J., S. Lafuerza, J. Garcia, and G. Subias. 2014. Structural properties in RFe_2O_4 compounds (R = Tm, Yb, and Lu). *Physical Review B* 90:094119.

56. Tanaka, M., H. Iwasaki, K. Siratori, and I. Shindo. 1989. Mossbauer Study on the Magnetic-Structure of $YbFe_2O_4$—a two-dimensional antiferromagnet on a triangular lattice. *Journal of the Physical Society of Japan* 58:1433–1440; Nakamura, S., H. Kito, and M. Tanaka. 1998. An approach to specify the spin configuration in the RFe_2O_4 (R = Y, Ho, Er, Tm, Yb, and Lu) family: Fe-57 Mossbauer study on a single crystal $LuFe_2O_4$. *Journal of Alloys and Compounds* 275:574–577.

57. Kuepper, K., M. Raekers, C. Taubitz, M. Prinz, C. Derks, M. Neumann, A. V. Postnikov, F. M. F. de Groot, C. Piamonteze, D. Prabhakaran, and S. J. Blundell. 2009. Charge order, enhanced orbital moment, and absence of magnetic frustration in layered multiferroic $LuFe_2O_4$. *Physical Review B* 80:220409.

58. Maglione, M., and M. A. Subramanian. 2008. Dielectric and polarization experiments in high loss dielectrics: A word of caution. *Applied Physics Letters* 93:032902; Park, J. Y., J. H. Park, Y. K. Jeong, and H. M. Jang. 2007. Dynamic magnetoelectric coupling in "electronic ferroelectric" $LuFe_2O_4$. *Applied Physics Letters* 91:152903.

59. Viana, D. S. F., R. A. M. Gotardo, L. F. Cotica, I. A. Santos, M. Olzon-Dionysio, S. D. Souza, D. Garcia, J. A. Eiras, and A. A. Coelho. 2011. Ferroic investigations in $LuFe_2O_4$ multiferroic ceramics. *Journal of Applied Physics* 110:034108.

60. Ruff, A., S. Krohns, F. Schrettle, V. Tsurkan, P. Lunkenheimer, and A. Loidl. 2012. Absence of polar order in $LuFe_2O_4$. *European Physical Journal B* 85:290.

61. Scott, J. F. 2008. Ferroelectrics go bananas. *Journal of Physics-Condensed Matter* 20:021001.

62. Lafuerza, S., J. Garcia, G. Subias, J. Blasco, K. Conder, and E. Pomjakushina. 2013. Intrinsic electrical properties of $LuFe_2O_4$. *Physical Review B* 88:085130.

63. Qin, Y. B., H. X. Yang, Y. Zhang, H. F. Tian, C. Ma, L. J. Zeng, and J. Q. Li. 2009. Suppression of the current leakage in charge ordered $Lu_2Fe_2Fe_{1-x}Mn_xO_7$ (0 < x < 0.86). *Applied Physics Letters* 95:072901.

64. Ren, P., Z. Yang, W. G. Zhu, C. H. A. Huan, and L. Wang. 2011. Origin of the colossal dielectric permittivity and magnetocapacitance in $LuFe_2O_4$. *Journal of Applied Physics* 109:074109.

65. Niermann, D., F. Waschkowski, J. de Groot, M. Angst, and J. Hemberger. 2012. Dielectric properties of charge-ordered $LuFe_2O_4$ revisited: The apparent influence of contacts. *Physical Review Letters* 109:016405.

66. Subramanian, M. A., T. He, J. Z. Chen, N. S. Rogado, T. G. Calvarese, and A. W. Sleight. 2006. Giant room-temperature magnetodielectric response in the electronic ferroelectric $LuFe_2O_4$. *Advanced Materials* 18:1737–1739.

67. Kambe, T., Y. Fukada, J. Kano, T. Nagata, H. Okazaki, T. Yokoya, S. Wakimoto, K. Kakurai, and N. Ikeda. 2013. Magnetoelectric effect driven by magnetic domain modification in $LuFe_2O_4$. *Physical Review Letters* 110:117602.

68. Fisher, B., J. Genossar, L. Patlagan, and G. M. Reisner. 2011. Electronic transport and I-V characteristics of polycrystalline $LuFe_2O_4$. *Journal of Applied Physics* 109:084111.

69. Verwey, E. J. W. 1939. Electronic conduction of magnetite (Fe_3O_4) and its transition point at low temperatures. *Nature* 144:327–328; Verwey, E. J. W., and P. W. Haayman. 1941. Electronic conductivity and transition point of magnetite ("Fe_3O_4"). *Physica* 8:979–987.

70. Garcia, J., and G. Subias. 2004. The Verwey transition—a new perspective. *Journal of Physics-Condensed Matter* 16:R145–R178.

71. Senn, M. S., J. P. Wright, and J. P. Attfield. 2012. Charge order and three-site distortions in the Verwey structure of magnetite. *Nature* 481:173–176.

72. Rado, G. T., and J. M. Ferrari. 1975. Electric-field dependence of magnetic-anisotropy energy in magnetite (Fe_3O_4). *Physical Review B* 12:5166–5174.

73. Kato, K., and S. Iida. 1982. Observation of ferroelectric hysteresis loop of Fe_3O_4 at 4.2-K. *Journal of the Physical Society of Japan* 51:1335–1336.

74. Alexe, M., M. Ziese, D. Hesse, P. Esquinazi, K. Yamauchi, T. Fukushima, S. Picozzi, and U. Gosele. 2009. Ferroelectric switching in multiferroic magnetite (Fe_3O_4) thin films. *Advanced Materials* 21:4452–4455.

75. Ziese, M., P. D. Esquinazi, D. Pantel, M. Alexe, N. M. Nemes, and M. Garcia-Hernandez. 2012. Magnetite (Fe_3O_4): A new variant of relaxor multiferroic? *Journal of Physics-Condensed Matter* 24:086007.

76. Schrettle, F., S. Krohns, P. Lunkenheimer, V. A. M. Brabers, and A. Loidl. 2011. Relaxor ferroelectricity and the freezing of short-range polar order in magnetite. *Physical Review B* 83:195109.

77. Cross, L. E. 1987. Relaxor Ferroelectrics. *Ferroelectrics* 76:241–267.

78. Takahashi, R., H. Misumi, and M. Lippmaa. 2012. Pyroelectric detection of spontaneous polarization in magnetite thin films. *Physical Review B* 86:144105.

79. Yamauchi, K., T. Fukushima, and S. Picozzi. 2009. Ferroelectricity in multiferroic magnetite Fe_3O_4 driven by noncentrosymmetric Fe^{2+}/Fe^{3+} charge-ordering: First-principles study. *Physical Review B* 79:212404.

80. Jeng, H. T., G. Y. Guo, and D. J. Huang. 2006. Charge-orbital ordering in low-temperature structures of magnetite: GGA+U investigations. *Physical Review B* 74:195115.

81. Verwey, E. J., P. W. Haayman, and F. C. Romeijn. 1947. Physical properties and cation arrangement of oxides with spinel structures II. Electronic conductivity. *Journal of Chemical Physics* 15:181–187.

82. Senn, M. S. 2012. Charge, orbital and magnetic ordering in transition metal oxides. *PhD diss.*, University of Edinburgh, UK.

83. Miles, P. A., W. B. Westphal, and A. Vonhippel. 1957. Dielectric spectroscopy of ferromagnetic semiconductors. *Reviews of Modern Physics* 29:279–307.

84. Yamauchi, K., and S. Picozzi. 2012. Orbital degrees of freedom as origin of magnetoelectric coupling in magnetite. *Physical Review B* 85:085131.

85. Arima, T. H. 2007. Ferroelectricity induced by proper-screw type magnetic order. *Journal of the Physical Society of Japan* 76:073702.

86. Scott, J. F., and R. Blinc. 2011. Multiferroic magnetoelectric fluorides: Why are there so many magnetic ferroelectrics? *Journal of Physics-Condensed Matter* 23:113202.

87. Yamauchi, K., and S. Picozzi. 2010. Interplay between charge order, ferroelectricity, and ferroelasticity: Tungsten bronze structures as a playground for multiferroicity. *Physical Review Letters* 105:107202.

88. Mezzadri, F., S. Fabbrici, E. Montanari, L. Righi, G. Calestani, E. Gilioli, F. Bolzoni, and A. Migliori. 2008. Structural properties and multiferroic phase diagram of $K_{0.6}Fe_{0.6}(II)Fe_{0.4}(III)F_3$ fluoride with TTB structure. *Physical Review B* 78:064111.

89. Valasek, J. 1921. Piezo-electric and allied phenomena in Rochelle salt. *Physical Review* 17:475–481.

90. Monceau, P., F. Y. Nad, and S. Brazovskii. 2001. Ferroelectric Mott-Hubbard phase of organic $(TMTTF)_2X$ conductors. *Physical Review Letters* 86:4080–4083.

91. Coulon, C., and S. S. P. Parkin. 1985. Structureless transition and strong localization effects in bis-tetramethyltetrahthiafulvalenium salts [(TMTTF)$_2$X]. *Physical Review B* 31:3583–3587.

92. Thorup, N., G. Rindorf, H. Soling, and K. Bechgaard. 1981. The structure of di(2,3,6,7-tetramethyl-1,4,5,8-tetraselenafulvalenium) hexafluorophosphate, (TMTSF)$_2$PF$_6$, the 1st superconducting organic-solid. *Acta Crystallographica Section B-Structural Science* 37:1236–1240.

93. Kohler, B., E. Rose, M. Dumm, G. Untereiner, and M. Dressel. 2011. Comprehensive transport study of anisotropy and ordering phenomena in quasi-one-dimensional (TMTTF)$_2$X salts (X = PF$_6$,AsF$_6$,SbF$_6$,BF$_4$,ClO$_4$, ReO$_4$). *Physical Review B* 84:035124.

94. de Souza, M., P. Foury-Leylekian, A. Moradpour, J. P. Pouget, and M. Lang. 2008. Evidence for lattice effects at the charge-ordering transition in (TMTTF)$_2$X. *Physical Review Letters* 101:216403.

95. Yoshioka, H., Y. Otsuka, and H. Seo. 2012. Theoretical studies on phase transitions in quasi-one-dimensional molecular conductors. *Crystals* 2:996–1016.

96. Foury-Leylekian, P., S. Petit, G. Andre, A. Moradpour, and J. P. Pouget. 2010. Neutron scattering evidence for a lattice displacement at the charge ordering transition of (TMTTF)$_2$PF$_6$. *Physica B-Condensed Matter* 405:S95–S97.

97. Coulon, C., A. Maaroufi, J. Amiell, E. Dupart, S. Flandrois, P. Delhaes, R. Moret, J. P. Pouget, and J. P. Morand. 1982. Anti-ferromagnetic and structural instabilities in tetramethyltetrathiafulvalene thiocyanate [(TMTTF)$_2$SCN]. *Physical Review B* 26:6322–6325.

98. Giovannetti, G., S. Kumar, J. P. Pouget, and M. Capone. 2012. Unraveling the polar state in TMTTF$_2$-PF$_6$ organic crystals. *Physical Review B* 85:205146.

99. Yasin, S., B. Salameh, E. Rose, M. Dumm, H. A. K. von Nidda, A. Loidl, M. Ozerov, G. Untereiner, L. Montgomery, and M. Dressel. 2012. Broken magnetic symmetry due to charge-order ferroelectricity discovered in (TMTTF)$_2$X salts by multifrequency ESR. *Physical Review B* 85:144428.

100. Seo, H., C. Hotta, and H. Fukuyama. 2004. Toward systematic understanding of diversity of electronic properties in low-dimensional molecular solids. *Chemical Reviews* 104:5005–5036.

101. Kakiuchi, T., Y. Wakabayashi, H. Sawa, T. Takahashi, and T. Nakamura. 2007. Charge ordering in alpha-(BEDT-TTF)$_2$I$_3$ by synchrotron x-ray diffraction. *Journal of the Physical Society of Japan* 76:113702.

102. Lunkenheimer, P., J. Muller, S. Krohns, F. Schrettle, A. Loidl, B. Hartmann, R. Rommel, M. de Souza, C. Hotta, J. A. Schlueter, and M. Lang. 2012. Multiferroicity in an organic charge-transfer salt that is suggestive of electric-dipole-driven magnetism. *Nature Materials* 11:755–758.

103. Yamamoto, K., S. Iwai, S. Boyko, A. Kashiwazaki, F. Hiramatsu, C. Okabe, N. Nishi, and K. Yakushi. 2008. Strong optical nonlinearity and its ultrafast response associated with electron ferroelectricity in an organic conductor. *Journal of the Physical Society of Japan* 77:074709.

104. Bender, K., I. Hennig, D. Schweitzer, K. Dietz, H. Endres, and H. J. Keller. 1984. Synthesis, structure and physical-properties of a two-dimensional organic metal, di[bis(ethylene-dithiolo)tetrathiofulvalene] triiodide, (BEDT-TTF)$^+_2$I$^-_3$. *Molecular Crystals and Liquid Crystals* 108:359–371.

105. Seo, H. 2000. Charge ordering in organic ET compounds. *Journal of the Physical Society of Japan* 69:805–820.

106. Yamamoto, K., A. A. Kowalska, and K. Yakushi. 2010. Direct observation of ferroelectric domains created by Wigner crystallization of electrons in alpha-[bis(ethylenedithio)tetrathiafulvalene]$_2$I$_3$. *Applied Physics Letters* 96:122901.

107. Itoh, H., K. Itoh, K. Goto, K. Yamamoto, K. Yakushi, and S. Iwai. 2014. Efficient terahertz-wave generation and its ultrafast optical modulation in charge ordered organic ferroelectrics. *Applied Physics Letters* 104:173302.

108. Abdel-Jawad, M., I. Terasaki, T. Sasaki, N. Yoneyama, N. Kobayashi, Y. Uesu, and C. Hotta. 2010. Anomalous dielectric response in the dimer Mott insulator kappa-(BEDT-TTF)$_2$Cu$_2$(CN)$_3$. *Physical Review B* 82:125119.

109. Hotta, C. 2010. Quantum electric dipoles in spin-liquid dimer Mott insulator kappa-ET$_2$Cu$_2$(CN)$_3$. *Physical Review B* 82:241104.

110. Sedlmeier, K., S. Elsasser, D. Neubauer, R. Beyer, D. Wu, T. Ivek, S. Tomic, J. A. Schlueter, and M. Dressel. 2012. Absence of charge order in the dimerized kappa-phase BEDT-TTF salts. *Physical Review B* 86:245103.

111. Tomic, S., M. Pinteric, T. Ivek, K. Sedlmeier, R. Beyer, D. Wu, J. A. Schlueter, D. Schweitzer, and M. Dressel. 2013. Magnetic ordering and charge dynamics in kappa-(BEDT-TTF)$_2$Cu[N(CN)$_2$]Cl. *Journal of Physics-Condensed Matter* 25:436004.

112. Itoh, K., H. Itoh, M. Naka, S. Saito, I. Hosako, N. Yoneyama, S. Ishihara, T. Sasaki, and S. Iwai. 2013. Collective excitation of an electric dipole on a molecular dimer in an organic dimer-Mott insulator. *Physical Review Letters* 110:106401.

113. Iguchi, S., S. Sasaki, N. Yoneyama, H. Taniguchi, T. Nishizaki, and T. Sasaki. 2013. Relaxor ferroelectricity induced by electron correlations in a molecular dimer Mott insulator. *Physical Review B* 87:075107.

114. Torrance, J. B., A. Girlando, J. J. Mayerle, J. I. Crowley, V. Y. Lee, P. Batail, and S. J. Laplaca. 1981. Anomalous nature of neutral-to-ionic phase-transition in tetrathiafulvalene-chloranil. *Physical Review Letters* 47:1747–1750.

115. Lecointe, M., M. H. Lemeecailleau, H. Cailleau, B. Toudic, L. Toupet, G. Heger, F. Moussa, P. Schweiss, K. H. Kraft, and N. Karl. 1995. Symmetry-breaking and structural-changes at the neutral-to-ionic transition in tetrathiafulvalene-p-chloranil. *Physical Review B* 51:3374–3386.

116. Kobayashi, K., S. Horiuchi, R. Kumai, F. Kagawa, Y. Murakami, and Y. Tokura. 2012. Electronic ferroelectricity in a molecular crystal with large polarization directing antiparallel to ionic displacement. *Physical Review Letters* 108:237601.

117. Dawber M. 2012. Viewpoint: Electrons weigh in on ferroelectricity. *Physics* 5:63.

118. Giovannetti, G., S. Kumar, A. Stroppa, J. van den Brink, and S. Picozzi. 2009. Multiferroicity in TTF-CA organic molecular crystals predicted through *ab initio* calculations. *Physical Review Letters* 103:266401.

119. Kagawa, F., S. Horiuchi, M. Tokunaga, J. Fujioka, and Y. Tokura. 2010. Ferroelectricity in a one-dimensional organic quantum magnet. *Nature Physics* 6:169–172.

120. Ravez, J., S. C. Abrahams, and R. Depape. 1989. Ferroelectric-ferroelastic properties of K$_3$Fe$_5$F$_{15}$ and the phase-transition at 490 K. *Journal of Applied Physics* 65:3987–3990.

121. Mitani, T., Y. Kaneko, S. Tanuma, Y. Tokura, T. Koda, and G. Saito. 1987. Electric-conductivity and phase-diagram of a mixed-stack charge-transfer crystal—tetrathiafulvalene-para-chloranil. *Physical Review B* 35:427–429.

122. Silveira, L. G. D., G. S. Dias, L. F. Cotica, J. A. Eiras, D. Garcia, J. A. Sampaio, F. Yokaichiya, and I. A. Santos. 2013. Charge carriers and small-polaron migration as the origin of intrinsic dielectric anomalies in multiferroic TbMnO$_3$ polycrystals. *Journal of Physics-Condensed Matter* 25:475401.

123. Kimura, T., T. Goto, H. Shintani, K. Ishizaka, T. Arima, and Y. Tokura. 2003. Magnetic control of ferroelectric polarization. *Nature* 426:55–58.

124. Lebeugle, D., D. Colson, A. Forget, M. Viret, P. Bonville, J. F. Marucco, and S. Fusil. 2007. Room-temperature coexistence of large electric polarization and magnetic order in $BiFeO_3$ single crystals. *Physical Review B* 76:024116.

125. Wernicke, R. 1978. Influence of kinetic processes on electrical-conductivity of donor-doped $BaTiO_3$ ceramics. *Physica Status Solidi A—Applied Research* 47:139–144.

126. Vonhippel, A. 1950. Ferroelectricity, domain structure, and phase transitions of barium titanate. *Reviews of Modern Physics* 22:221–237.

127. Seo, H., J. Merino, H. Yoshioka, and M. Ogata. 2006. Theoretical aspects of charge ordering in molecular conductors. *Journal of the Physical Society of Japan* 75:051009.

128. Morin, F. J. 1959. Oxides which show a metal-to-insulator transition at the Neel temperature. *Physical Review Letters* 3:34–36.

4

Single-Phase Type-II Multiferroics
Frustrated Magnetism-Triggered Ferroelectricity

Shuai Dong and Jun-Ming Liu

Contents

4.1 Introduction

In this chapter, type-II multiferroics are discussed [1–3]. They are often referred to as magnetic ferroelectrics or magnetic multiferroics, in which the ferroelectricity is directly generated by some type of frustrated spin order. Khomskii [4] viewed these materials as the most exciting development in the field of multiferroics.

Compared with type-I multiferroics, type-II multiferroics possess intrinsically stronger magnetoelectric coupling, since their ferroelectricity is directly triggered by magnetism. Complete control of electric polarization by magnetic fields has been demonstrated in various type-II multiferroics. This makes them more desirable for potential applications. In contrast, it is rather challenging to control electric polarization magnetically in type-I multiferroics despite their superior ferroelectric properties, which is almost independent of magnetism. In addition, the mechanisms of magnetoelectric coupling in type-II multiferroics are more novel and fascinating compared with, for example, the mechanisms of the interfacial strain-mediated coupling in ferroelectric–ferromagnetic composites [5].

This chapter is organized as follows: In Section 4.2, selected examples of type-II multiferroics are introduced, mostly from an experimental perspective. Thanks to the fast development in the past decade, many new type-II multiferroics have been discovered in compounds involving almost every $3d$ transition metals. In Section 4.3, the underlying mechanisms are discussed, mostly from atheoretical perspective. Both the phenomenological theories based on symmetry consideration and microscopic theories at the quantum mechanical level are covered. In Section 4.4, the latest progress in this field is introduced. Some novel materials have been discovered in recent years, with new structures, new elements, new physics, and most importantly, better performance. A brief summary and perspective are given in Section 4.5.

4.2 Examples of Magnatic Multiferroics

4.2.1 Manganites

4.2.1.1 $TbMnO_3$ and $DyMnO_3$

The manganite family has attracted much attention since the 1990s, mostly with regard to the colossal magnetoresistence effect [6–9]. In 2003, Kimura et al. [10] discovered that the pyroelectric polarization of $TbMnO_3$ can be rotated between the c- and a-axis by magnetic fields. This observation opened a new era of research on multiferroics. In this sense, $TbMnO_3$ was the first recognized magnetic ferroelectric, that is, the first type-II multiferroic.

$TbMnO_3$ has a highly distorted perovskite structure due to the small size of Tb^{3+}, as shown in Figure 4.1a. Upon cooling, it changes first from the paramagnetic state to an incommensurate modulated sinusoidal antiferromagnetic state at the Néel temperature ($T_N = 40$ K), then to a cycloid type spiral order with all spin moments lying in the b–c plane at the so-called lock-in temperature ($T_{lock} = 28$ K), as shown in Figure 4.1b. The magnetic modulation is always along the b-axis, with its wave vector q_{Mn} changes continuously in the sinusoidal antiferromagnetic state but is locked to a value of 0.28 in the spiral state. The ferroelectric polarization along the c-axis emerges below T_{lock}, as shown in Figure 4.1c, suggesting a direct coupling between the ferroelectricity and spiral spin order [11,12]. The polarization measured by the pyroelectric method reaches about 600–800 $\mu C/m^2$ [10,13], which is only ~0.1% of the saturation polarization of $BiFeO_3$ [14,15]. An external magnetic field up to several Tesla along the b-axis can rotate the polarization to the a-axis [10,13], as shown in Figure 4.2, suggesting an intrinsic strong magnetoelectric coupling.

Upon reducing the A-site cation size, for instance, by partially or fully replacing Tb^{3+} with Dy^{3+}, the wave vector q_{Mn} gradually increases. For $DyMnO_3$, the value of q_{Mn} is 0.38 below T_{lock}. The improper ferroelectric polarization of $DyMnO_3$ can reach ~2000 $\mu C/cm^2$, much larger than that of $TbMnO_3$, although the T_{lock} decreases to 18 K, as shown in Figure 4.3a [13].

The magnetic field–driven polarization rotation from the c- to a-axis is also observed in DyMnO3, as shown in Figure 4.3b [13]. One more interesting point is that its polarization along the c-axis has a sudden decrease below 8 K, which is the ordering temperature of Dy^{3+} magnetic moments. This complex temperature dependence of ferroelectric polarization is a unique character of DyMnO3, not observed in other multiferroic manganites. Despite this, other magnetoelectric features are very similar between $DyMnO_3$ and $TbMnO_3$, implying the unified physics in these two materials.

More information about the spiral spin order in manganites and other systems can be found in the excellent review by Kimura [16].

4.2.1.2 Orthorhombic $HoMnO_3$ and $YMnO_3$

In 2006, the manganite family contributed two new members to type-II multiferroics: orthorhombic $HoMnO_3$ and $YMnO_3$. Since Ho^{3+} and Y^{3+} are very small, the orthorhombic perovskite structure becomes metastable, while the hexagonal structure is more stable. Interestingly, both the orthorhombic

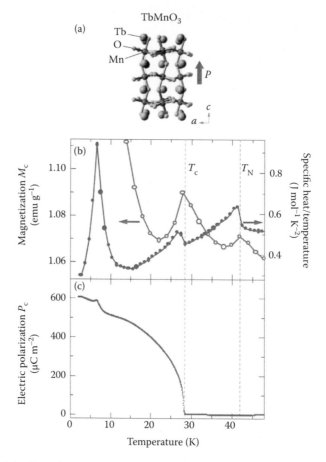

FIGURE 4.1 Ferroelectricity accompanied by magnetic transitions in TbMnO₃. (a) Side view of the crystal structure of TbMnO₃. The electric polarization is along the *c*-axis. (b) Magnetic susceptibility, specific heat divided by temperature, and (c) electric polarization along the *c* axis in TbMnO₃ as a function of temperature [16]. (Reproduced from Kimura, T., *Annu. Rev. Mater. Res.*, 37, 387–413, 2007.)

and hexagonal HoMnO₃/YMnO₃ are multiferroics, although they belong to different types: type-I for the hexagonal ones and type-II for the orthorhombic ones. In this chapter, we focus on the orthorhombic type-II multiferroic phases.

The magnetic order in orthorhombic HoMnO₃ is different from the cycloidal one in TbMnO₃. It forms a collinear zigzag-type order, leading to the so-called E-type antiferromagnetism [18,19]. In 2006, Sergienko et al. [20] predicted that this E-type antiferromagnetism could induce a large improper polarization up to 1–2 μC/cm² along the *a*-axis. Soon after the prediction, Lorenz et al. [21] synthesized orthorhombic HoMnO₃ and YMnO₃ using a high-pressure method and confirmed their multiferroicity. The ferroelectric Curie temperature T_C equals to the magnetic lock-in temperature: 26 K for

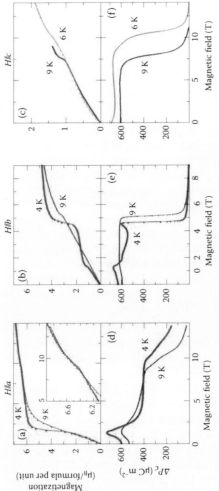

FIGURE 4.2 Effects of magnetic field on various properties of TbMnO$_3$ crystals, obtained upon increasing the magnetic field. (a–c) Magnetization and (d–f) changes in electric polarization along the c axis, under external magnetic fields parallel to a, b, and c axes, respectively. The inset of (a) shows a magnified view of the high-magnetic-field region [16]. (Reproduced from Kimura, T., *Annu. Rev. Mater. Res.*, 37, 387–413, 2007.)

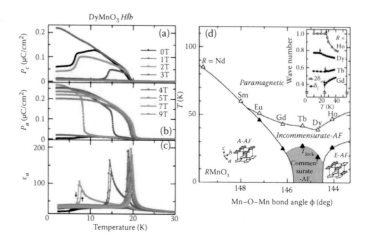

FIGURE 4.3 (a) and (b) Temperature and magnetic-field dependences of electric polarization along the *c* and *a* axes, respectively, and (c) dielectric constant along the *a*-axis at different magnetic fields for DyMnO$_3$ crystals. A magnetic field is applied along the *b*-axis. (d) Magnetic phase diagram of *R*MnO$_3$ as a function of Mn–O–Mn bond angle. The Néel and lock-in transition temperatures are denoted by open and closed triangles, respectively. Inset: Temperature profiles of the wave numbers of modulated crystallographic (δ_l) or magnetic (δ_m) structures. The lock-in transition temperatures are indicated by arrows. Left and right lower insets show schematic illustrations of the A-type and the E-type antiferromagnetic structures, respectively. The commensurate-antiferromagnetic state (gray area) has the spiral spin order, while the incommensurate-antiferromagnetic state above the gray area has the sinusoidal-type spin density wave [17]. (Reproduced from Goto, T. et al., *Phys. Rev. Lett.*, 92, 257201, 2004.)

HoMnO$_3$ and 28 K for YMnO$_3$. As a fingerprint of type-II multiferroics, the ferroelectric polarization of HoMnO$_3$ is significantly suppressed by magnetic field. In contrast, the polarization of YMnO$_3$, although also induced by magnetism, is almost unaffected by magnetic field up to 7 Tesla [21]. The sizes of Ho^{3+} and Y^{3+} are similar. The difference lies in their magnetic moments: Ho^{3+} has a large local moment, while Y^{3+} is nonmagnetic.

The measured pyroelectric polarization is less than 100 μC/m^2 for the polycrystalline HoMnO$_3$ sample [21], two orders of magnitude smaller than the theoretical prediction [20,23]. The difference has been gradually reduced by improving sample quality. For example, Ishiwata et al. [22] synthesized high-quality polycrystalline samples of *R*MnO$_3$ (R = Eu$_{1-x}$Y$_x$, Y$_{1-y}$Lu$_y$, Dy, Ho, Er, Tm, and Yb), which showed polarization of up to 800 μC/m^2 for the E-type antiferromagnets, as shown in Figure 4.4. By multiplying a calibration factor of 6, the value for corresponding single crystals was estimated to be ~5000 μC/m^2, approaching the theoretical prediction. Using a larger poling field (37.5 kV/cm), Pomjakushin et al. [24] obtained a larger polarization of up to 1500 μC/m^2 in polycrystalline TmMnO$_3$ (also with the E-type antiferromagnetism), which

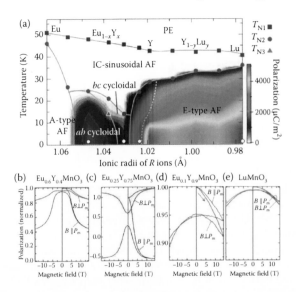

FIGURE 4.4 (a) Contour plot of (corrected) magnitude of polarization in the phase diagram of o-$RMnO_3$ with nonmagnetic R ($Eu_{1-x}Y_x$, $Y_{1-y}Lu_y$). Polarization at 2 K under magnetic fields for (b) $Eu_{0.6}Y_{0.4}MnO_3$, (c) $Eu_{0.25}Y_{0.75}MnO_3$, (d) $Eu_{0.1}Y_{0.9}MnO_3$, and (e) $LuMnO_3$. The polarization values are normalized by that at 2 K [22]. (Reproduced from Ishiwata, S. et al., *Phys. Rev. B*, 81, 100411, 2010.)

seemed to be unsaturated yet. Single crystals of $YMnO_3$ grown under high pressure showed an unsaturated polarization of up to 2200 $\mu C/m^2$ at 2 K under a 10 kV/cm poling field [25]. The largest polarization obtained so far was from $YMnO_3$ thin film grown on the $YAlO_3$ substrate, whose in-plane polarization reached 8000 $\mu C/m^2$ at 4 K under a 10 kV/cm poling field [26,27].

4.2.1.3 A-Site Doping Effect in $RMnO_3$

The size of R^{3+} (R = rare earth ion or Y) affects the ground state of $RMnO_3$, which changes from A-type antiferromagnetic to the spiral spin order and then to E-type antiferromagnetic with decreasing size of R^{3+}. The magnetic field–driven polarization rotation from c-axis to a-axis is also observed in $DyMnO_3$, as shown in Figure 4.3d. As discussed previously, the latter two states are multiferroic, with ferroelectric polarizations driven by particular magnetic orders. The A-type antiferromagnetic state is nonferroelectric.

The phase diagram can be extended by combining two Rs at the A-site, for example, $Eu_{1-x}Y_xMnO_3$. $EuMnO_3$ is located at the A-type antiferromagnetic side, while $YMnO_3$ is located at the E-type antiferromagnetic side. By combining Eu and Y in the $Eu_{1-x}Y_xMnO_3$ alloy, one can tune the average size of R continuously, and reproduce the phase diagram. Furthermore, the spiral spin order in $Eu_{1-x}Y_xMnO_3$ is nontrivial. The easy plane of spiral is the a–b plane when x is small, and flips to the b–c plane when x is large. This spin-flop behavior does not occur in $Tb_{1-x}Dy_xMnO_3$, in which the ground state spiral is always in the $b-c$ plane. And in $Eu_{1-x}Y_xMnO_3$, the ferroelectric polarization of the $a-b$ plane spiral state is significantly larger than that of the $b-c$ plane one.

Based on previous studies on colossal magnetoresistive manganites, phase separation is naturally expected at the first-order phase transition boundaries when the quenching disorder exists, rendering a phase-coexisting system, which is sensitive to external stimuli. This phenomenon also exists in multiferroic manganites. The phase transition from spiral spin order to E-type antiferromagnetism is a first-order one, and the quenching disorder from the alloy mixture of Eu^{3+} and Y^{3+} may induce a coexistence of spiral and E-type order, as shown in Figure 4.4 [22]. Such phase coexistence has also been observed in $Tb_{1-x}Ho_xMnO_3$ [28].

Besides the size effect, the A-site ion can also modulate the multiferroic properties via the magnetic coupling if it is magnetic. Many rare earth ions, for example, Tb^{3+}, Dy^{3+}, and Ho^{3+}, show large $4f$ magnetic moments, which become ordered at very low temperatures, typically <10 K. However, even above their ordering temperatures, these magnetic rare earth ions can interact with Mn^{3+}. In particular, neutron studies have revealed the coexistence of magnetic modulations from Dy^{3+} and Mn^{3+} between 8 and 28 K [29,30], which enhances the ferroelectric polarization that happens due to the exchange striction between Dy^{3+} and Mn^{3+} [31]. In addition, a sinusoidal incommensurate magnetic order has been observed in single crystal $HoMnO_3$, which is different from the E-type antiferromagnetic order [32]. Ferroelectric polarization was observed along the c-axis due to the exchange striction between Ho^{3+} and Mn^{3+}, while it should be along the a-axis for the E-type antiferromagnetic order [32].

The difference between $Eu_{1-x}Y_xMnO_3$ and $TbMnO_3/DyMnO_3$ can thus be understood from the A-site magnetism. Both Eu^{3+} and Y^{3+} are nonmagnetic, giving rising to pure Mn-based magnetic ferroelectricity. $DyMnO_3$ is more complex, with at least two sources of ferroelectric polarization, both of which are driven by magnetic orders. For more information on multiferroic perovskite manganites, readers can refer to our recent review and the references mentioned there [33].

4.2.1.4 RMn_2O_5

Besides the most studied $RMnO_3$, Mn-oxides can also take other more complex structures. In 2004, Hur et al. [34] found a magnetic-induced electric polarization in $TbMn_2O_5$ as shown in Figure 4.5. The ferroelectric Curie temperature of $TbMn_2O_5$ is 40 K [34], coinciding with its magnetic Néel temperature, a little higher than that of $TbMnO_3$ but remains very low. The pyroelectric polarization is about 400 $\mu C/m^2$, smaller than that of $TbMnO_3$. Most interestingly, the polarization can be reversed by a magnetic field, which is a desirable function not common even in type-II multiferroics [34].

Further studies on RMn_2O_5 (R = Tb, Dy, Ho, and Bi) revealed a complex scenario. Mn^{3+} and Mn^{4+} coexist in such a system, occupying the oxygen octahedron and tetrahedron, respectively. The spins of Mn^{3+} and Mn^{4+} are noncollinear at low temperatures. With decreasing temperatures, several magnetic phases appear, inducing polarization and the complex behavior

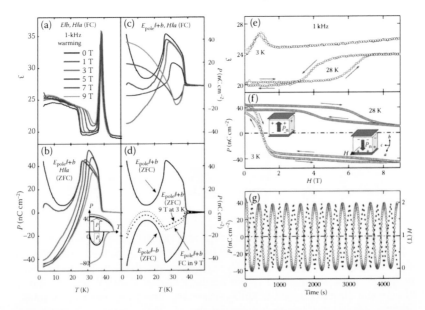

FIGURE 4.5 (a–d) Temperature and magnetic-field (along the *a*-axis) dependences of ferroelectric properties and memory effect. (a) The dielectric constants along the *b*-axis measured during warming. (b) The total electric polarization along the *b*-axis under various magnetic fields, measured after zero magnetic-field cooling. Inset: the total polarization may be the sum of positive and negative components. (c) Total electric polarization under the zero magnetic field after magnetic-field cooling. (d) Total electric polarization measured under zero magnetic field after cooling under four different conditions, as indicated next the curves. The magnetic field leaves a permanent imprint in the polarization, implying a memory effect. (e–g) Reproducible polarization reversal by magnetic fields. (e) The dielectric constant as a function of applied magnetic field at 3 and 28 K. (f) Change of total electric polarization by applied magnetic fields at 3 and 28 K, which was calculated from the magnetoelectric current measured after zero magnetic-field cooling. Insets: the reversal of polarization. (g) Polarization reversal at 3 K by linearly varying magnetic field from 0 to 2 T, which clearly displays highly reproducible polarization switching by magnetic fields [34]. (Reproduced from Hur, N. et al., *Nature*, 429, 392–395, 2004.)

of compensation and reversal of net polarization [35,36]. In this sense, the RMn_2O_5 system is more likely a magnetic ferrielectric family.

4.2.2 Other Type-II Multiferroics

There are other type-II multiferroics besides manganites. It fact, it seems possible to find type-II multiferroics in compounds with any $3d$ magnetic ions. This implies common physical mechanisms to be discussed in Section 4.2.3. In the following, we briefly introduce some other typical type-II multiferroics.

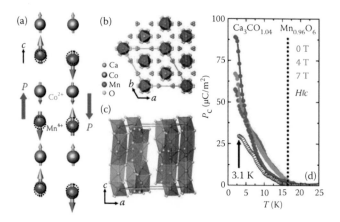

FIGURE 4.6 (a) Co–Mn–Co–Mn chains with Ising-type ↑↑↓↓ spin order in Ca_3CoMnO_6. The ionic displacements (from the original dashed positions) are also indicated, which give rise to the net electric polarization. The two possible magnetic configurations lead to the positive/negative polarizations. (b) and (c) The crystal structure of Ca_3CoMnO_6. The green boxes represent the crystallographic unit cell. (d) Pyroelectric polarization of single crystal $Ca_3Co_{1.04}Mn_{0.96}O_6$ along the chain direction (c-axis). The samples were poled upon cooling from 40 to 2 K (filled circles), and to 3.1 K (open circles) before the measurement. For the 2 K poled sample, various magnetic fields are applied during the measurements to illustrate the magnetoelectric coupling effect [37]. (Reproduced from Choi, Y. et al., *Phys. Rev. Lett.*, 100, 047601, 2008.)

4.2.2.1 Ca_3CoMnO_6

In 2008, Choi et al. [37] reported a weak magnetic ferroelectricity ($\sim$90 μC/m²) in Ca_3CoMnO_6 below 16.5 K, as shown in Figure 4.6. Ca_3CoMnO_6 possesses quasi-one-dimensional $-Co^{2+}-Mn^{4+}-Co^{2+}-Mn^{4+}-$ spin chains. Within each chain, the spin order is ↑↑↓↓. By applying a strong magnetic field to change it to ↑↑↑↓ or even a fully ferromagnetic state, the electric polarization will be suppressed [38]. Further experiments revealed that non-stoichiometry played an important role in $Ca_3Co_{2-x}Mn_xO_6$ ($x \approx 1$). The ↑↑↓↓ magnetic structure and its affiliated ferroelectricity disappear when the ratio of Co/Mn is very close to one, which is against intuition. The multiferroicity emerges only when the x of $Ca_3Co_{2-x}Mn_xO_6$ deviates a little from 1 [39,40].

Further density functional theory (DFT) calculations also revealed some contradictions, for instance, the calculated ferroelectric polarization is much larger than the experimental one [41,42]. And the ↑↑↓↓ magnetic order is not so robust as the ground state for the ideal stoichiometric system [41,42]. Further investigations are needed to better understand this system.

4.2.2.2 Organic Magnetic Multiferroics

Type-II multiferroic materials exist not only in various oxides but also in organic materials. In 2010, Kagawa et al. [43] reported an organic charge-transfer salt tetrathiafulvalene-p-bromanil (TTFCBA) to be a magnetic ferroelectric material.

As schematically shown in Figure 4.7a and b, TTF-BA contains ionic TTF donor (D^+) and BA acceptor (A^-) molecules, which form $D^+A^-D^+A^-$ mixed chains along the a and b axes, respectively. The dimerization occurs in both chains below 53 K. TTF and BA molecules in TTF-BA are almost ionic in the whole temperature region (the ionicity is as high as 0.95); thus, the D^+A^- chain can be regarded as an one-dimensional Heisenberg chain with spin $-1/2$, similar to the inorganic Ca_3CoMnO_6.

Due to the spin-Peierls transition, that is, the dimerization between D^+A^-, ferroelectricity emerges below 53 K. The anomalies of spin susceptibility (Figure 4.7c) and infrared reflectivity (Figure 4.7d) coincide with the sharp peak of dielectric constant (Figure 4.7e) and the emerging point of spontaneous polarization (Figure 4.7f) along the b-axis.

Besides these typical type-II multiferroics, there are many others that have been discovered in the past decade, such as $CuFeO_2$ [44], $MnWO_4$ [45,46], $Ni_3V_2O_8$ [47], $CoCr_2O_4$ [48], $LiCu_2O_2$ [49], $LiCuVO_4$ [50], $RFeO_3$ [51,52], and

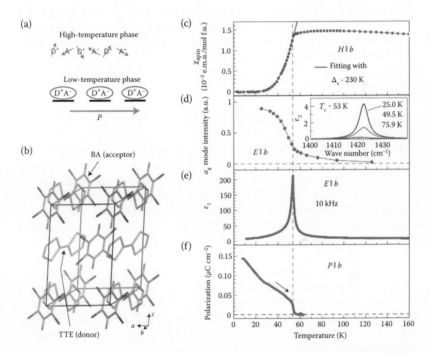

FIGURE 4.7 Ferroelectric spin-Peierls transition in TTF-BA. (a) Schematic of ionic donor (D^+) and acceptor (A^-) in high- and low-temperature phases of TTF-BA. Spins are denoted by arrows. Ellipsoids with underlines represent dimers. (b) Crystal structure of TTF-BA. Temperature dependence of the physical properties. (c) Spin susceptibility. (d) Normalized spectral weight of the infrared a_g mode at 1422 cm^{-1} as a fingerprint of local D^+A^- dimerization. Inset: The temperature dependence of the a_g mode spectra. (e) Dielectric constant at 10 kHz. (f) Ferroelectric polarization along the b axis [43]. (Reproduced from Kagawa, F. et al., *Nature Phys.*, 6, 2010.)

possible $RNiO_3$ [53]. Most of them show complex magnetic orders but the underlying mechanisms of multiferroicity are similar, which will be clarified in the following section. Readers interested in these materials can refer to the comprehensive reviews [2,11] and the original papers for more details.

4.3 Mechanisms of Spin Order–Induced Ferroelectricity

4.3.1 Symmetry Consideration

The order parameters of magnetism and ferroelectricity are magnetization (M) and polarization (P), which break the time-reversal and space-inversion symmetries, respectively. In the nonrelativistic limit, time and space are independent. Thus, it is expected that M and P are decoupled. In fact, in most materials, they seem to be mutually exclusive.

According to the Landau theory of phase transition, the free energy of a system close to a phase transition can be expressed as the Taylor series expansion of an order parameter. For example, for a ferromagnet/ferroelectric without external magnetic/electric fields, the free energy can be written as

$$F_m = F_m^0 + a_m \mathbf{M}^2 + b_m \mathbf{M}^4 + \cdots \tag{4.1}$$

$$F_p = F_p^0 + a_p \mathbf{P}^2 + b_p \mathbf{P}^4 + \cdots \tag{4.2}$$

where F^0 is the "original" free energy without magnetization/polarization, and α and β are corresponding coefficients. The absence of odd power orders of M and P is because the energy as a scalar quantity should remain unchanged under both time-reversal and space-inversion operations. For example, a component like M^3 will break the time-reversal symmetry; thus F_m will change if time sequence is reversed, which is physically forbidden. If an external magnetic field H (electric field E) is applied, components like $\mathbf{M \cdot H}$ ($\mathbf{P \cdot E}$) will not violate the symmetry requirement and thus can appear in the free energy expression.

Therefore, restricted by symmetry requirements, the lowest power order of magnetoelectric coupling can be expressed as $M^2 P^2$ in the free energy. It is clear that such a coupling, being in the fourth order, is rather weak and most likely indirect (e.g., via strains). However, great progress has been made in the past decade and it is now recognized that, although a single magnetic moment only breaks the time-reversal symmetry, a collection of many can break the space-inversion symmetry when forming some particular magnetic orders $M(r)$ and thus generate ferroelectricity. Until now, most of the special magnetic orders can be classified into two categories.

4.3.1.1 Chiral Structure

Chirality is ubiquitous in nature. For example, the morning glories always follow the right-hand rule, namely, wind in the anticlockwise direction.

This right-hand rule applies in most winding plants, with only a few exceptions. This winding chirality also exists in DNA structure, hurricane, as well as our solar system. In the quantum world, if spins form a spiral order, with clockwise or anticlockwise chirality, the system can break the space-inversion symmetry. A space-inversion operation ($r \leftrightarrow -r$) will change the chirality between clockwise rotation and anticlockwise rotation, corresponding to the switching between P and $-P$.

Mathematically, Mostovoy [54] proposed a phenomenological equation to describe the spiral spin order–induced electric polarization. The free energy of magnetoelectric coupling term reads as

$$F_{em} = \gamma P \cdot \left[M(\nabla \cdot M) - (M \cdot \nabla)M + \cdots \right] \tag{4.3}$$

where γ is the coefficient. Since the operation ∇ breaks the space-inversion symmetry, the above expression will not violate the space-inversion or time-reversal symmetry. Such a coupling is in the cubic power order, lower than the fourth order as in Equation 4.2, and therefore should be intrinsically stronger.

If the quadratic term of the polarization energy $P^2/2\varepsilon$ (ε is the dielectric constant) is considered in the total free energy, it is straightforward to obtain the induced polarization as following:

$$P = \varepsilon\gamma \left[M(\nabla \cdot M) - (M \cdot \nabla)M \right] \tag{4.4}$$

According to Equation 4.4, if magnetic moments are uniformly aligned, namely, in the ideal ferromagnetic order, the derivation of M is zero, leading to a nonferroelectric state. Only when the magnetism is frustrated, namely, the magnetic moments are spatially modulated, a nonzero P can be obtained.

Let us apply Equation 4.4 to noncollinear spin orders. Consider a general mathematic form of spiral/conical spin order, which is as follows:

$$M(r) = M_x \cos(Q \cdot r)x + M_y \sin(Q \cdot r)y + M_z z, \tag{4.5}$$

where Q is the propagation vector in the reciprocal space and r is the position vector in the real space. M_x, M_y, and M_z are the magnitudes of magnetic moment along the x, y, and z axes, respectively. If $M_z = 0$, it is a coplaner spiral order. If M_z is nonzero, it is a three-dimensional conical order with a net magnetic moment along the z axis. Using Equation 4.4, we can obtain the average polarization as

$$\langle P \rangle = \varepsilon\gamma M_x M_y [z \times Q] \tag{4.6}$$

This equation clearly indicates that a noncollinear spin order can induce an electric polarization perpendicular to the propagation vector Q and the spiral axis z. In this sense, a cycloid spiral, with $z \perp Q$, can induce a nonzero <P>, while a screw spiral, with $z \| Q$, is incapable for such an effect.

For a conical spin order, although the M_z component does not explicitly appear in Equation 4.6, it provides a degree of freedom to tune the induced polarization by magnetic field h. The Zeeman energy $h_z M_z$ can tune the conical spiral plane, as well as the magnitudes of M_x and M_y.

4.3.1.2 Parity Structure

Besides noncollinear spin orders, certain special collinear spin orders can also break the space-inversion symmetry. Let us take the simplest one-dimensional Ising spin chain along the x axis as an example to illustrate how it works. As shown in Figure 4.8b, spins form a ↑↑↓↓ type order. For any site in this chain, its left and right neighbors are inequivalent. In other words, the parity symmetry is broken in this type of magnetic structure. A parity-related order parameter for each site i (Ψ_i) is defined as $(M_{i-1} - M_{i+1})$. Then a time-reversal invariance can be constructed as

$$\Omega = \Psi_1 \cdot \Psi_4 = \Psi_2 \cdot \Psi_3 = (M_1 - M_3) \cdot (M_2 - M_4) \quad (4.7)$$

Then a shift of spin pattern by one step, giving ↑↓↓↑ (or ↓↑↑↓), will reverse the Ψ_3 (or Ψ_2) but keep the Ψ_2 (or Ψ_3) invariant. Thus, the sign of Ω will be reversed by the shifting operation but invariant under the time-reversal operation.

A space-inversion and time-reversal invariant free energy term can be constructed as

$$F_{em} = [\alpha(r) \cdot P]\Omega \quad (4.8)$$

where α is a coefficient containing odd power orders of space vector r. For example, α can be $\partial J / \partial x$, where J is the exchange interaction between nearest-neighbor sites. Then the induced polarization is

$$P = -\varepsilon\alpha(r)\Omega \quad (4.9)$$

in which the direction of P is governed by $\alpha(r)$ instead of the magnetic order.

This mechanism applies to the aforementioned E-type antiferromagnetic as well as the ↑↑↓↓ orders.

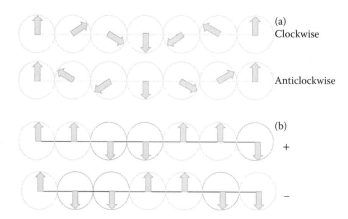

(a) Clockwise

Anticlockwise

(b)

+

−

FIGURE 4.8 Sketch of two types of symmetry-broken spin orders. (a) The clockwise versus anticlockwise cycloidal spirals. (b) ↑↑↓↓ (parity +) vs. ↑↓↓↑ (parity −). The parity is only definable when one-unit translation is forbidden; otherwise the translational symmetry will make these two patterns identical.

4.3.1.3 Other Possibilities

Following the previous discussion, it is clear that electric polarization can be induced by magnetism via other combinations of the magnetic and dipole orders. For example, a "good" combination could be

$$F_{em} = [P \cdot \beta(r)][M \cdot \gamma(r)]^2 \tag{4.10}$$

where β and γ are two functions of space vector r. β must be in the odd power order of r. There is no special requirement for γ, which can even be a plain function $\gamma = 1$. Similar to the parity mechanism, the orientation of induced P is determined by the nonmagnetic $\beta(r)$, as

$$P = -\beta(r)[M \cdot \gamma(r)]^2 \tag{4.11}$$

The coordinate vector need not be the only one to be incorporated into the free energy expression; time can also be included for the dynamic process. For example, by considering the dynamics of polarization, the coupling can be

$$F_{em} = \frac{dP}{dt} \cdot \nabla \cdot M \tag{4.12}$$

which is in the second power order.

In short, any combinations of M and P that are space-reversal and time-inversion invariant, can be the source of magnetoelectric coupling. Further studies may reveal more types of magnetoelectric couplings.

4.3.2 Quantum-Level Microscopic Mechanisms

The previous analyses are based on the phenomenological Landau theory and symmetry consideration. In this subsection, the quantum-level microscopic physics will be employed to understand the magnetoelectric coupling in type-II multiferroics.

4.3.2.1 Dzyaloshinskii–Moriya Interaction

In 1958, Dzyaloshinskii [55] proposed a thermodynamic theory to explain the weak ferromagnetism in Cr_2O_3, which was illustrated using the quantum perturbation theory by Moriya in 1960 [56]. The origin of the Dzyaloshinskii–Moriya interaction is the relativistic spin-orbit coupling, and it plays a crucial role not only in many multiferroics (not limited to type-II ones) but also in many subfields of magnetism, for example, the skyrmion quasiparticle [57].

Let us use perovskite oxides (ABO_3) as examples to show how Dzyaloshinskii–Moriya interaction couples magnetism and polarization. In perovskite oxides with ideal cubic structure, the B–O–B bonds are straight with 180° bond angles. Such bonds possess rotation symmetry with respect to the B–B axis. However, in most real perovskites, due to the size mismatch between A and B ions, the oxygen octahedra will tilt and rotate (usually described using the Glazer notation) [58–60]. The middle oxygen ion may move away from the original position, bending the B–O–B bond and breaking the rotation symmetry. This bending will induce the Dzyaloshinskii–Moriya

interaction as a relativistic correction to the superexchange between magnetic
B ions. The Hamiltonian can be expressed as [56,61]

$$H_{DM} = \mathbf{D}_{ij} \cdot (\mathbf{S}_i \times \mathbf{S}_j) \qquad (4.13)$$

where $\mathbf{D}_{ij}$ is the coefficient of Dzyaloshinskii–Moriya interaction between two
spins $\mathbf{S}_i$ and $\mathbf{S}_j$. In perovskite structures with bent B–O–B bonds, the vector
$\mathbf{D}_{ij}$ must be perpendicular to the B–O–B plane, determined by the symmetry
conditions, as shown in Figure 4.9a [56]. In the first-order approximation, the
intensity of $\mathbf{D}_{ij}$ is proportional to the displacement of oxygen ion away from
the original point $\mathbf{d}_O$.

$$\mathbf{D}_{ij} = \zeta \mathbf{e}_{ij} \times \mathbf{d}_O \qquad (4.14)$$

where ζ is the coefficient and $\mathbf{e}_{ij}$ is the vector pointing from site i to site j. Thus,
in the cubic limit with the highest symmetry, $\mathbf{D}_{ij} = 0$. Due to the cooperative
rotation or tilting of oxygen octahedra, the direction of $\mathbf{D}_{ij}$ should be reversed
between nearest-neighbor B–O–B bonds if all O–B–O bonds are 180° (rigid
octahedra).

Based on the Hamiltonian, it is straightforward to see that the role of
Dzyaloshinskii–Moriya interaction is to induce a noncollinear spin pattern
and determine the easy magnetic plane. For example, let us consider the simplest Hamiltonian for a one-dimensional spin chain:

$$H = J \sum_{\langle ij \rangle} (\mathbf{S}_i \cdot \mathbf{S}_j) + \sum_{ij} \mathbf{D}_{ij} \cdot (\mathbf{S}_i \times \mathbf{S}_j) \qquad (4.15)$$

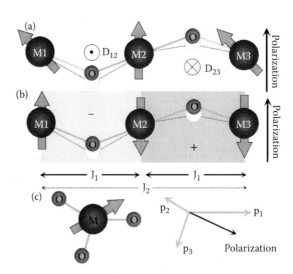

FIGURE 4.9 Sketch of three mechanisms to generate electric polarizations
via magnetic orders. Here metal (M)–oxygen (O) bonds are used for example,
though the anion is not limited to oxygen. (a) The (inverse) Dzyaloshinskii–Moriya
interaction. (b) Exchange striction. (c) Spin-dependent metal–ligand hybridization.
In (a) and (b), the positions of oxygen ions are unidirectionally shifted.

where J is the standard exchange. If $D_{i,i+1}$ is uniform, for instance, along one direction and J is negative (positive), then a spiral spin order is the ground state, with the nearest-neighbor spin angle being $\arctan[-|D|/J]$ ($\pi - \arctan[|D|/J]$).

The converse effect of Dzyaloshinskii–Moriya interaction is to generate a uniform bias of D (and the corresponding uniform oxygen shift) when a spiral spin order is present. The process can be described by the following Hamiltonian equation:

$$H = \left(\zeta e_{ij} \times d_O \right) \cdot \left(S_i \times S_j \right) + \frac{\kappa}{2} d_O^2 \qquad (4.16)$$

where the second term is the elastic energy with κ as the stiffness. By minimizing the total energy, the displacement d_O is $-(\zeta / \kappa) e_{ij} \times (S_i \times S_j)$. Since $S_i \times S_j$ is uniform in a spiral magnet with fixed chirality, the induced ionic displacements are also parallel, giving a macroscopic electric polarization, as shown in Figure 4.9a. This scenario was firstly proposed by Sergienko and Dagotto [61] to explain the origin of ferroelectricity in $TbMnO_3$.

Even without the ionic displacement, a shift of electron can also be induced by a noncollinear spin pair. By applying the quantum perturbation theory to the Hubbard model with spin-orbit coupling, Katsura et al. [62] proved that the induced charge dipole is proportional to $P \sim e_{ij} \times (S_i \times S_j)$. Although the theoretical derivation is complex, this conclusion is elegant and in agreement with the mechanism based on the Dzyaloshinskii–Moriya interaction. In this sense, these two theories are syngeneic, although the route is a little different: the first one is based on the shift of ions, while the second one is based on electrons. Both contribute together to the total ferroelectric polarization.

4.3.2.2 Spin-Dependent Metal–Ligand Hybridization

According to previous analysis, a screw-type spiral spin order $[e_{ij} \times (S_i \times S_j)]$ cannot generate any polarization. However, in some materials, for example, $CuFeO_2$, the spiral spin order is the in-plane screw type under a magnetic field. And experiments have revealed a finite polarization, which cannot be understood under the previous scenario based on the Dzyaloshinskii–Moriya interaction [44]. There must be other mechanisms at work. For example, Arima [63] proposed an alternative mechanism based on spin-orbit coupling. For each transition metal ion, the local polarization can be expressed as

$$P = \sum \left(e_i \cdot S \right)^2 e_i \qquad (4.17)$$

where the summation is over all bonds and e denotes the bond direction, as shown in Figure 4.9c.

Different from the Dzyaloshinskii–Moriya interaction involving two magnetic sites, here only a single magnetic site is enough to generate local polarization. The underlying mechanism is the spin-orbit coupling-induced perturbation of the hybridization between metal d- and anion p-orbitals. Although this mechanism was originally proposed for $CuFeO_2$, it was ruled out later by careful analysis of its crystal symmetry. However, this mechanism is responsible for the polarization observed in $Ba_2CoGe_2O_7$ [64].

4.3.2.3 Exchange Striction

For transition metals with active $3d$ electrons, the spin-orbit coupling is intrinsically weak. Thus, the coefficient ζ is a small quantity, giving a very small dipole moment for the noncollinear spin pair. To overcome this drawback, Sergienko et al. [20] proposed an alternative microscopic mechanism, namely, the exchange striction, to generate polarization in orthorhombic $HoMnO_3$. As discussed previously, the ↑↑↓↓ magnetic order may break the space-inversion symmetry if it is coupled with a coordinate-related coefficient $\alpha(r)$. Let us take the one-dimensional ↑↑↓↓ spin chain as an example to illustrate the microscopic mechanism, while leaving the more complex E-type antiferromagnetism in $HoMnO_3$ for the readers to explore (partially illustrated in Figure 4.9b). The exchange coefficient J between nearest-neighbor sites depends on the bond length. In the first-order approximation, the Tylor expansion of $J(x)$ around the equilibrium length can be written as: $J(\delta) = J_0 + \left.\frac{\partial J}{\partial x}\right|_{x_0} \delta$ where x is the bond length, x_0 is the equilibrium one, J_0 is the exchange coefficient at x_0, and $\delta = x - x_0$. By assuming a periodic $(\delta, -\delta, \delta, -\delta)$ type displacement mode, the Hamiltonian for a given ↑↑↓↓ spin chain is

$$H = \sum_{\langle i,j \rangle} \left[J_{ij}(\delta_{ij})(\mathbf{S}_i \cdot \mathbf{S}_j) + \frac{\kappa}{2}(x_{ij} - x_0)^2 \right] \tag{4.18}$$

$$= \sum_{n=4i} \left[4\left.\frac{\partial J}{\partial x}\right|_{x_0} \delta + 2\kappa\delta^2 \right] \tag{4.19}$$

By minimizing the energy, we can obtain the equilibrium displacement δ as $-\frac{1}{\kappa}\left.\frac{\partial J}{\partial x}\right|_{x_0}$. This displacement may induce a polarization if the charge at site i is not identical to that at site $i + 1$.

4.3.3 A New Unified Model

Recently, Xiang et al. [65,66] proposed a phenomenological framework to unify the various models for magnetism-driven polarization, which also accounts for those phenomena that cannot be well explained by previous models.

In this unified model, the total polarization is separated into two parts:

$$P = P_e(S_1, S_2, \ldots, S_m; U = 0) + P_{ion}(U) \tag{4.20}$$

where the first part P_e is the pure electronic contribution and the second part P_{ion} comes from the ionic displacements. S_i denotes the individual spin. $U = (u_1, u_2, \ldots, u_n)$ denote the ionic displacements from their centrosymmetric positions, which may also depend on spins.

As in proper ferroelectrics, $P_{ion}(U)$ can be obtained by summing all the local dipoles:

$$P_{ion}(U) = \sum_i [u_i Z_i] \tag{4.21}$$

where Z_i is the Born effective charge. The P_e item is a bit more complex, which can be expressed as

$$P_e = \sum_{i,\alpha\beta} P_{i,\alpha\beta} S_{i\alpha} S_{i\beta} + \sum_{\langle i,j \rangle} P_{es}^{ij} S_i S_j + \sum_{\langle i,j \rangle} \vec{M}^{ij} \cdot (S_i \times S_j) \tag{4.22}$$

where the first term is the single-site contribution through the aforementioned p–d hybridization, which is responsible for the ferroelectricity in $Ba_2CoGe_2O_7$ [64] and Cu_2OSeO_3 [67], but not important in most magnetic multiferroics. The second term is the exchange striction contribution. The last one is the general spin-current term, with $\vec{M}_{ij}$ being a 3×3 matrix:

$$\vec{M}_{ij} = \begin{vmatrix} \left(P_{ij}^{yz}\right)_x & \left(P_{ij}^{zx}\right)_x & \left(P_{ij}^{xy}\right)_x \\ \left(P_{ij}^{yz}\right)_y & \left(P_{ij}^{zx}\right)_y & \left(P_{ij}^{xy}\right)_y \\ \left(P_{ij}^{yz}\right)_z & \left(P_{ij}^{zx}\right)_z & \left(P_{ij}^{xy}\right)_z \end{vmatrix} \tag{4.23}$$

The Katsura–Nagaosa–Balatsky spin-current model discussed previously is a special case of such a general model, for example, when the i–j bond is along the x direction and only $\left(P_{ij}^{zx}\right)_z = -\left(P_{ij}^{xy}\right)_y$ are nonzero elements, Equation 4.23 becomes the Katsura–Nagaosa–Balatsky one. This general model can explain the ferroelectricity of $CuFeO_2$, for which the Katsura–Nagaosa–Balatsky model fails.

4.3.4 Origin of the Special Magnetic Orders Responsible for Ferroelectricity

The previous discussion focuses on the origin of electric polarization in type-II multiferroics, namely, how the magnetic orders generate electric dipoles. Equally important is the question of how to obtain such special magnetic orders in real materials.

To answer this question, we first introduce an order parameter defined as $< S_i \cdot S_j >/N_c$, where the spins are normalized unit vectors, N_c is the average coordination number, and $< >$ represents the average overall nearest-neighbor pairs. Perfect ferromagnetism and antiferromagnetism are thus described by order parameters of +1 and −1, respectively. All the special spin orders described previously can be considered transition states between ferromagnetism and antiferromagnetism, since the values of corresponding order parameter are between +1 and −1. It is well recognized now that such transition states can be obtained by exchange frustration.

Exchange frustration exists in many magnetic systems [68]. A simple illustration is shown is Figure 4.10. For a classical spin system, the Hamiltonian can be written as

$$H = \sum_{i,j} J_{i,j} \left(S_i \cdot S_j\right) \tag{4.24}$$

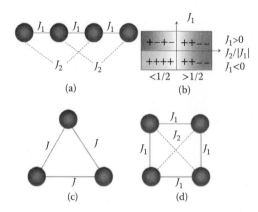

FIGURE 4.10 Illustration of exchange frustration. (a) A one-dimensional spin chain with nearest-neighbor exchange J_1 and next-nearest-neighbor exchange J_2. (b) Ground state phase diagram for one-dimensional Ising spin chain. (c) Frustration in a triangular lattice. (d) Frustration in a square lattice with nearest-neighbor exchange J_1 and next-nearest-neighbor exchange J_2.

where $J_{i,j}$ denotes the exchange between spins S_i and S_j, and i/j denotes the site index. For a one-dimensional classical spin chain, when there is only the nearest-neighbor exchange J_1, the ground state is ferromagnetic if $J_1 < 0$ or antiferromagnetic ($\uparrow\downarrow\uparrow\downarrow$) if $J_1 > 0$. However, when there is the next-nearest-neighbor exchange J_2, exchange frustration occurs and the ground state may change. For an Ising-type spin chain, the ground state becomes $\uparrow\uparrow\downarrow\downarrow$ if $J_2/|J_1| > 0.5$ regardless of the sign of J_1, as shown in Figure 4.10b. For a Heisenberg-type spin chain, the ground state becomes a spiral order, with a wavelength of $2\pi/\arccos(-J_1/4J_2)$ if $J_2 > (-J_1/4)$.

Exchange frustration can also occur in any two-dimensional lattice. For example, in a triangular lattice as shown in Figure 4.10c, if all nearest-neighbor exchanges are uniform and positive, the three spins will be frustrated, giving rise to the $\uparrow\downarrow\downarrow$ or $\uparrow\uparrow\downarrow$ pattern for Ising-type spins, or 120° noncollinear pattern for Heisenberg-type spins. This kind of frustration is not due to the competing next-nearest-neighbor interaction, but the geometry. For a square lattice, as shown in Figure 4.10d, the next-nearest-neighbor exchanges may also bring frustration into the system, as in the one-dimensional case.

Let us take the orthorhombic $RMnO_3$ family as an example to show how the exchange frustration tunes the phase diagram. In the beginning, Kimura et al. [69] presented a simulated phase diagram (Figure 4.11a) of a Heisenberg spin model with $J_1-J_2-J_3$ exchanges. With increasing J_2, the ground state changes from A-type antiferromagnetism to E-type antiferromagnetism. In the intermediate region with $J_2 \sim -J_1$, incommensurate modulated spin order is obtained at finite temperatures. Later, Dong et al. [70] studied the two-orbital double-exchange model with both the nearest-neighbor and next-nearest-neighbor exchanges. The Jahn–Teller distortion was found to be important to obtain the desired wave numbers for TbMnO$_3$ and DyMnO$_3$, as shown in Figure 4.11b. Monte Carlo

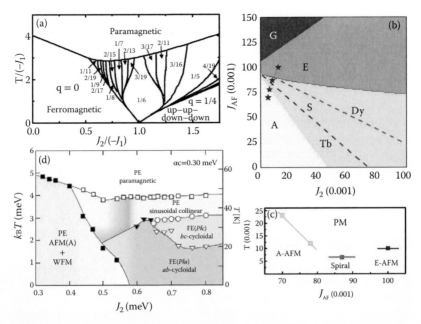

FIGURE 4.11 Phase diagrams of $RMnO_3$. (a) Mean-field result of the two-dimensional J_1–J_2–J_3 model. J_1 is the nearest-neighbor exchange, J_2 and J_3 are the next-nearest-neighbor exchange along the b and a axes, respectively. $J_3/J_1 = 0.01$. Each phase is characterized by a wave number q of the spin structure along the b axis. $q = 0$ and $1/4$ correspond to the A- and E-type antiferromagnetic states in $RMnO_3$, respectively [69]. (Reproduced from Kimura, T. et al., *Phys. Rev. B*, 68, 060403, 2003.) (b) Zero-temperature phase diagram of the two-orbital double-exchange model with Jahn–Teller distortions. The possible values of J_{AF}–J_{2b} for realistic wave numbers in $TbMnO_3$ and $DyMnO_3$ are shown with dashed lines. The four asterisks represent the set of couplings, which have been confirmed in the finite temperature Monte Carlo simulations as shown in [70] (c). (Reproduced from Dong, S. et al., *Phys. Rev. B*, 78, 155121, 2008.) (d) A typical finite temperature multiferroic phase diagram of the Mochizuki–Furukawa model as a function of J_2. AFM + WFM is the A-type antiferromagnetic phase with weak ferromagnetism due to spin canting. PE/FE denotes paraelectric–ferroelectric phase, respectively [77]. (Reproduced from Mochizuki, M., and Furukawa, N., *Phys. Rev. B*, 80, 134416, 2009.)

simulation also confirmed the phase evolution. In such a quantum model, the required next-nearest-neighbor interaction is small, about 10% of the nearest-neighbor superexchange, which is more realistic considering the orthorhombic structure.

In 2009, Mochizuki et al. constructed a classical Heisenberg spin model with many interactions; this led to a phase diagram close to the actual one, as shown in Figure 4.11d. Using this comprehensive model, Mochizuki et al. [71–76] studied many properties of $RMnO_3$, for example, magnetic switching of polarization, multiferroic excitation (electromagnons). An interesting discovery is that the exchange striction also contributes to the total polarization in the a–b plane spiral phase, together with the contribution from the Dzyaloshinskii–Moriya interaction. This can explain the experimental observation that the polarization along the a-axis for the a–b plane spiral is larger than that along the c-axis for the b–c plane spiral in $Eu_{1-x}Y_xMnO_3$.

With the help of the DFT-based calculations, the exchange interactions, as well as the Dzyaloshinskii–Moriya interaction and magnetic anisotropic energy, can be extracted by mapping the magnetic systems using the classical Heisenberg spin models. Xiang et al. [78] developed a general and efficient method to quantitatively calculate the coefficients of these interactions and their derivatives.

4.4 Pursuit of Better Performance

In the previous sections, we have summarized the progress in both the experimental and theoretical investigations of type-II multiferroics during the last decade. Despite the success in understanding the underlying physics, serious drawbacks remain for the type-II multiferroics from the view point of applications. These include low working temperatures (mostly below 40 K), very small ferroelectric polarization (mostly below 1000 $\mu C/m^2$, only 0.1% of that for typical perovskite ferroelectrics), and very weak magnetization (mostly antiferromagnetic). Furthermore, even though the polarization is directly driven by magnetism in these systems, the magnetoelectric coupling may be weak, leading to small response within low field region. To overcome these issues, various strategies have been explored and new type-II multiferroics with better performance have been developed.

4.4.1 CuO

In general, the critical temperature (T_C) is proportional to the exchange interaction (J): $T_C \propto J$. It is thus straightforward to look for high temperature type-II multiferroics in materials with a strong exchange interaction. Despite the conceptual difference between superconductivity and multiferroicity, the search for high-T_C superconductors has helped the magnetoelectric community to develop multiferroics with higher T_C.

In 2008, Kimura et al. [81] investigated CuO, since previous studies on cuprates suggested that it possessed a very strong exchange interaction (Figure 4.12). CuO has two magnetic transitions at 213 K and 230 K, respectively. Above 230 K, it is paramagnetic, and below 213 K, it is collinear antiferromagnetic. Between 213 and 230 K, a noncollinear spiral spin order develops [81]. Similar to that observed in spiral magnets, CuO shows a ferroelectric polarization within this narrow temperature window. Although the polarization is small (~160 $\mu C/m^2$), the working temperature is significantly higher. Later, DFT calculations further confirmed that the high T_C originates from the strong exchange interaction [82,83].

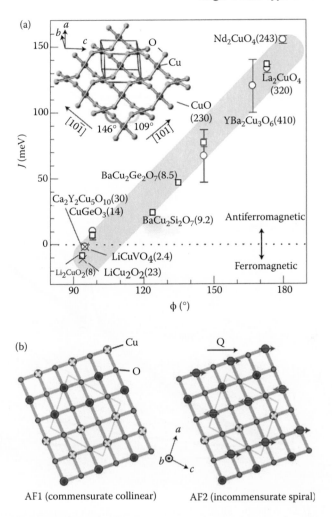

FIGURE 4.12 Multiferroicity of CuO. (a) Superexchange interaction J for various cuprates. ϕ is the Cu–O–Cu bond angle. CuO has an intermediate magnitude of J among the cuprates. The plot of J–ϕ for cuprates was originally presented in References [79,80]. $LiCu_2O_2$ and $LiCuVO_4$ are multiferroics with the spiral magnetic order. Numbers in parentheses denote magnetic ordering temperatures of the respective compounds. Inset: A schematic diagram of the crystal structure of CuO. (b) Sketch of the magnetic structures: the commensurate collinear (AF1) and the incommensurate spiral (AF2) phases. (c) Magnetic susceptibility as a function of temperature (d) Dielectric constant measured along the b-axis at various frequencies. Inset: A magnified view around T_{N1} at 100 kHz. (e) Electric polarization along the b axis as a function of temperature, poled under electric fields of +117 (black) and −117 (gray) kV/m above T_{N2} and cooled to the lower boundary of the AF2 phase (220 K) [81]. (Reproduced from Kimura, T. et al., *Nat. Mater.*, 7, 291–294, 2008.)

(Continued)

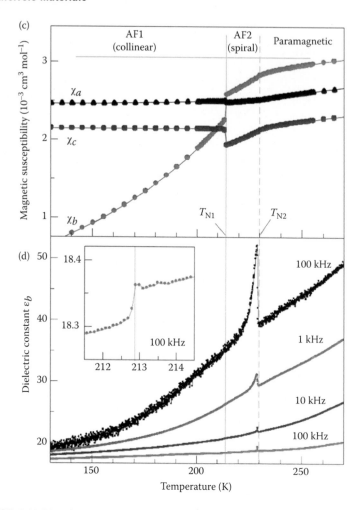

FIGURE 4.12 (*Continued*) Multiferroicity of CuO. (a) Superexchange interaction *J* for various cuprates. φ is the Cu–O–Cu bond angle. CuO has an intermediate magnitude of *J* among the cuprates. The plot of *J*–φ for cuprates was originally presented in References [79,80]. $LiCu_2O_2$ and $LiCuVO_4$ are multiferroics with the spiral magnetic order. Numbers in parentheses denote magnetic ordering temperatures of the respective compounds. Inset: A schematic diagram of the crystal structure of CuO. (b) Sketch of the magnetic structures: the commensurate collinear (AF1) and the incommensurate spiral (AF2) phases. (c) Magnetic susceptibility as a function of temperature (d) Dielectric constant measured along the *b*˙-axis at various frequencies. Inset: A magnified view around T_{N1} at 100 kHz. (e) Electric polarization along the *b*˙ axis as a function of temperature, poled under electric fields of +117 (black) and −117 (gray) kV/m above T_{N2} and cooled to the lower boundary of the AF2 phase (220 K) [81]. (Reproduced from Kimura, T. et al., *Nat. Mater.*, 7, 291–294, 2008.)

(Continued)

(e)

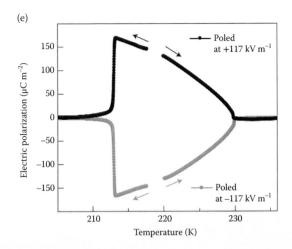

FIGURE 4.12 (*Continued*) Multiferroicity of CuO. (a) Superexchange interaction J for various cuprates. ϕ is the Cu–O–Cu bond angle. CuO has an intermediate magnitude of J among the cuprates. The plot of J–ϕ for cuprates was originally presented in References [79,80]. $LiCu_2O_2$ and $LiCuVO_4$ are multiferroics with the spiral magnetic order. Numbers in parentheses denote magnetic ordering temperatures of the respective compounds. Inset: A schematic diagram of the crystal structure of CuO. (b) Sketch of the magnetic structures: the commensurate collinear (AF1) and the incommensurate spiral (AF2) phases. (c) Magnetic susceptibility as a function of temperature (d) Dielectric constant measured along the b^*-axis at various frequencies. Inset: A magnified view around T_{N1} at 100 kHz. (e) Electric polarization along the b^* axis as a function of temperature, poled under electric fields of +117 (black) and −117 (gray) kV/m above T_{N2} and cooled to the lower boundary of the AF2 phase (220 K) [81]. (Reproduced from Kimura, T. et al., *Nat. Mater.*, 7, 291–294, 2008.)

4.4.2 Hexaferrite

Fe ions of various valences also show large $3d$ magnetic moments, which may provide opportunities to find magnetic ferroelectrics. In 2005, Kimura et al. [85] reported the control of electric polarization by magnetic fields in a Y-type hexaferrite $Ba_{0.5}Sr_{1.5}Zn_2Fe_{12}O_{22}$. Although it is a nonferroelectric helimagnetic insulator at the zero-field ground state, the system undergoes successive metamagnetic transitions under external magnetic fields, and shows a concomitant ferroelectric polarization in some of the magnetic-induced phases with long-wavelength magnetic structures. The polarization can be rotated 360° by external magnetic fields. This function opens up the potential not only for room temperature magnetoelectric devices but also for devices based on the magnetically controlled electro-optical response.

In 2008, Ishiwata et al. [84] reported low-magnetic-field control of electric polarization vector in a Y-type hexaferrite $Ba_2Mg_2Fe_{12}O_{22}$ (Figure 4.13).

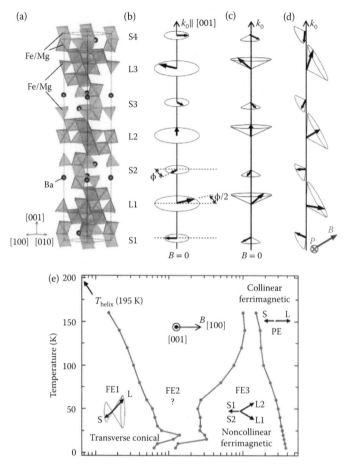

FIGURE 4.13 Y-type hexaferrite $Ba_2Mg_2Fe_{12}O_{22}$. (a) Schematic crystal structure. The magnetic structure consists of alternating stacks of L blocks (brown, large magnetic moment) and S blocks (green, small magnetic moment). (b–d) Illustrations of helicoidal spin structures: (b) proper screw type when the temperature is between 50 and 195 K; (c) longitudinal conical type below 50 K; (d) slanted conical type below 195 K under a small magnetic field (~30 mT). (e) Magnetoelectric phase diagram. (f) Magnetization and (g) ferroelectric polarization as a function of magnetic field at 5 K. Inset: The temperature-dependent magnetization measured when heating under zero-field after field cooling under 5 T magnetic field. Plausible spin configurations within each phase are shown. Phase boundaries in (f) and (g) are indicated by dashed lines. In (e–g), the magnetic field is applied along the [001] axis. (h) Schematic drawing of change of the spin cone and helicity under an oscillating magnetic field [84]. (Reproduced from Ishiwata, S. et al., *Science*, 319, 1643–1646, 2008.)

(Continued)

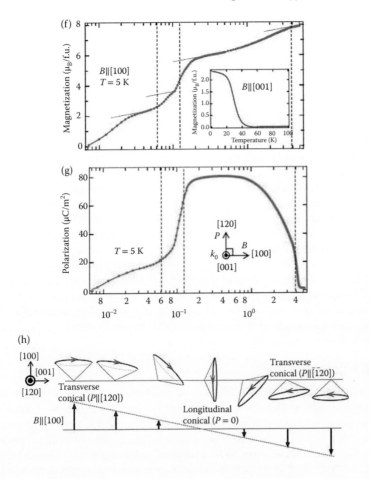

FIGURE 4.13 (*Continued*) Y-type hexaferrite $Ba_2Mg_2Fe_{12}O_{22}$. (a) Schematic crystal structure. The magnetic structure consists of alternating stacks of L blocks (brown, large magnetic moment) and S blocks (green, small magnetic moment). (b–d) Illustrations of helicoidal spin structures: (b) proper screw type when the temperature is between 50 and 195 K; (c) longitudinal conical type below 50 K; (d) slanted conical type below 195 K under a small magnetic field (~30 mT). (e) Magnetoelectric phase diagram. (f) Magnetization and (g) ferroelectric polarization as a function of magnetic field at 5 K. Inset: The temperature-dependent magnetization measured when heating under zero-field after field cooling under 5 T magnetic field. Plausible spin configurations within each phase are shown. Phase boundaries in (f) and (g) are indicated by dashed lines. In (e–g), the magnetic field is applied along the [001] axis. (h) Schematic drawing of change of the spin cone and helicity under an oscillating magnetic field [84]. (Reproduced from Ishiwata, S. et al., *Science*, 319, 1643–1646, 2008.)

Its spin structure is helimagnetic with the propagation vector k_0 parallel to [001] axis. The external magnetic field–induced transverse conical spin structure carries the P vector perpendicular to both the magnetic field and k_0, consistent with the spin current model discussed previously. The oscillating magnetic field produces the cyclic displacement current by continuously changing the magnetic cone axis.

The advantages of $Ba_2Mg_2Fe_{12}O_{22}$ are twofold. First, the magnetoelectric response is very sensitive. Different from most type-II multiferroics, which usually require large magnetic fields up to several Tesla to reverse/rotate/suppress the electric polarization, the polarization of $Ba_2Mg_2Fe_{12}O_{22}$ responses to magnetic field as small as 0.03 Tesla. Second, the ferroelectric Curie temperature can be relatively high. The proper screw spin structure emerges below 195 K. Although it is nonferroelectric, it can be easily tuned continuously to the ferroelectric conical spin order by the magnetic field [84].

There are other types of hexaferrites. For example, Kitagawa et al. [86] reported the low-field magnetoelectric effect at room temperature in Z-type hexaferrite $Sr_3Co_2Fe_{24}O_{41}$. The magnetic transition temperatures are 510 K and 670 K, well above the room temperature. At room temperature, it shows a large magnetic moment 6–8 μ_B per formula unit. A small magnetic field of 0.2 Tesla can induce a polarization of up to 20 $\mu C/m^2$ at room temperature. For more information about the progress of multiferroic hexaferrites, readers can refer to a recent review by Kimura [87].

4.4.3 $CaMn_7O_{12}$

Even in manganites, the multiferroic performance can be improved. In 2009, Dong et al. [88] predicted a spin-orthogonal stripe state in quarter-hole-doped perovskite manganites, which should be multiferroic due to the noncollinear spin order as well as the exchange striction. The estimated critical temperature may reach 100 K, three to four times of those of $RMnO_3$. The underlying mechanism to stabilize the noncollinear spin structure is not the traditional exchange frustration involving the next-nearest-neighbor exchanges, as in $TbMnO_3$. Instead, the electronic self-organization is responsible for such spin-orthogonal stripes. Further studies extended the spin-orthogonal stripe state to other doping concentrations, for example, 1/2, 1/3, 1/5, 1/6, ..., 1/∞ [89].

Then how can we realize it? According to the theoretical phase diagram, it requires the following parameters to be fulfilled: (1) a very small lattice, (2) quarter doping, (3) zero or weak Jahn-Teller distortion, (4) insulating, and (5) zero or weak disorder [88]. It seems almost impossible to fulfill all these conditions in a typical manganite. However, there is a special case: quadruple perovskite $CaMn_7O_{12}$, which can also be written as $(CaMn_3)Mn_4O_{12}$ [90]. Three Mn^{3+} ions occupy the A-site together with one Ca^{2+} ion, which are fully ordered, as shown in Figure 4.14a and b. The average valence of the remaining four Mn ions is +3.75, exactly at the quarter doping. The lattice constants are extremely small when compared with normal manganites, but with quite a weak Jahn-Teller distortion [91]. And this system is insulating in nature. All conditions required seem to be satisfied!

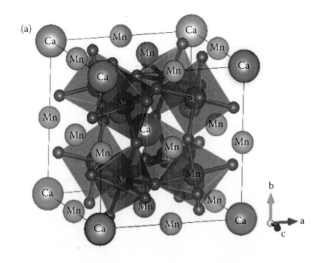

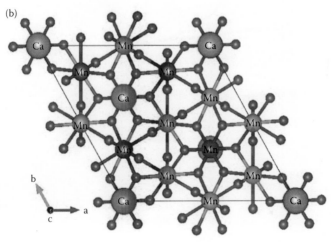

FIGURE 4.14 CaMn$_7$O$_{12}$. (a) Cubic (*Im$\overline{3}$*) crystal structure at high temperatures (>440 K). (b) Rhombohedral (*R$\overline{3}$*) crystal structure at low temperatures (<440 K) [33]. (Reproduced from Dong, S., and Liu, J.-M., *Mod. Phys. Lett. B*, 26, 1230004, 2012.) (c) Polarization of CaMn$_7$O$_{12}$ along the hexagonal *c* axis (black) and the three pseudocubic <100> axes. Inset: A small anomaly at T_{N2} = 48 K. (d) The magnetic susceptibility parallel and perpendicular to the hexagonal *c* axis, measured under a 500 Oe magnetic field, in both zero field-cooled and field-cooled (500 Oe) conditions. Inset: The in-plane magnetic susceptibility at T_{N1}. (e) Temperature dependence of incommensurate magnetic propagation along the *c*-axis. (f–e) Magnetic structure of CaMn$_7$O$_{12}$ in the *ab* and *ac* planes, respectively. Mn1, Mn2, and Mn3 are shown in red, black, and yellow, respectively. The magnetic moments rotate in the *ab* plane with a circular envelope as depicted in [92] (g). (Reproduced from Johnson, R. et al., *Phys. Rev. Lett.*, 108, 067201, 2012.)

(Continued)

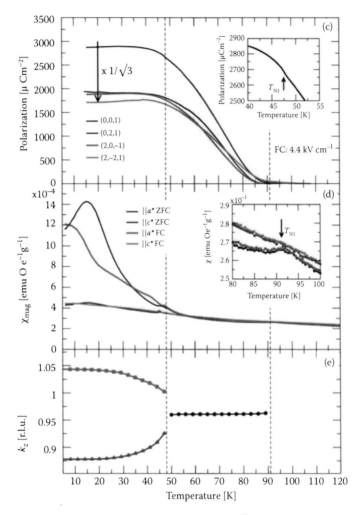

FIGURE 4.14 (Continued) CaMn$_7$O$_{12}$. (a) Cubic (*Im*$\bar{3}$) crystal structure at high temperatures (>440 K). (b) Rhombohedral (*R*$\bar{3}$) crystal structure at low temperatures (<440 K) [33]. (Reproduced from Dong, S., and Liu, J.-M., *Mod. Phys. Lett. B*, 26, 1230004, 2012.) (c) Polarization of CaMn$_7$O$_{12}$ along the hexagonal *c* axis (black) and the three pseudocubic <100> axes. Inset: A small anomaly at T_{N2} = 48 K. (d) The magnetic susceptibility parallel and perpendicular to the hexagonal *c* axis, measured under a 500 Oe magnetic field, in both zero field-cooled and field-cooled (500 Oe) conditions. Inset: The in-plane magnetic susceptibility at T_{N1}. (e) Temperature dependence of incommensurate magnetic propagation along the *c*-axis. (f–e) Magnetic structure of CaMn$_7$O$_{12}$ in the *ab* and *ac* planes, respectively. Mn1, Mn2, and Mn3 are shown in red, black, and yellow, respectively. The magnetic moments rotate in the *ab* plane with a circular envelope as depicted in [92] (g). (Reproduced from Johnson, R. et al., *Phys. Rev. Lett.*, 108, 067201, 2012.)

(Continued)

(f)

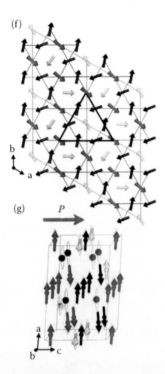

(g)

FIGURE 4.14 (*Continued*) $CaMn_7O_{12}$. (a) Cubic ($Im\bar{3}$) crystal structure at high temperatures (>440 K). (b) Rhombohedral ($R\bar{3}$) crystal structure at low temperatures (<440 K) [33]. (Reproduced from Dong, S., and Liu, J.-M., *Mod. Phys. Lett. B*, 26, 1230004, 2012.) (c) Polarization of $CaMn_7O_{12}$ along the hexagonal c axis (black) and the three pseudocubic <100> axes. Inset: A small anomaly at T_{N2} = 48 K. (d) The magnetic susceptibility parallel and perpendicular to the hexagonal c axis, measured under a 500 Oe magnetic field, in both zero field-cooled and field-cooled (500 Oe) conditions. Inset: The in-plane magnetic susceptibility at T_{N1}. (e) Temperature dependence of incommensurate magnetic propagation along the c-axis. (f–e) Magnetic structure of $CaMn_7O_{12}$ in the ab and ac planes, respectively. Mn1, Mn2, and Mn3 are shown in red, black, and yellow, respectively. The magnetic moments rotate in the ab plane with a circular envelope as depicted in [92] (g). (Reproduced from Johnson, R. et al., *Phys. Rev. Lett.*, 108, 067201, 2012.)

Zhang et al. [93] synthesized polycrystalline $CaMn_7O_{12}$ samples and found a magnetic-related electric polarization below 90 K. There are two magnetic transitions at 90 K and 48 K, respectively. Later, Johnson et al. [92] confirmed the magnetic ferroelectricity using a single crystal and resolved the magnetic order between 48 K and 90 K, as shown in Figure 4.14c through g. The polarization can reach 2870 $\mu C/m^2$, larger than that of $DyMnO_3$ [92]. Johnson et al. [92] proposed a new mechanism for the ferroelectricity in $CaMn_7O_{12}$, coupled to the ferroaxial component of the crystal structure. Alternatively, based on first-principle calculations, Lu et al. [94]

proposed that the combination of the Dzyaloshinskii–Moriya interaction and exchange striction is responsible for the giant ferroelectric polarization and strong magnetoelectric response. The exchange striction contributes most of the polarization, while the Dzyaloshinskii–Moriya interaction controls the direction of polarization.

4.4.4 BaFe$_2$Se$_3$

Besides cuprates, iron-based pnictides and chalcogenides have attracted much attention in the condensed matter community since 2008 because of their superconducting properties [95–97]. Recently, Dong et al. [98] predicted that the iron-selenide BaFe$_2$Se$_3$ is a magnetic ferrielectric with attractive performance.

The crystal structure of BaFe$_2$Se$_3$ is shown in Figure 4.15a and b, which has two iron ladders (labeled as A and B) in each unit. Long-range block-type antiferromagnetic order (Figure 4.15b and c) is established below 256 K, confirmed by both neutron studies and first-principles calculations [99–102].

The block antiferromagnetic order is a variant of ↑↑↓↓ spin chain; therefore, it induces exchange striction. Indeed, as revealed by neutron studies, the

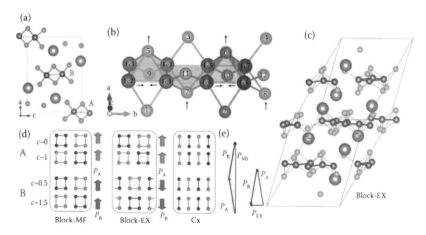

FIGURE 4.15 BaFe$_2$Se$_3$. (a) Side view of the crystal structure along the *b*-axis. (b) A Fe–Se ladder along the *b*-axis and its magnetic order. Partial ionic displacements driven by the exchange striction are marked (black arrows). (c) A unit cell considering the antiferromagnetic magnetic order. (d) Spin structures. Left: Block-MF; middle: Block-EX; right: Cx. The side arrows denote the local polarizations of the ladders. In (b–d), the spins (↑/↓) of Fe ions are distinguished by colors. (e) Vector addition of ferroelectric polarizations of ladders A and B. For the Block-MF case, the two ferroelectric polarizations are nearly parallel giving a large polarization P_{MF}. For the Block-EX case, the two ferroelectric polarizations are nearly antiparallel, giving a small polarization P_{MF} induced by the tilting of the iron ladders [98]. (Reproduced from Dong, S. et al., *Phys. Rev. Lett.* 113, 187204, 2014.)

iron displacements are prominent [99,101,103,104]: the nearest-neighbor distance between Fe(↑)–Fe(↑) [or Fe(↓)–Fe(↓)] at 200 K becomes 2.593 Å, shorter than the Fe(↑)–Fe(↓) distance of 2.840 Å [104]. As shown in Figure 4.15b, the Se(5) and Se(7) positions do not need to be antisymmetric with respect to the Fe ladder plane anymore. The same mechanism works for the edge Se's, for example, Se(1) and Se(11). As a consequence, the Se atomic positions break the space-inversion symmetry, generating a local ferroelectric polarization perpendicular to the iron ladders plane.

According to neutron studies [99], the block antiferromagnetic pattern has a $\pi/2$-phase shift between the nearest-neighbor A–B ladders but a π-phase shift between the nearest-neighbor A–A ladders (and nearest-neighbor B–B ladders), as in the Block-EX state shown in Figure 4.15d. The $\pi/2$- phase shift between A–B ladders will induce (nearly) opposite ferroelectric polarizations, as sketched in Figure 4.15d and e. A full cancellation does not occur due to a small canting between the ladder A and B planes (see Figure 4.15a), leading to a residual ferroelectric polarization (P_{EX}) pointing almost along the c-axis (Figure 4.15e). If the $\pi/2$-phase shift between ladders A and B is eliminated, the magnetic structure becomes the Block-MF state. Then, the magnetism-induced ferroelectric polarizations of all ladders will coherently produce a combined polarization P_{MF} pointing along the a-axis (Figure 4.15e).

DFT calculations confirm the above predictions. Both the Block-EX and Block-MF states are found to be multiferroic. P_{MF} is large and mostly along the a-axis ($\sim$2–3 $\mu C/cm^2$), among the largest values reported for type-II multiferroics. The net polarization of Block-EX state is mostly along the c-axis and its amplitude is about 0.19 $\mu C/cm^2$ when using generalized gradient approximation (GGA) in the calculation, which is one order of magnitude smaller than P_{MF} as expected.

4.5 Summary and Perspective

Since the discovery of $TbMnO_3$ in 2003, the field of type-II multiferroic has flourished in the past decade. More and more magnetic ferroelectrics have been discovered. The theoretical framework has been established from both phenomenological and quantum mechanical points of view. On one hand, type-II multiferroics share common features with other correlated electron materials (e.g., high-T_C superconducting cuprates and colossal magnetoresistive manganites). The interplay between electrons and the cross-couplings among charge-spin-orbital-lattice degrees of freedom lead to many fascinating emergent phenomena. On the other hand, type-II multiferroics are also unique. While high-T_C superconducting cuprates and colossal magnetoresistive manganites are restricted to particular transition metal ions with certain crystal structures, type-II multiferroics can be found in various crystal structures with various 3d metal ions. The studies on type-II multiferroics will continue. Further investigation on known physical mechanisms and more intensive search for new materials with better performances will lead to more exciting discoveries.

Acknowledgment

This work was supported by the National Natural Science Foundation of China under Grants 11274060 and 51322206.

References

1. Cheong, S. W., and M. Mostovoy. 2007. Multiferroics: A magnetic twist for ferroelectricity. *Nature Materials* 6:13–20.
2. Wang, K. F., J. M. Liu, and Z. F. Ren. 2009. Multiferroicity: The coupling between magnetic and polarization orders. *Advances in Physics* 58:321–448.
3. Picozzi, S., and C. Ederer. 2009. First principles studies of multiferroic materials. *Journal of Physics-Condensed Matter* 21:303201.
4. Khomskii, D. 2009. Classifying multiferroics: Mechanisms and effects. *Physics* 2:20.
5. Tokura, Y. 2006. Materials science—Multiferroics as quantum electromagnets. *Science* 312:1481–1482.
6. Tokura, Y. 2000. *Colossal Magnetoresistive Oxides*. London: Gordon and Breach Science Publishers.
7. Dagotto, E., T. Hotta, and A. Moreo. 2001. Colossal magnetoresistant materials: The key role of phase separation. *Physics Reports—Review Section of Physics Letters* 344:1–153.
8. Dagotto, E. 2002. *Nanoscale Phase Separation and Colossal Magnetoresistance*. Berlin: Springer.
9. Tokura, Y. 2006. Critical features of colossal magnetoresistive manganites. *Reports on Progress in Physics* 69:797–851.
10. Kimura, T., T. Goto, H. Shintani, K. Ishizaka, T. Arima, and Y. Tokura. 2003. Magnetic control of ferroelectric polarization. *Nature* 426:55–58.
11. Kenzelmann, M., A. Harris, S. Jonas, C. Broholm, J. Schefer, S. Kim, C. Zhang, S.-W. Cheong, O. Vajk, and J. Lynn. 2005. Magnetic inversion symmetry breaking and ferroelectricity in $TbMnO_3$. *Physical Review Letters* 95:087206.
12. Arima, T., A. Tokunaga, T. Goto, H. Kimura, Y. Noda, and Y. Tokura. 2006. Collinear to spiral spin transformation without changing the modulation wavelength upon ferroelectric transition in $Tb_{1-x}Dy_xMnO_3$. *Physical Review Letters* 96:097202.
13. Kimura, T., G. Lawes, T. Goto, Y. Tokura, and A. Ramirez. 2005. Magnetoelectric phase diagrams of orthorhombic $RMnO_3$ (R = Gd, Tb, and Dy). *Physical Review B* 71:224425.
14. Wang, J., J. B. Neaton, H. Zheng, V. Nagarajan, S. B. Ogale, B. Liu, D. Viehland, V. Vaithyanathan, D. G. Schlom, U. V. Waghmare, N. A. Spaldin, K. M. Rabe, M. Wuttig, and R. Ramesh. 2003. Epitaxial $BiFeO_3$ multiferroic thin film heterostructures. *Science* 299:1719–1722.
15. Choi, T., S. Lee, Y. J. Choi, V. Kiryukhin, and S. W. Cheong. 2009. Switchable ferroelectric diode and photovoltaic effect in $BiFeO_3$. *Science* 324:63–66.
16. Kimura, T. 2007. Spiral magnets as magnetoelectrics. *Annual Review of Materials Research* 37:387–413.
17. Goto, T., T. Kimura, G. Lawes, A. Ramirez, and Y. Tokura. 2004. Ferroelectricity and giant magnetocapacitance in perovskite rare-earth manganites. *Physical Review Letters* 92:257201.
18. Munoz, A., M. Casáis, J. Alonso, M. Martínez-Lope, J. Martinez, and M. Fernandez-Diaz. 2001. Complex magnetism and magnetic structures of the metastable $HoMnO_3$ perovskite. *Inorganic Chemistry* 40:1020–1028.

19. Hotta, T., M. Moraghebi, A. Feiguin, A. Moreo, S. Yunoki, and E. Dagotto. 2003. Unveiling new magnetic phases of undoped and doped manganites. *Physical Review Letters* 90:247203.

20. Sergienko, I. A., C. Şen, and E. Dagotto. 2006. Ferroelectricity in the magnetic E-phase of orthorhombic perovskites. *Physical Review Letters* 97:227204.

21. Lorenz, B., Y.-Q. Wang, and C.-W. Chu. 2007. Ferroelectricity in perovskite $HoMnO_3$ and $YMnO_3$. *Physical Review B* 76:104405.

22. Ishiwata, S., Y. Kaneko, Y. Tokunaga, Y. Taguchi, T.-h. Arima, and Y. Tokura. 2010. Perovskite manganites hosting versatile multiferroic phases with symmetric and antisymmetric exchange strictions. *Physical Review B* 81:100411.

23. Picozzi, S., K. Yamauchi, B. Sanyal, I. A. Sergienko, and E. Dagotto. 2007. Dual nature of improper ferroelectricity in a magnetoelectric multiferroic. *Physical Review Letters* 99:227201.

24. Pomjakushin, V. Y., M. Kenzelmann, A. Dönni, A. Harris, T. Nakajima, S. Mitsuda, M. Tachibana, L. Keller, J. Mesot, and H. Kitazawa. 2009. Evidence for large electric polarization from collinear magnetism in $TmMnO_3$. *New Journal of Physics* 11:043019.

25. Ishiwata, S., Y. Tokunaga, Y. Taguchi, and Y. Tokura. 2011. High-pressure hydrothermal crystal growth and multiferroic properties of a perovskite $YMnO_3$. *Journal of the American Chemical Society* 133:13818–13820.

26. Nakamura, M., Y. Tokunaga, M. Kawasaki, and Y. Tokura. 2011. Multiferroicity in an orthorhombic $YMnO_3$ single-crystal film. *Applied Physics Letters* 98:082902.

27. Wadati, H., J. Okamoto, M. Garganourakis, V. Scagnoli, U. Staub, Y. Yamasaki, H. Nakao, Y. Murakami, M. Mochizuki, and M. Nakamura. 2012. Origin of the large polarization in multiferroic $YMnO_3$ thin films revealed by soft-and hard-X-ray diffraction. *Physical Review Letters* 108:047203.

28. Lu, C., S. Dong, K. Wang, and J.-M. Liu. 2010. Enhanced polarization and magnetoelectric response in $Tb_{1-x}Ho_xMnO_3$. *Applied Physics A* 99:323–331.

29. Schierle, E., V. Soltwisch, D. Schmitz, R. Feyerherm, A. Maljuk, F. Yokaichiya, D. Argyriou, and E. Weschke. 2010. Cycloidal order of 4f moments as a probe of chiral domains in $DyMnO_3$. *Physical Review Letters* 105:167207.

30. Feyerherm, R., E. Dudzik, O. Prokhnenko, and D. Argyriou. 2010. Rare earth magnetism and ferroelectricity in $RMnO_3$. *Journal of Physics: Conference Series* 200:012032.

31. Zhang, N., Y. Guo, L. Lin, S. Dong, Z. Yan, X. Li, and J.-M. Liu. 2011. Ho substitution suppresses collinear Dy spin order and enhances polarization in $DyMnO_3$. *Applied Physics Letters* 99:102509.

32. Lee, N., Y. Choi, M. Ramazanoglu, I. W Ratcliff, V. Kiryukhin, and S.-W. Cheong. 2011. Mechanism of exchange striction of ferroelectricity in multiferroic orthorhombic $HoMnO_3$ single crystals. *Physical Review B* 84:020101.

33. Dong, S., and J.-M. Liu. 2012. Recent progress of multiferroic perovskite manganites. *Modern Physics Letters B* 26:1230004.

34. Hur, N., S. Park, P. A. Sharma, J. S. Ahn, S. Guha, and S. W. Cheong. 2004. Electric polarization reversal and memory in a multiferroic material induced by magnetic fields. *Nature* 429:392–395.

35. Kim, J. W., S. Haam, Y. Oh, S. Park, S.-W. Cheong, P. Sharma, M. Jaime, N. Harrison, J. H. Han, and G.-S. Jeon. 2009. Observation of a multiferroic critical end point. *Proceedings of the National Academy of Sciences* 106:15573–15576.

36. Zhao, Z., M. Liu, X. Li, L. Lin, Z. Yan, S. Dong, and J.-M. Liu. 2014. Experimental observation of ferrielectricity in multiferroic $DyMn_2O_5$. *Scientific reports* 4:3984.

37. Choi, Y., H. Yi, S. Lee, Q. Huang, V. Kiryukhin, and S.-W. Cheong. 2008. Ferroelectricity in an Ising chain magnet. *Physical Review Letters* 100:047601.

38. Jo, Y., S. Lee, E. Choi, H. Yi, I. W Ratcliff, Y. Choi, V. Kiryukhin, S. Cheong, and L. Balicas. 2009. 3:1 magnetization plateau and suppression of ferroelectric polarization in an Ising chain multiferroic. *Physical Review B* 79:012407.

39. Kiryukhin, V., S. Lee, I. W Ratcliff, Q. Huang, H. Yi, Y. Choi, and S. Cheong. 2009. Order by static disorder in the Ising chain magnet $Ca_3Co_{2-x}Mn_xO_6$. *Physical Review Letters* 102:187202.

40. Ding, P., L. Li, Y. Guo, Q. He, X. Gao, and J. Liu. 2010. Influence of Co: Mn ratio on multiferroicity of $Ca_3Co_{2-x}Mn_xO_6$ around x~ 1. *Applied Physics Letters* 97:032901.

41. Wu, H., T. Burnus, Z. Hu, C. Martin, A. Maignan, J. Cezar, A. Tanaka, N. Brookes, D. Khomskii, and L. Tjeng. 2009. Ising magnetism and ferroelectricity in Ca_3CoMnO_6. *Physical Review Letters* 102:026404.

42. Zhang, Y., H. Xiang, and M.-H. Whangbo. 2009. Interplay between Jahn-Teller instability, uniaxial magnetism, and ferroelectricity in Ca_3CoMnO_6. *Physical Review B* 79:054432.

43. Kagawa, F., S. Horiuchi, M. Tokunaga, J. Fujioka, and Y. Tokura. 2010. Ferroelectricity in a one-dimensional organic quantum magnet. *Nature Physics* 6:169–172.

44. Kimura, T., J. Lashley, and A. Ramirez. 2006. Inversion-symmetry breaking in the noncollinear magnetic phase of the triangular-lattice antiferromagnet $CuFeO_2$. *Physical Review B* 73:220401.

45. Taniguchi, K., N. Abe, T. Takenobu, Y. Iwasa, and T. Arima. 2006. Ferroelectric polarization flop in a frustrated magnet $MnWO_4$ induced by a magnetic field. *Physical Review Letters* 97:097203.

46. Heyer, O., N. Hollmann, I. Klassen, S. Jodlauk, L. Bohaty, P. Becker, J. A. Mydosh, T. Lorenz, and D. Khomskii. 2006. A new multiferroic material: $MnWO_4$. *Journal of Physics-Condensed Matter* 18:L471–L475.

47. Lawes, G., A. Harris, T. Kimura, N. Rogado, R. Cava, A. Aharony, O. Entin-Wohlman, T. Yildirim, M. Kenzelmann, and C. Broholm. 2005. Magnetically driven ferroelectric order in $Ni_3V_2O_8$. *Physical Review Letters* 95:087205.

48. Yamasaki, Y., H. Sagayama, N. Abe, T. Arima, K. Sasai, M. Matsuura, K. Hirota, D. Okuyama, Y. Noda, and Y. Tokura. 2008. Cycloidal spin order in the a-axis polarized ferroelectric phase of orthorhombic perovskite manganite. *Physical Review Letters* 101:097204.

49. Park, S., Y. Choi, C. Zhang, and S. Cheong. 2007. Ferroelectricity in an S = 1/2 chain cuprate. *Physical Review Letters* 98:057601.

50. Naito, Y., K. Sato, Y. Yasui, Y. Kobayashi, Y. Kobayashi, and M. Sato. 2007. Ferroelectric transition induced by the incommensurate magnetic ordering in $LiCuVO_4$. *Journal of the Physical Society of Japan* 76:023708.

51. Tokunaga, Y., N. Furukawa, H. Sakai, Y. Taguchi, T.-h. Arima, and Y. Tokura. 2009. Composite domain walls in a multiferroic perovskite ferrite. *Nature Materials* 8:558–562.

52. Lee, J.-H., Y. K. Jeong, J. H. Park, M.-A. Oak, H. M. Jang, J. Y. Son, and J. F. Scott. 2011. Spin-canting-induced improper ferroelectricity and spontaneous magnetization reversal in $SmFeO_3$. *Physical Review Letters* 107:117201.

53. Giovannetti, G., S. Kumar, D. Khomskii, S. Picozzi, and J. van den Brink. 2009. Multiferroicity in rare-earth nickelates $RNiO_3$. *Physical Review Letters* 103:156401.

54. Mostovoy, M. 2006. Ferroelectricity in spiral magnets. *Physical Review Letters* 96:067601.

55. Dzyaloshinsky, I. 1958. A thermodynamic theory of "weak" ferromagnetism of antiferromagnetics. *Journal of Physics and Chemistry of Solids* 4:241–255.

56. Moriya, T. 1960. Anisotropic superexchange interaction and weak ferromagnetism. *Physical Review* 120:91.

57. Iwasaki, J., M. Mochizuki, and N. Nagaosa. 2013. Universal current-velocity relation of skyrmion motion in chiral magnets. *Nature Communications* 4:1463.

58. Glazer, A. 1972. The classification of tilted octahedra in perovskites. *Acta Crystallographica Section B: Structural Crystallography and Crystal Chemistry* 28:3384–3392.

59. Woodward, P. M. 1997. Octahedral tilting in perovskites. I. Geometrical considerations. *Acta Crystallographica Section B: Structural Science* 53:32–43.

60. Repeated Author. 1997. Octahedral tilting in perovskites. II. structure stabilizing forces. *Acta Crystallographica Section B: Structural Science* 53:44–66.

61. Sergienko, I. A., and E. Dagotto. 2006. Role of the Dzyaloshinskii-Moriya interaction in multiferroic perovskites. *Physical Review B* 73:094434.

62. Katsura, H., N. Nagaosa, and A. V. Balatsky. 2005. Spin current and magnetoelectric effect in noncollinear magnets. *Physical Review Letters* 95:057205.

63. Arima, T.-h. 2007. Ferroelectricity induced by proper-screw type magnetic order. *Journal of the Physical Society of Japan* 76:073702.

64. Murakawa, H., Y. Onose, S. Miyahara, N. Furukawa, and Y. Tokura. 2010. Ferroelectricity induced by spin-dependent metal-ligand hybridization in $Ba_2CoGe_2O_7$. *Physical Review Letters* 105:137202.

65. Xiang, H., E. Kan, Y. Zhang, M.-H. Whangbo, and X. Gong. 2011. General theory for the ferroelectric polarization induced by spin-spiral order. *Physical Review Letters* 107:157202.

66. Xiang, H., P. Wang, M.-H. Whangbo, and X. Gong. 2013. Unified model of ferroelectricity induced by spin order. *Physical Review B* 88:054404.

67. Yang, J.-H., Z.-L. Li, X. Lu, M.-H. Whangbo, S.-H. Wei, X. Gong, and H. Xiang. 2012. Strong Dzyaloshinskii-Moriya interaction and origin of ferroelectricity in Cu_2OSeO_3. *Physical Review Letters* 109:107203.

68. Ramirez, A. 1994. Strongly geometrically frustrated magnets. *Annual Review of Materials Science* 24:453–480.

69. Kimura, T., S. Ishihara, H. Shintani, T. Arima, K. Takahashi, K. Ishizaka, and Y. Tokura. 2003. Distorted perovskite with $e_g(1)$ configuration as a frustrated spin system. *Physical Review B* 68:060403.

70. Dong, S., R. Yu, S. Yunoki, J.-M. Liu, and E. Dagotto. 2008. Origin of multiferroic spiral spin order in the $RMnO_3$ perovskites. *Physical Review B* 78:155121.

71. Mochizuki, M., and N. Furukawa. 2009. Mechanism of lattice-distortion-induced electric-polarization flop in the multiferroic perovskite manganites. *Journal of the Physical Society of Japan* 78:053704.

72. Mochizuki, M., N. Furukawa, and N. Nagaosa. 2010. Theory of electromagnons in the multiferroic Mn perovskites: the vital role of higher harmonic components of the spiral spin order. *Physical Review Letters* 104:177206.

73. Mochizuki, M., N. Furukawa, and N. Nagaosa. 2010. Spin model of magnetostrictions in multiferroic Mn perovskites. *Physical Review Letters* 105:037205.

74. Mochizuki, M., and N. Nagaosa. 2010. Theoretically predicted picosecond optical switching of spin chirality in multiferroics. *Physical Review Letters* 105:147202.

75. Mochizuki, M., and N. Furukawa. 2010. Theory of magnetic switching of ferroelectricity in spiral magnets. *Physical Review Letters* 105:187601.

76. Mochizuki, M., N. Furukawa, and N. Nagaosa. 2011. Theory of spin-phonon coupling in multiferroic manganese perovskites $RMnO_3$. *Physical Review B* 84:144409.

77. Mochizuki, M., and N. Furukawa. 2009. Microscopic model and phase diagrams of the multiferroic perovskite manganites. *Physical Review B* 80:134416.

78. Xiang, H., E. Kan, S.-H. Wei, M.-H. Whangbo, and X. Gong. 2011. Predicting the spin-lattice order of frustrated systems from first principles. *Physical Review B* 84:224429.

79. Shimizu, T., T. Matsumoto, A. Goto, K. Yoshimura, and K. Kosuge. 2003. Magnetic dimensionality of the antiferromagnet CuO. *Journal of the Physical Society of Japan* 72:2165–2168.

80. Mizuno, Y., T. Tohyama, S. Maekawa, T. Osafune, N. Motoyama, H. Eisaki, and S. Uchida. 1998. Electronic states and magnetic properties of edge-sharing Cu-O chains. *Physical Review B* 57:5326.

81. Kimura, T., Y. Sekio, H. Nakamura, T. Siegrist, and A. Ramirez. 2008. Cupric oxide as an induced-multiferroic with high-T_C. *Nature Materials* 7:291–294.

82. Giovannetti, G., S. Kumar, A. Stroppa, J. van den Brink, S. Picozzi, and J. Lorenzana. 2011. High-T_C ferroelectricity emerging from magnetic degeneracy in cupric oxide. *Physical Review Letters* 106:026401.

83. Jin, G., K. Cao, G.-C. Guo, and L. He. 2012. Origin of ferroelectricity in high-Tc magnetic ferroelectric CuO. *Physical Review Letters* 108:187205.

84. Ishiwata, S., Y. Taguchi, H. Murakawa, Y. Onose, and Y. Tokura. 2008. Low-magnetic-field control of electric polarization vector in a helimagnet. *Science* 319:1643–1646.

85. Kimura, T., G. Lawes, and A. Ramirez. 2005. Electric polarization rotation in a hexaferrite with long-wavelength magnetic structures. *Physical Review Letters* 94:137201.

86. Kitagawa, Y., Y. Hiraoka, T. Honda, T. Ishikura, H. Nakamura, and T. Kimura. 2010. Low-field magnetoelectric effect at room temperature. *Nature Materials* 9:797–802.

87. Kimura, T. 2012. Magnetoelectric hexaferrites. *Annu. Rev. Condens. Matter Phys.* 3:93–110.

88. Dong, S., R. Yu, J.-M. Liu, and E. Dagotto. 2009. Striped multiferroic phase in double-exchange model for quarter-doped manganites. *Physical Review Letters* 103:107204.

89. Liang, S., M. Daghofer, S. Dong, C. Şen, and E. Dagotto. 2011. Emergent dimensional reduction of the spin sector in a model for narrow-band manganites. *Physical Review B* 84:024408.

90. Vasil'ev, A., and O. Volkova. 2007. New functional materials $AC_3B_4O_{12}$ (Review). *Low Temperature Physics* 33:895–914.

91. Przenioslo, R., I. Sosnowska, E. Suard, A. Hewat, and A. Fitch. 2004. Charge ordering and anisotropic thermal expansion of the manganese perovskite $CaMn_7O_{12}$. *Physica B: Condensed Matter* 344:358–367.

92. Johnson, R., L. Chapon, D. Khalyavin, P. Manuel, P. Radaelli, and C. Martin. 2012. Giant improper ferroelectricity in the ferroaxial magnet $CaMn_7O_{12}$. *Physical Review Letters* 108:067201.

93. Zhang, G., S. Dong, Z. Yan, Y. Guo, Q. Zhang, S. Yunoki, E. Dagotto, and J.-M. Liu. 2011. Multiferroic properties of $CaMn_7O_{12}$. *Physical Review B* 84:174413.

94. Lu, X., M.-H. Whangbo, S. Dong, X. Gong, and H. Xiang. 2012. Giant ferroelectric polarization of $CaMn_7O_{12}$ induced by a combined effect of Dzyaloshinskii-Moriya interaction and exchange striction. *Physical Review Letters* 108:187204.

95. Johnston, D. C. 2010. The puzzle of high temperature superconductivity in layered iron pnictides and chalcogenides. *Advances in Physics* 59:803–1061.

96. Stewart, G. 2011. Superconductivity in iron compounds. *Reviews of Modern Physics* 83:1589.

97. Dagotto, E. 2013. Colloquium: The unexpected properties of alkali metal iron selenide superconductors. *Reviews of Modern Physics* 85:849–867.

98. Dong, S., J.-M. Liu, and E. Dagotto. 2014. $BaFe_2Se_3$: A high T_C magnetic multiferroic with large ferrielectric polarization. *Physical Review Letters* 113:187204.

99. Caron, J., J. Neilson, D. Miller, A. Llobet, and T. McQueen. 2011. Iron displacements and magnetoelastic coupling in the antiferromagnetic spin-ladder compound $BaFe_2Se_3$. *Physical Review B* 84:180409.

100. Saparov, B., S. Calder, B. Sipos, H. Cao, S. Chi, D. J. Singh, A. D. Christianson, M. D. Lumsden, and A. S. Sefat. 2011. Spin glass and semiconducting behavior in one-dimensional $BaFe_{2-\delta}Se_3$ ($\delta \approx 0.2$) crystals. *Physical Review B* 84:245132.

101. Nambu, Y., K. Ohgushi, S. Suzuki, F. Du, M. Avdeev, Y. Uwatoko, K. Munakata, H. Fukazawa, S. Chi, and Y. Ueda. 2012. Block magnetism coupled with local distortion in the iron-based spin-ladder compound $BaFe_2Se_3$. *Physical Review B* 85:064413.

102. Medvedev, M. V., I. A. Nekrasov, and M. V. Sadovskii. 2012. Electronic and magnetic structure of presumable superconductor $BaFe_2Se_3$. *Journal of Experimental and Theoretical Physics Letters* 95:37–41.

103. Krzton-Maziopa, A., E. Pomjakushina, V. Pomjakushin, D. Sheptyakov, D. Chernyshov, V. Svitlyk, and K. Conder. 2011. The synthesis, and crystal and magnetic structure of the iron selenide $BaFe_2Se_3$ with possible superconductivity at Tc = 11 K. *Journal of Physics: Condensed Matter* 23:402201.

104. Caron, J., J. Neilson, D. Miller, K. Arpino, A. Llobet, and T. McQueen. 2012. Orbital-selective magnetism in the spin-ladder iron selenides $Ba_{1-x}K_xFe_2Se_3$. *Physical Review B* 85:180405.

Multiferroics for Spintronics

Xiaoli Lu, Heng Li, Xin Li, Jiwen Zhang,
Jincheng Zhang, Yue Hao, and Marin Alexe

Contents

5.1 Introduction to Conventional Charge-Based Electronics and the Emerging Spintronics

By scaling law, the essential component of modern integrated circuits, the Si-based complementary metal-oxide semiconductor (CMOS) field effect transistor, has become much smaller in size now to about 10 nm with reduction in the number of features [1–3]. Yet this "cheap way to do electronics," as pointed out by Moore [4], has many challenges in sight such as large leakage current [5,6], sizeable parasitic resistance [7], and exponential increase of circuit power density in chips with billions of transistors as shown in Figure 5.1 [8]. To combat these challenges, many emerging technologies arise

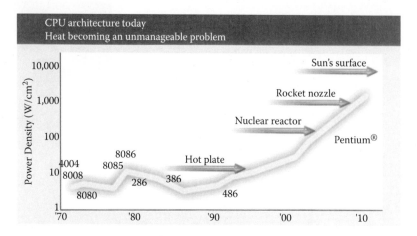

FIGURE 5.1 Power density trend for Intel CPUs. (Courtesy of Pat Gelsinger, Inter Developer Forum, Spring 2004.)

with potential improvements over conventional Si-based transistors. From the materials aspect, efforts have been focused on other channel materials with better transport properties such as Ge [9], III-V compounds [10], carbon nanotube [11], and so on. From the device engineering aspect, new designs such as three-dimensional FinFET [12], high dielectric constant gate dielectrics [13], quantum electronic devices (QED) [14], and single electron transistors (SET) [15] have been proposed.

Besides conventional electronic devices, spin-based electronics, that is, spintronics, has gained increasing attention [16]. Unlike conventional devices, where the charge of electron is used to transfer and store information [17], electron spin is employed in spintronic devices, which greatly enhances data-processing efficiency [18]. More importantly, with the intrinsic nonvolatility, high speed, and low power consumption, spin-based devices can be a game changer for the microelectronic industry [19].

In fact, spin is not a totally new concept for the information technology industry. Magnetization in ferromagnetic materials has long been used to store information on hard disk drives [20]. Yet in this traditional approach, detection and manipulation of local spins are accomplished by using the magnetic field, which limits the data storage density and increases energy consumption. The discovery of giant magnetoresistance (GMR) [21] and tunneling magnetoresistance (TMR) [18] opened new ways to use spin-dependent charge transport in magnetic multilayers. Employing these techniques in the read head (Figure 5.2) of hard disk drives greatly increased the data recording density [22].

Motivated by the success of TMR reading head, more efforts have been devoted to developing a magnetic random access memory (MRAM) based on this technology. The crossbar architecture as shown in Figure 5.3, where

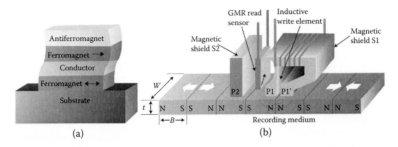

FIGURE 5.2 Schematic diagrams of (a) a spin valve and (b) the GMR head introduced in 1991 by IBM for its hard disk drives ((a) From Wolf, S. A. et al., *Science,* 294, 1488–1495, 2001. Reprinted with permission of AAAS. (b) Reprinted by permission from Macmillan Publishers Ltd. *Nat. Mater.* Chappert, C. et al., The emergence of spin electronics in data storage, 6:813–823, copyright 2007.)

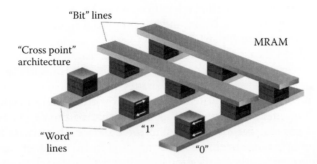

FIGURE 5.3 Schematic working principle of MRAM with cross-point architecture. (Reprinted by permission from Macmillan Publishers Ltd. *Nat. Mater.* Chappert, C. et al., The emergence of spin electronics in data storage, 6:813–823, copyright 2007.)

TMR-based memory cells record the data, shows great promise because of high data storage density, high read/write speed, and good endurance [20,23]. The main issue is related to the dependence on magnetic field for "writing," which leads to high energy consumption and poses a great challenge for heat management [24]. Although some new concepts have been proposed, such as heat-assisted recording [25] and resonance frequency microwave excitation–assisted switching [26], they still require a magnetic field for the "writing" process.

In an attempt to eliminate the need for magnetic field, "spin transfer torque" (STT)-induced magnetization switching has been investigated. This phenomenon was discovered while investigating spin injection across the interface between different materials. Initial attempts focused on forming Ohmic contact between a ferromagnetic metal and a semiconductor, where heavily doped semiconductor usually led to spin-flip scattering and eliminated the most injected spin polarization [27]. Later on, spin injection in the ballistic regime was proposed, which can produce high spin-polarized current (up to 40%) [28]. At the cost of overall efficiency, "hot" electrons with energies much greater than Fermi energy can be used in ballistic injection and the spin polarization value can be further enhanced to 90% [29]. Further experiments revealed that the spin-polarized current that flows through a nonmagnetic layer to another ferromagnetic layer could excite strong, uniform spin-wave due to the spin-dependent scattering of the polarized current. The spin-dependent scattering can also lead to the reversal of the magnetic moment in a ferromagnetic layer, which represents a pure electric current–induced spin transfer process and is denoted as STT [30,31].

STT opens a new pathway to practical MRAM devices (as shown in Figure 5.4), but at the cost of a high writing power. Since the transfer torque per unit area is proportional to the spin current density, very narrow STT stacks (normally nanowires) are required [32]. In addition to the heat management issue, high current density may introduce other issues such as compatibility with CMOS transistor, thermal reliability, and so on. Further investigations are needed to resolve these issues, and one promising approach is to combine heat-assisted recording with STT [33].

Although STT-based MRAM does not use magnetic field for "writing," it is nevertheless a current-based technique that leads to high energy consumption and heat management issues in high-density data storage devices. In this aspect, multiferroic materials with a strong magnetoelectric coupling effect have emerged as an alternative choice [34,35]. They hold great promise for low-power spintronic devices, multistate memory [36] and even neuromorphic computing [37]. In this chapter, we review recent progresses on the application of multiferroic materials in spintronic devices.

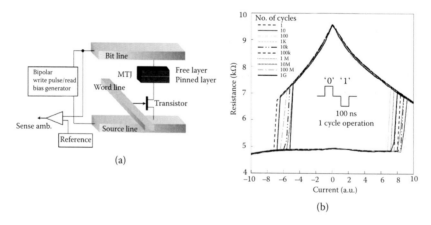

FIGURE 5.4 (a) Schematic architecture of an STT-based MRAM cell, and (b) stability of an MRAM cell with increasing number of writing cycles. (Reprinted from Kawahara, T. et al. *Digest of Technical Papers*, 480–617, 2007 IEEE International Solid-State Circuits Conference, San Francisco, CA, 11–15, February 2007, © 2007 IEEE.)

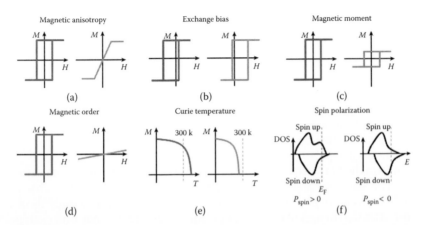

FIGURE 5.5 Magnetic properties that can be affected by magnetoelectric coupling [42]. (Reprinted from Fusil, S. et al., *Annu. Rev. Mater. Res.*, 44, 91–116, 2014. Copyright 2014 Annual Reviews.)

5.2 Spin Manipulation with Magnetoelectric Coupling

The electric-field-control of magnetization in multiferroic materials is characterized by the converse magnetoelectric coupling effect [38,39]. Currently, single-phase multiferroic materials suffer from either small net magnetization (type-I multiferroics) or low transition temperatures (type-II multiferroics), and cannot be used alone to build spintronic devices. Practically, multilayer heterostructures are normally used to transfer electric field–induced magnetic response into adjacent ferromagnetic layer for spin valves or magnetic tunnel junctions (MTJs). In such heterostructures, the couplings are mostly interfacial effects, where different mechanisms such as field effect (through the accumulation or depletion of charge carriers in the ferromagnetic layer), interface orbital hybridization, magnetic exchange interaction, and strain transfer (via a combination of piezoelectric and magnetistrictive effects) [40,41] can be at work. In principle, with magnetoelectric coupling, various magnetic properties including magnetic anisotropy, exchange bias, magnetic moment, and spin polarization can all be affected by the electric field (as shown in Figure 5.5).

5.2.1 Ferroelectric Field Effect

Using phenomenological calculations, Hu et al. [43] reported that there is a critical thickness of the ferromagnetic layer, below which field effect–induced charge accumulation/depletion will dominate in a ferromagnetic/ferroelectric multiferroic heterostructure. Thus, nonvolatile control of the magnetization using this mechanism requires the ferromagnetic layer to be very thin (a few nanometers). Since most ferroelectric materials are perovskite oxides, lattice-matching oxides such as $La_{1-x}A_xMnO_3$ (A = Ca, Sr), $CoFe_2O_4$, $SrRuO_3$, and Fe_3O_4 are common choices for the magnetic layer.

Vaz et al. [44] have studied a $Pb(Zr_{0.2}Ti_{0.8})O_3/La_{0.8}Sr_{0.2}MnO_3$ multiferroic heterostructure, where the $La_{0.8}Sr_{0.2}MnO_3$ layer is only 4–5 nm thick. Transport and magnetic behaviors of the $La_{0.8}Sr_{0.2}MnO_3$ layer clearly correlates with the polarization direction of $Pb(Zr_{0.2}Ti_{0.8})O_3$. As shown in Figure 5.6a, the metal insulator–transition temperature is shifted from 188 K in the depletion state to 226 K in the accumulation state. Clear change in the saturation magnetization is also observed depending on the polarization direction of $Pb(Zr_{0.2}Ti_{0.8})O_3$ (Figure 5.6b). X-ray absorption near edge spectroscopy (XANES) measurement confirms that the change in Mn valence state tracks the switching of the $Pb(Zr_{0.2}Ti_{0.8})O_3$ polarization (Figure 5.6c). From XANES spectra, the average change in the Mn valence was estimated to be around 0.1/Mn between the two remnant states of $Pb(Zr_{0.2}Ti_{0.8})O_3$, and the value is in agreement with change in the surface charge due to polarization of $Pb(Zr_{0.2}Ti_{0.8})O_3$. This study clearly correlates the changes of electron population in Mn 3d e_g bands with charge-based interfacial effects, and shows an effective pathway to enhance the coupling in complex oxide heterostructures by taking advantage of the delicate balance among the various order parameters in strongly correlated systems.

Yin et al. [45] reported the carrier density modulation–induced metal–insulator transition in $La_{0.5}Ca_{0.5}MnO_3$ through ferroelectric polarization switching. In a $La_{0.7}Sr_{0.3}MnO_3/BaTiO_3/La_{0.5}Ca_{0.5}MnO_3/La_{0.7}Sr_{0.3}MnO_3$ multiferroic tunnel junction, the $BaTiO_3$ and $La_{0.5}Ca_{0.5}MnO_3$ layers were maintained at a few nanometers thick. The composition of $La_{0.5}Ca_{0.5}MnO_3$ was chosen at the boundary between ferromagnetic metallic and antiferromagnetic insulating phases. When the polarization of $BaTiO_3$ is reversed, electron accumulation or depletion modulates the doping level of $La_{0.5}Ca_{0.5}MnO_3$, and ferroelectric-induced magnetic phase transition was achieved. With this transition, tunneling electroresistance effect (TER) of the junction was greatly enhanced (up to 10,000%) when compared to a reference device without $La_{0.5}Ca_{0.5}MnO_3$ (Figure 5.7). Combining the TER and TMR effects, a four-state memory device was demonstrated (Figure 5.7c).

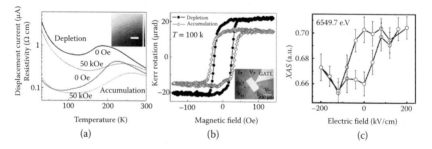

FIGURE 5.6 (a) Resistivity versus temperature curves of $La_{0.8}Sr_{0.2}MnO_3$ in a $Pb(Zr_{0.2}Ti_{0.8})O_3/La_{0.8}Sr_{0.2}MnO_3$ heterostructure for the two polarization states of $Pb(Zr_{0.2}Ti_{0.8})O_3$ measured under zero and 50 kOe magnetic field, and (b) the corresponding in-plane magnetic hysteresis loops measured at 100 K. (c) XAS signal as a function of the electric field applied to $Pb(Zr_{0.2}Ti_{0.8})O_3$. (Reprinted with permission from Vaz, C.A.F. et al., *Phys. Rev. Lett.*, 104, 127202. Copyright 2010 by the American Physical Society.)

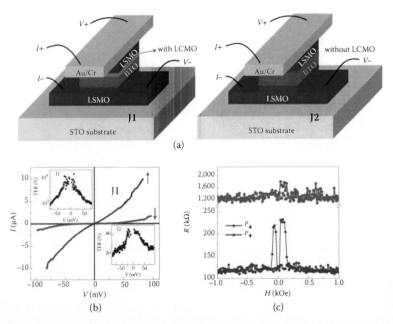

FIGURE 5.7 (a) Schematic diagrams of the tunnel junctions with and without $La_{0.5}Ca_{0.5}MnO_3$ layer. (b) Current–voltage curves of junction 1 at 5 K after $BaTiO_3$ is poled upward (blue) and downward (red). (c) TMR of junction 1 at the polarization up and down states. (Reprinted by permission from Macmillan Publishers Ltd. *Nat. Mater.* Yin, Y. W. et al., Enhanced tunnelling electroresistance effect due to a ferroelectrically induced phase transition at a magnetic complex oxide interface, 12:397–402, copyright 2013.)

Pantel et al. [46] have investigated a different multiferroic tunnel junction, $Co/PbZr_{0.2}Ti_{0.8}O_3/La_{0.7}Sr_{0.3}MnO_3$, in which the $PbZr_{0.2}Ti_{0.8}O_3$ is as thin as 3 nm (Figure 5.8). With the combined TER and TMR effects, four distinct nonvolatile resistance states were observed at 50 K. More interestingly, switching the polarization of $PbZr_{0.2}Ti_{0.8}O_3$ not only changes the resistance of the junction (TER effect), but also switches the TMR from negative (−3%) to positive (4%, as shown in Figure 5.8b). And this polarization-induced TMR sign switching is reversible. The microscopic mechanism of this effect is not yet clear, but it could be a pure electronic effect derived either from orbital hybridization at the interface or from spin-dependent screening in the ferromagnetic layer. This study demonstrates the possibility of complete remnant control of spin polarization with strong magnetoelectric coupling, which certainly benefits spintronic applications.

Despite the progress on interfacial charge-mediated spin control in multiferroic heterostructures, issues such as how to maintain a stable and switchable remnant state at nanometer thicknesses, and how to extend the sizeable coupling effect to above room temperature are still challenging.

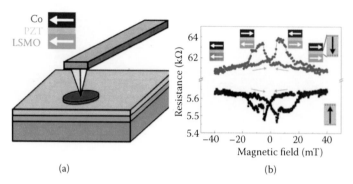

(a)   (b)

FIGURE 5.8 (a) Structure of the multiferroic Co/PZT/LSMO tunnel junction. (b) TMR curves of the junction measured at 50 K in the as-grown state (black squares, polarization up) and after poling with a 3 V electrical bias (red circles). (Reprinted by permission from Macmillan Publishers Ltd. *Nat. Mater.* Pantel, D. et al., Reversible electrical switching of spin polarization in multiferroic tunnel junctions, 11:289–293, copyright 2012.)

5.2.2 Strain-Mediated Coupling

When ferromagnetic films are grown on ferroelectric substrates such as $BaTiO_3$, $Pb(Zr,Ti)O_3$, or $(PbMg_{1/3}Nb_{2/3}O_3)_{1-x}–(PbTiO_3)_x$ (PMN–PT) single crystals, strain-mediated coupling can be achieved. This indirect coupling mechanism was found to be a simple yet effective way to control magnetic properties using electric field. Geprägs et al. [47] reported an increase in saturation magnetic field by up to 200 mT in $Ni/BaTiO_3$ under electric field. And if the $BaTiO_3$ substrate was replaced by $Pb(Zr,Ti)O_3$, an electric field–induced magnetic anisotropy change (90° in plane) can be achieved at room temperature (Figure 5.9) [48]. However, these results were all found to be volatile due to their piezoelectricity-driven nature, which is not ideal for spintronic applications. To achieve nonvolatile spin manipulation, one needs to find a remnant polarization/strain-related coupling mechanism [49].

Another possible pathway to nonvolatile spin control is ferroelastic domain switching–induced strain effect [51]. Unlike the piezoelectric-induced strain, different remnant strain states can be achieved with ferroelastic domain switching, leading to hysteresis loop in the strain–electric field (S–E) curve instead of the butterfly-like curve as shown in Figure 5.10 [52].

Zhao et al. [52] reported a nonvolatile electric field–induced magnetization change at room temperature in a $Co_{40}Fe_{40}B_{20}$/PMN–PT heterostructure (Figure 5.11a). The amorphous $Co_{40}Fe_{40}B_{20}$ ferromagnetic layer was chosen because of its soft ferromagnetic nature and sensitivity to strain. The thickness of the $Co_{40}Fe_{40}B_{20}$ layer is around 20 nm; thus, the contributions from electric field–induced charge accumulation/depletion and/or other interface effect cannot account for the large change in magnetization observed. After an initial poling (8 kV/cm) along PMN–PT [001], in-plane magnetic measurements were carried out along [–110] with a varying electric field. Clear room temperature electric field dependence was observed in the saturation magnetization, which was attributed to the polarization switching in PMN–PT (Figure 5.11b). This result is certainly different from the piezoelectric strain–driven effect,

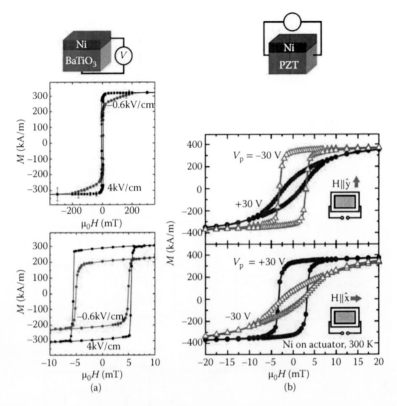

FIGURE 5.9 Piezoelectric strain-induced magnetization and magnetic anisotropy variations in: (a) Ni/BTO. (Reprinted with permission from Pantel, D. et al., *Nat. Mater.*, 11, 289–293. Copyright 2010, American Institute of Physics.) (b) Ni/PZT heterostructures. (Gepraegs, S. et al., *Appl. Phys. Lett.*, 96, 142509, 2010, Copyright 2009 Institute of Physics.)

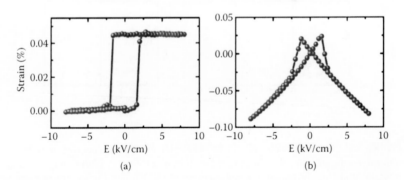

FIGURE 5.10 Electric field–induced (a) remnant and (b) piezoelectric strain behaviors. (Reprinted by permission from Macmillan Publishers Ltd. *Sci. Rep.* Yang, L. et al., Bipolar loop-like non-volatile strain in the (001)-oriented Pb(Mg1/3Nb2/3)O3-PbTiO3 single crystals, 4, 4591, Copyright 2014.)

Multiferroic Materials

and it was suggested that 109° ferroelastic domain switching plays a significant role. As we know, the electric field applied along [001] direction can induce 71°, 109°, or 180° domain switching in PMN–PT (Figure 5.11c through e), but only the 109° domain switching leads to remnant strain, that is, net strain at zero electric field. Quantitative x-ray reciprocal space mapping (RSM) analysis revealed that 109° switching only accounts for about 26% of the total switching events, which may limit the application potential of the $Co_{40}Fe_{40}B_{20}$/ PMN–PT heterostructure. Furthermore, local ferroelastic switching may produce a high-energy domain state, leading to the relaxation to a more stable ferroelastic domain arrangement [53]. This would lead to reliability problems in devices. In this respect, Eom et al. [54] have demonstrated both selective control and stabilization of 71° ferroelastic switching in $BiFeO_3$ by suppressing

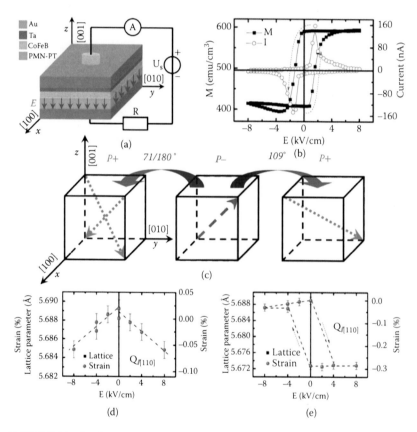

FIGURE 5.11 (a) Schematic of the CoFeB/PMN–PT heterostructure and experimental configuration. (b) Electric field tuning of the in-plane magnetization (square) and the corresponding polarization current (open circle). (c) Schematic illustration of the 71°/180° and 109° polarization switching process, and (d and e) the corresponding changes in lattice parameters and strains observed experimentally. (Reprinted from Zhang, S. et al., *Phys. Rev. Lett.*, 108, 137203. Copyright 2012 by the American Physical Society.)

the back switching in island structures. In a 3.4 × 3.4 μm² island, 71° ferroelastic switching was stable for more than 26 h without relaxation, whereas a domain with a similar size in a continuous film was fully relaxed only after 2 h. The enhanced stability of ferroelastic switching relies on eliminating the constraint imposed by the surrounding high-energy domain walls in the isolated island structure, and opens a new avenue for the device design to achieve nonvolatile electric field–controlled spintronics.

5.2.3 Tuning the Magnetic Exchange Effect

In the $Co_{40}Fe_{40}B_{20}$/PMN–PT heterostructure discussed previously, the $Co_{40}Fe_{40}B_{20}$ is a soft ferromagnetic material with no magnetocrystalline anisotropy, which was considered an essential requirement for the large magnetoelectric coupling effect. Yet in practical applications, highly anisotropic ferromagnetic materials are normally used for data storage or logic operation. Thus, the ferroelastic switching–induced strain may not be large enough for effective magnetization tuning (e.g., the 109° domain switching in (001) PMN–PT can only induce a lattice distortion of ~0.3%) [52]. Thus, other mechanisms for strong nonvolatile coupling at room temperature are required.

Since most type-I multiferroics are antiferromagnetic, even if the magnetoelectric coupling is strong enough to induce antiferromagnetic domain switching under the electric field, the absence of net magnetization makes it undetectable. However, by depositing a ferromagnetic layer on top and employing the exchange bias effect [55], electric-field control of magnetization may be obtained. Ideally, an electric field–induced reversal or suppression of exchange bias, resulting in the reversal of magnetization, is desirable.

Chu et al. [56,57] have observed a clear one-to-one correlation between the ferroelectric domains in $BiFeO_3$ and ferromagnetic domains in $Co_{0.9}Fe_{0.1}$ in a $BiFeO_3$/$Co_{0.9}Fe_{0.1}$ heterostructure, which was believed to be mediated by the exchange coupling between the two layers. Due to the strong coupling between ferroelectric polarization and antiferromagnetic easy plane in $BiFeO_3$, when an in-plane electric field is applied to the (001)-oriented $BiFeO_3$ film and induces 71° domain switching, the corresponding antiferromagnetic easy plane rotates from $(1\bar{1}1)$ to $(\bar{1}11)$ as shown in Figure 5.12c. Due to the exchange bias between $BiFeO_3$ and $Co_{0.9}Fe_{0.1}$, magnetization in the adjacent $Co_{0.9}Fe_{0.1}$ layer may also be affected. Experimentally, the combination of piezoelectric force microscopy (PFM) and x-ray magnetic circular dichroism photoemission electron microscopy (XMCD–PEEM) indeed confirmed the correlation (Figure 5.12d and e). The in-plane electric field applied to $BiFeO_3$ leads to a reversible 90° rotation of the magnetization in $Co_{0.9}Fe_{0.1}$ as revealed by the contrast change in XMCD–PEEM image (Figure 5.12e). Such functionality can be used in the next-generation, low-power electric field–controlled MRAM.

5.2.4 Other Systems and Effects

Besides the mechanisms discussed in previous sections, other effects have also been explored to achieve electric-field control of magnetization [38]. For example, under large in-plane compressive strain, rhombohedral phase (R-phase)

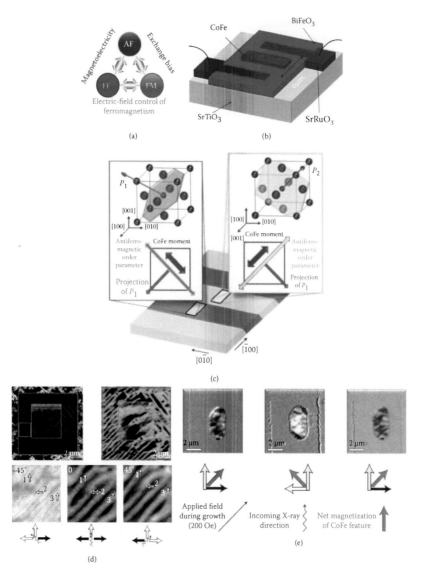

FIGURE 5.12 (a) Schematic diagram showing the combination of magnetoelectric coupling in $BiFeO_3$, and exchange bias leads to electric-field control of magnetization in a $BiFeO_3/Co_{0.9}Fe_{0.1}$ heterostructure. (b) The device structure used to demonstrate the concept. (c) Polarization switching–induced easy magnetic plane rotation in $BiFeO_3$. (d) In-plane PFM images of $BiFeO_3$ and the corresponding XMCD-PEEM images of the same location. The bottom images are obtained with different incoming x-ray. (e) PEEM images of $Co_{0.9}Fe_{0.1}$ taken before and after switching the polarization in $BiFeO_3$ demonstrate the electric-field control of local magnetization. (Reprinted by permission from Macmillan Publishers Ltd. *Nat. Mater.* Chu, Y.-H. et al., Electric-field control of local ferromagnetism using a magnetoelectric multiferroic, 7:478–482, copyright 2008.)

$BiFeO_3$ can be transformed into a tetragonal-like phase (T-phase) [58]. When the strain is partially released, a nanoscale mixture of T- and R-phase will be formed (Figure 5.13). He et al. [59] found that in this mixed-phase system, the highly distorted R-phase has enhanced magnetization than that of the bulk-canted antiferromagnetic phase. Since this R to T phase transition can be induced by electric field, it enables the electric-field control of magnetization at room temperature. As demonstrated in Figure 5.13c and d, the magnetization can be erased and regenerated with the application of electric fields.

Another example is shown in Figure 5.14, though it is in principle a strain effect in a different form [60]. Following the original work of Zheng et al.

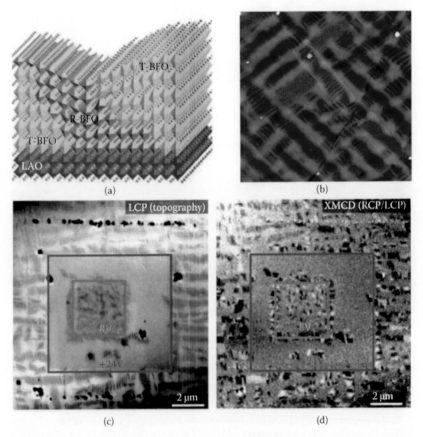

FIGURE 5.13 (a) Schematic illustration of the boundaries between tetragonal-like and rhombohedral-like phases in highly strained $BiFeO_3$ thin films on the $LaAlO_3$ substrate. (b) Topography images of mixed phase $BiFeO_3$ thin films obtained by atomic force microscope. (c and d) The topography and corresponding XMCD images reveal that the rhombohedral-like phase with larger magnetization can be erased and regenerated by the electric field. (Reprinted by permission from Macmillan Publishers Ltd. *Nat. Commun.* He, Q. et al., Electrically controllable spontaneous magnetism in nanoscale mixed phase multiferroics, 2:225, copyright 2011.)

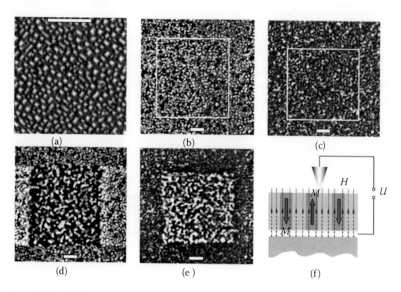

FIGURE 5.14 (a) Topography of a $BiFeO_3$–$CoFe_2O_4$ vertically aligned-nanocomposite film. (b) and (c) Magnetic force microscopy (MFM) images of the film after being magnetized in 20 kOe fields, respectively. (d and e) MFM images of the same areas as in (b and c) after electrical poling at −16 V. (f) The proposed electric field–assisted magnetic recording concept, where an electric field is used to soften the magnetic medium during the "write" process resulting in a device with low power consumption and high recording density. (Reprinted with permission from Zavaliche, F. et al., *Nano Lett.*, 7, 1586–1590. Copyright 2007 American Chemical Society.)

[60], Zavaliche et al. [61] fabricated thick epitaxial film of vertically aligned nanocomposition, $BiFeO_3$–$CoFe_2O_4$ system. The system undergoes a spontaneous phase separation during pulsed laser deposition, forming a nanopillars ($CoFe_2O_4$) embedded in matrix ($BiFeO_3$) nanostructure (Figure 5.14a). In this vertically aligned nanocomposite, strong interface coupling can be maintained by reducing the substrate clamping effect and increasing the interfacial area [62], leading to large magnetoelectric coupling. Due to the shape and strain effects, an out-of-plane magnetic anisotropy was found in the $CoFe_2O_4$ nanopillars. The sample was initially poled into a single domain state under 20 kOe magnetic field along the film normal direction (Figure 5.14b and c). After the magnetic poling, an electric bias of −16 V was applied and led to unidirectional magnetization reversal in many nanopillars as shown by the contrast change in the magnetic force microscopy (MFM) images (Figure 5.14d and e). This process can be explained as follows: under the electric field, the strain derived from ferroelectric polarization switching in $BiFeO_3$ matrix is transferred to $CoFe_2O_4$ and reduces its magnetic anisotropy. Therefore, a spin reorientation from out-of-plane to in-plane may occur when the electric field is applied. Upon removal of the applied field, the initial out-of-plane magnetic anisotropy is restored and magnetization may flip either up or down with equal probabilities. To achieve deterministic control of magnetization, a small magnetic field,

which could be much smaller than the coercive field, is needed to break the symmetry. Following the heat-assisted magnetic recording concept, an electric field–assisted magnetic recording can thus be proposed, where electric field is used to soften the magnetic medium during the "write" process resulting in a device with low power consumption and high recording density.

It has been reported that the ferroelectric domain walls in $BiFeO_3$, where both time-reversal and space-inversion symmetries are broken, are conducting and possess net magnetization [63]. It is thus possible to manipulate the domain wall transport via a magnetic field, and magnetization via an electric field. Alexe et al. [64] have investigated the effect of magnetic field on the domain wall conductivity in Li-doped $BiFeO_3$ thin films. The clearly different conductivities of the fully (negative and positive) and partially switched films (Figure 5.15b) suggest contribution from domain walls. When a magnetic field is applied at a certain angle θ with respect to the film's normal direction, fully and partially switched films again showed different angle dependence (Figure 5.15a and c). The clear hysteresis in the half-switched $BiFeO_3$ film is likely associated with the ferromagnetic order at domain walls, in contrast to the antiferromagnetic order of the domain bulk. Although these results were obtained at low temperatures (below 100 K), it nevertheless opens up new possibilities to utilize single-phase multiferroic $BiFeO_3$ in spintronic devices.

5.3 Spintronics with Multiferroics: Prototype Devices

With the progress in the understanding of magnetoelectric coupling and electric-field control of magnetization, various prototype devices have been proposed and tested. For example, Lei et al. [65] demonstrated that the domain wall propagation in a magnetic $Co_{40}Fe_{40}B_{20}$ nanowire can be controlled by an electric field applied to the $PbZr_{0.5}Ti_{0.5}O_3$ substrate. Logic device and racetrack memory can be designed based on this functionality.

As shown in Figure 5.16a and b, without voltages applied to the $PbZr_{0.5}Ti_{0.5}O_3$ substrate, domain walls can propagate freely in the nanowire. However, the propagation can be blocked by a local stress induced through a voltage applied to the substrate. This is due to an increase of coercivity of the $Co_{40}Fe_{40}B_{20}$ layer under stress (Figure 5.16c), which increases the energy barrier for domain wall motion. Based on this functionality, logic devices such as NOR gate can be achieved as shown in Figure 5.16d. Control A drives domain wall motion, and the writing line is used to generate domain walls. The $PbZr_{0.5}Ti_{0.5}O_3$ gates B0 and B1 act as inputs and the magnetic tunnel junction C serves as the output. Initially, the output "C" would be set to the low resistance state "0." A domain wall is then nucleated in the nanowire, and if either or both the gate voltages "B0" and "B1" are applied (i.e., in the "1" state), the induced anisotropy change at the gate leads to a local pinning of the domain wall, leaving the output "C" in the "0" state. However, if both gate voltages are switched off (inputs "B0" and "B1" set to "0"), the generated domain wall propagates freely along the wire, driven by the spin-polarized current controlled by "A." Eventually, this leads to magnetization reversal at

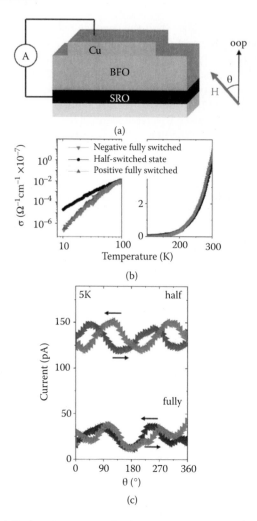

FIGURE 5.15 (a) Device structure to check the magnetoelectric coupling related to domain walls in BiFeO$_3$. (b) Temperature dependence of the effective conductivity of BiFeO$_3$ in the fully switched (red and blue) and partially switched (black) states, revealing contributions from domain walls. (c) Current measured with a magnetic field applied at different angles from the film's normal direction again reveals the difference between fully and partially switched samples. (Lee, J.H. et al.: Spintronic functionality of BiFeO$_3$ domain walls. *Adv. Mater.* 26. 7078–7082. 2005. Copyright Wiley-VCH Verlag GmbH & Co. KGaA. Reproduced with permission.)

the output "C" subsequently switching its state to "1." Thus, the NOR operation table can be obtained. Besides the logic option, an individually address-able racetrack memory can also be realized with the domain wall controlling gates, as shown in Figure 5.16e. If a new state "1 0 0 0" is going to be written, the voltages at BL0 and BL2 are switched on and that at BL1 is switched off, combined with the spin polarized current from control A, then the final state

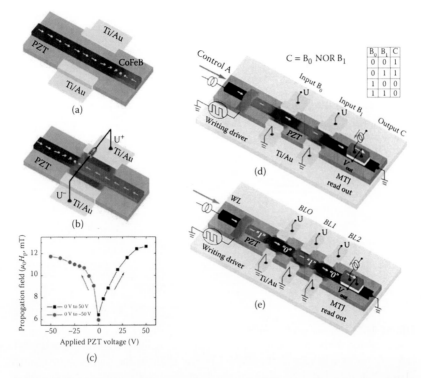

FIGURE 5.16 Domain wall propagation in a CoFeB nanowire (a) without and (b) with electric field applied to the PbZr$_{0.5}$Ti$_{0.5}$O$_3$ substrate. (c) The propagation field is tuned by the electric field applied to the substrate. (d) NOR logic function can be achieved using multiple voltage control as inputs and magnetic tunnel junction response as the output. (e) A racetrack memory can also be designed based on this concept. (Reprinted by permission from Macmillan Publishers Ltd. *Nat. Commun.* Lei, N. et al., Strain-controlled magnetic domain wall propagation in hybrid piezoelectric/ferromagnetic structures, 4:1378, copyright 2013.)

"1 0 0 0" can be achieved. This would be much faster than the word-by-word writing process in the conventional racetrack memory.

A strong magnetoelectric coupling could also benefit the emerging MRAM technology. Nan et al. [35] designed a simple MRAM architecture with polycrystalline Ni spin valve on (011) PMN–PT as the basic building block (Figure 5.17a). Using a combination of phase-field modeling and micromagnetic simulation, a time-dependent electric field–induced magnetization switching in a 64 × 64 × 5 nm^3 Ni-free layer in the spin valve was obtained as shown in Figure 5.17b. By operating the (011) PMN–PT layer below its ferroelectric coercive field, bistable in-plane strains can be obtained and they can induce a nonvolatile 90° in-plane magnetization rotation in Ni. By this electric field–induced magnetization switching in the free layer, a hysteresis in the device resistance can be obtained (Figure 5.17c). These results enable the design of a strain-mediated magnetoelectric random access memory (Figure 5.17d). Taking the 64 × 64 nm^2 cell as an

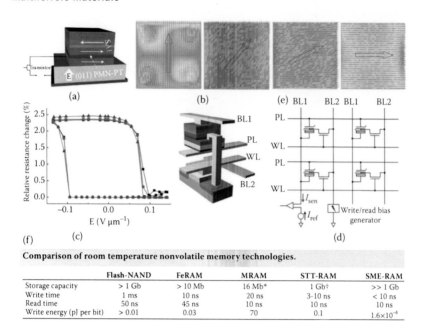

Comparison of room temperature nonvolatile memory technologies.

	Flash-NAND	FeRAM	MRAM	STT-RAM	SME-RAM
Storage capacity	> 1 Gb	> 10 Mb	16 Mb*	1 Gb†	>> 1 Gb
Write time	1 ms	10 ns	20 ns	3-10 ns	< 10 ns
Read time	50 ns	45 ns	10 ns	10 ns	10 ns
Write energy (pJ per bit)	> 0.01	0.03	70	0.1	1.6×10^{-4}

FIGURE 5.17 (a) Schematic building block of the MRAM proposed by Nan et al., where a spin valve or magnetic tunnel junction is integrated onto a PMN–PT layer. Applying a voltage to the PMN–PT layer changes the tunneling resistance of the cell. (b) Snapshots of the magnetization vector diagrams in 64 × 64 × 5 nm³ Ni-free layers at (from left to right) $t = 0.13$, 0.64, 0.90, and 1.02 ns, respectively. (c) Hysteric loops of relative junction resistance change upon the perpendicular electric fields applied to the (011) PMN–PT layer, accompanied by the magnetization switching in the Ni-free layers with thicknesses of 35 nm (squares), 15 nm (circles), and 5 nm (triangles), respectively. (d) Architecture of the 1-T(transistor)/1-magnetoresistive (MR) memory unit cell and (e) cell array of the voltage-controlled MRAM device. BL, PL, and WL stand for bit-line, plate-line, and word-line, respectively. (f) Comparison with other nonvolatile memory technologies. (Reprinted by permission from Macmillan Publishers Ltd. *Nat. Commun.* Hu, J.-M. et al., High-density agnetoresistive random access memory operating at ultralow voltage at room temperature, 2:553, copyright 2011.)

example, the switching (writing) field is around 0.26 V and the writing energy is estimated to be 0.16 fJ per bit (Pr ~ 30 μC/cm²), which is drastically lower than both conventional MRAM (70 pJ/bit) and STT MRAM (0.1 pJ/bit). The operation sequence of the proposed device is quite simple. For writing, the voltage bias is applied between the bit-line (BL2) and the plate-line (PL), which is controlled by the word-line (WL) connecting to the gate of transistor (Figure 5.17e). To read, the sensing current (I_{sen}) is sent to the memory cell from BL2 to BL1, with the two bistable resistance states of the cell, and R_{high} and R_{low}, different output voltages compared to the reference voltage (V_R) can be obtained. In Figure 5.17f, the performance of various nonvolatile memories is compared with the proposed technology, clearly demonstrating its potential.

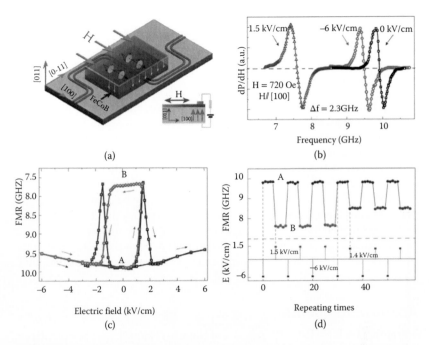

FIGURE 5.18 (a) Schematic of ferromagnetic resonance (FMR) measurement and device structure. The sample is laid face down on an S-shaped co-planar wave-guide. Magnetic fields are applied in the [100] direction and electric fields are applied along the [011] direction. (b) Electric field dependence of the FMR field in frequency sweeping mode. (c) FMR frequency responses under unipolar (red) and bipolar (blue) sweeping of electric fields at room temperature. (d) Voltage impulse–induced nonvolatile switching of FMR frequency. (Liu, M. et al.: Voltage-impulse-induced non-volatile ferroelastic switching of FMR for reconfigurable magnetoelectric microwave devices. *Adv. Mater.* 25. 4886–4892. 2013. Copyright Wiley-VCH Verlag GmbH & Co. KGaA. Reproduced with permission.)

As a simple yet efficient way to tune magnetization, strain-mediated magnetoelectric coupling effect has also been introduced in magnetic microwave devices such as tunable filters, resonators, phase shifters, and so on. For example, Liu et al. [66] reported a voltage-driven nonvolatile tuning of FMR up to 2.3 GHz in amorphous $Fe_{60}Co_{20}B_{20}$ films deposited on (011) PMN–PT single crystal. The device structure is schematically shown in Figure 5.18a. The $Fe_{60}Co_{20}B_{20}$/PMN–PT heterostructure is placed face down on an S-shaped co-planar waveguide with magnetic field applied in the [100] direction and electric field applied in the [011] direction. In the frequency sweeping mode, electric field–induced FMR frequency shift was clearly observed and attributed to the tensile strain generated in the PMN–PT substrate due to the piezoelectric effect (Figure 5.18b). With the 71° and 109° ferroelastic switching of PMN–PT, two stable and reversible FMR remnant states A and B were obtained. By choosing a proper voltage pulse, resonance frequency of any value between A and B is achievable, and this provides a low power alternative for linear FMR modulators used in aircraft, satellite, and portable communication systems.

5.4 Conclusion and Outlook

To conclude, magnetoelectric coupling effects have found promising applications in many areas, and various prototype devices have been proposed and tested. However, strong coupling effects are mostly limited to bulk composites where strain across the interface is crucial. Further work is still needed for thin-film heterostructures and nanostructures, where other coupling mechanisms are at play [38]. In this aspect, atomic-scale controllable growth for high interface quality is very important. For use in applications at room temperature and above, more efforts are needed to search for suitable multiferroic and magnetic materials, which may extend outside the complex oxides family [67]. Calculations have suggested that compounds with competing magnetic interactions and critical temperatures above 300 K are promising candidates.

Acknowledgments

The authors acknowledge the support of National Natural Science Foundation of China (Grant Nos: 51202176, 51572211).

References

1. Moore, G. E. 1975. Progress in digital electronics. *IEDM Technical Digest* 21: 11–13.
2. Thompson, S. E. and S. Parthasarathy. 2006. Moore's law: The future of Si microelectronics. *Materials Today* 9:20–25.
3. Markov, I. L. 2014. Limits on fundamental limits to computation. *Nature* 512:147–154.
4. Moore, G. E. 1965. Cramming more components onto integrated circuits. *Electronics* 38:114–177.
5. Thompson, S. E., M. Armstrong, C. Auth, M. Alavi, M. Buehler, R. Chau, S. Cea, T. Ghani, G. Glass, T. Hoffman, C. H. Jan, C. Kenyon, J. Klaus, K. Kuhn, Z. Y. Ma, B. McIntyre, K. Mistry, A. Murthy, B. Obradovic, R. Nagisetty, P. Nguyen, S. Sivakumar, R. Shaheed, L. Shiften, B. Tufts, S. Tyagi, M. Bohr, and Y. El-Mansy. 2004. A 90-nm logic technology featuring strained-silicon. *IEEE Transactions on Electron Devices* 51:1790–1797.
6. Kent, J. P. and J. Prasad. 2008. Microelectronics for the real world: "Moore" versus "More than Moore". *Proceedings of the IEEE 2008 Custom Integrated Circuits Conference*, San Jose, CA.
7. Dixit, A., A. Kottantharayil, N. Collaert, M. Goodwin, M. Jurezak, and K. De Meyer. 2005. Analysis of the parasitic S/D resistance in multiple-gate FETs. *IEEE Transactions on Electron Devices* 52:1132–1140.
8. Grove, A. 2002. Changing vectors of Moore's law. *International Electron Devices Meeting 2002*, San Francisco, CA.

9. Lee, M. L., E. A. Fitzgerald, M. T. Bulsara, M. T. Currie, and A. Lochtefeld. 2005. Strained Si, SiGe, and Ge channels for high-mobility metal-oxide-semiconductor field-effect transistors. *Journal of Applied Physics* 97:011101.

10. Ye, P. D., G. D. Wilk, B. Yang, J. Kwo, S. N. G. Chu, S. Nakahara, H. J. L. Gossmann, J. P. Mannaerts, M. Hong, K. K. Ng, and J. Bude. 2003. GaAs metal-oxide-semiconductor field-effect transistor with nanometerthin dielectric grown by atomic layer deposition. *Applied Physics Letters* 83:180–182.

11. Wind, S. J., J. Appenzeller, R. Martel, V. Derycke, and P. Avouris. 2002. Vertical scaling of carbon nanotube field-effect transistors using top gate electrodes. *Applied Physics Letters* 80:3817–3819.

12. Hisamoto, D., W. C. Lee, J. Kedzierski, H. Takeuchi, K. Asano, C. Kuo, E. Anderson, T. J. King, J. Bokor, and C. M. Hu. 2000. FinFET—A self-aligned double-gate MOSFET scalable to 20 nm. *IEEE Transactions on Electron Devices* 47:2320–2325.

13. Robertson, J. 2006. High dielectric constant gate oxides for metal oxide Si transistors. *Reports on Progress in Physics* 69:327–396.

14. Rauschenbeutel, A., G. Nogues, S. Osnaghi, P. Bertet, M. Brune, J. M. Raimond, and S. Haroche. 1999. Coherent operation of a tunable quantum phase gate in cavity QED. *Physical Review Letters* 83:5166–5169.

15. Park, J., A. N. Pasupathy, J. I. Goldsmith, C. Chang, Y. Yaish, J. R. Petta, M. Rinkoski, J. P. Sethna, H. D. Abruna, P. L. McEuen, and D. C. Ralph. 2002. Coulomb blockade and the Kondo effect in single-atom transistors. *Nature* 417:722–725.

16. Fabian, J., A. Matos-Abiague, C. Ertler, P. Stano, and I. Zutic. 2007. Semiconductor spintronics. *Acta Physica Slovaca* 57:565–907.

17. Frank, D. J., R. H. Dennard, E. Nowak, P. M. Solomon, Y. Taur, and H. S. P. Wong. 2001. Device scaling limits of Si MOSFETs and their application dependencies. *Proceedings of the IEEE* 89:259–288.

18. Ikeda, S., J. Hayakawa, Y. M. Lee, F. Matsukura, Y. Ohno, T. Hanyu, and H. Ohno. 2007. Magnetic tunnel junctions for spintronic memories and beyond. *IEEE Transactions on Electron Devices* 54:991–1002.

19. Zutic, I., J. Fabian, and S. Das Sarma. 2004. Spintronics: Fundamentals and applications. *Reviews of Modern Physics* 76:323–410.

20. Chappert, C., A. Fert, and F. N. Van Dau. 2007. The emergence of spin electronics in data storage. *Nature Materials* 6:813–823.

21. Daughton, J. M. 1999. GMR applications. *Journal of Magnetism and Magnetic Materials* 192:334–342.

22. Wolf, S. A., D. D. Awschalom, R. A. Buhrman, J. M. Daughton, S. von Molnár, M. L. Roukes, A. Y. Chtchelkanova, and D. M. Treger. 2001. Spintronics: A spin-based electronics vision for the future. *Science* 294:1488–1495.

23. Tehrani, S., J. M. Slaughter, E. Chen, M. Durlam, J. Shi, and M. DeHerrera. 1999. Progress and outlook for MRAM technology. *IEEE Transactions on Magnetics* 35:2814–2819.

24. Gallagher, W. J. and S. S. P. Parkin. 2006. Development of the magnetic tunnel junction MRAM at IBM: From first junctions to a 16-Mb MRAM demonstrator chip. *IBM Journal of Research and Development* 50:5–23.

25. Daughton, J. M., and A. V. Pohm. 2003. Design of Curie point written magneto-resistance random access memory cells. *Journal of Applied Physics* 93:7304–7306.

26. Moriyama, T., R. Cao, J. Q. Xiao, J. Lu, X. R. Wang, Q. Wen, and H. W. Zhang. 2007. Microwave-assisted magnetization switching of $Ni_{80}Fe_{20}$ in magnetic tunnel junctions. *Applied Physics Letters* 90:152503.

27. Gardelis, S., C. G. Smith, C. H. W. Barnes, E. H. Linfield, and D. A. Ritchie. 1999. Spin-valve effects in a semiconductor field-effect transistor: A spintronic device. *Physical Review B* 60:7764–7767.

28. Upadhyay, S. K., R. N. Louie, and R. A. Buhrman. 1999. Spin filtering by ultra-thin ferromagnetic films. *Applied Physics Letters* 74:3881–3883.

29. Rippard, W. H., and R. A. Buhrman. 2000. Spin-dependent hot electron transport in Co/Cu thin films. *Physical Review Letters* 84:971–974.

30. Katine, J. A., F. J. Albert, R. A. Buhrman, E. B. Myers, and D. C. Ralph. 2000. Current-driven magnetization reversal and spin-wave excitations in Co/Cu/Co pillars. *Physical Review Letters* 84:3149–3152.

31. Stiles, M. D. and A. Zangwill. 2002. Anatomy of spin-transfer torque. *Physical Review B* 66:014407.

32. Koyama, T., D. Chiba, K. Ueda, K. Kondou, H. Tanigawa, S. Fukami, T. Suzuki, N. Ohshima, N. Ishiwata, Y. Nakatani, K. Kobayashi, and T. Ono. 2011. Observation of the intrinsic pinning of a magnetic domain wall in a ferromagnetic nanowire. *Nature Materials* 10:194–197.

33. Zhao, W., J. Duval, J.-O. Klein, and C. Chappert. 2011. A compact model for magnetic tunnel junction (MTJ) switched by thermally assisted Spin transfer torque (TAS plus STT). *Nanoscale Research Letters* 6:368.

34. Zhao, T., A. Scholl, F. Zavaliche, K. Lee, M. Barry, A. Doran, M. P. Cruz, Y. H. Chu, C. Ederer, N. A. Spaldin, R. R. Das, D. M. Kim, S. H. Baek, C. B. Eom, and R. Ramesh. 2006. Electrical control of antiferromagnetic domains in multiferroic $BiFeO_3$ films at room temperature. *Nature Materials* 5:823–829.

35. Hu, J.-M., Z. Li, L.-Q. Chen, and C.-W. Nan. 2011. High-density magnetoresistive random access memory operating at ultralow voltage at room temperature. *Nature Communications* 2:553.

36. Gajek, M., M. Bibes, S. Fusil, K. Bouzehouane, J. Fontcuberta, A. Barthelemy, and A. Fert. 2007. Tunnel junctions with multiferroic barriers. *Nature Materials* 6:296–302.

37. Jo, S. H., T. Chang, I. Ebong, B. B. Bhadviya, P. Mazumder, and W. Lu. 2010. Nanoscale memristor device as synapse in neuromorphic systems. *Nano Letters* 10:1297–1301.

38. Vaz, C. A. F., J. Hoffman, C. H. Anh, and R. Ramesh. 2010. Magnetoelectric coupling effects in multiferroic complex oxide composite structures. *Advanced Materials* 22:2900–2918.

39. Ma, J., J. Hu, Z. Li, and C.-W. Nan. 2011. Recent progress in multiferroic magnetoelectric composites: from bulk to thin films. *Advanced Materials* 23:1062–1087.

40. Huang, W., J. Zhu, H. Z. Zeng, X. H. Wei, Y. Zhang, and Y. R. Li. 2006. Strain induced magnetic anisotropy in highly epitaxial $CoFe_2O_4$ thin films. *Applied Physics Letters* 89:262506.

41. Lu, X. L., Y. S. Kim, S. Goetze, X. G. Li, S. N. Dong, P. Werner, M. Alexe, and D. Hesse. 2011. Magnetoelectric coupling in ordered arrays of multilayered heteroepitaxial $BaTiO_3/CoFe_2O_4$ nanodots. *Nano Letters* 11:3202–3206.

42. Fusil, S., Garcia, V., Barthélémy, A., and Bibes, M. 2014. Magnetoelectric devices for spintronics. *Annual Review of Materials Research* 44:91–116.

43. Hu, J.-M., C.-W. Nan, and L.-Q. Chen. 2011. Size-dependent electric voltage controlled magnetic anisotropy in multiferroic heterostructures: Interface-charge and strain comediated magnetoelectric coupling. *Physical Review B* 83:134408.

44. Vaz, C. A. F., J. Hoffman, Y. Segal, J. W. Reiner, R. D. Grober, Z. Zhang, C. H. Ahn, and F. J. Walker. 2010. Origin of the magnetoelectric coupling effect in $Pb(Zr_{0.2}Ti_{0.8})O_3/La_{0.8}Sr_{0.2}MnO_3$ multiferroic heterostructures. *Physical Review Letters* 104:127202.

45. Yin, Y. W., J. D. Burton, Y. M. Kim, A. Y. Borisevich, S. J. Pennycook, S. M. Yang, T. W. Noh, A. Gruverman, X. G. Li, E. Y. Tsymbal, and Q. Li. 2013. Enhanced tunnelling electroresistance effect due to a ferroelectrically induced phase transition at a magnetic complex oxide interface. *Nature Materials* 12:397–402.

46. Pantel, D., S. Goetze, D. Hesse, and M. Alexe. 2012. Reversible electrical switching of spin polarization in multiferroic tunnel junctions. *Nature Materials* 11:289–293.

47. Gepraegs, S., A. Brandlmaier, M. Opel, R. Gross, and S. T. B. Goennenwein. 2010. Electric field controlled manipulation of the magnetization in $Ni/BaTiO_3$ hybrid structures. *Applied Physics Letters* 96:142509.

48. Weiler, M., A. Brandlmaier, S. Gepraegs, M. Althammer, M. Opel, C. Bihler, H. Huebl, M. S. Brandt, R. Gross, and S. T. B. Goennenwein. 2009. Voltage controlled inversion of magnetic anisotropy in a ferromagnetic thin film at room temperature. *New Journal of Physics* 11:013021.

49. Lahtinen, T. H. E., K. J. A. Franke, and S. van Dijken. 2012. Electric-field control of magnetic domain wall motion and local magnetization reversal. *Scientific Reports* 2:258.

50. Yang, L., Zhao, Y, Zhang, S., Li, P., Gao, Y., Yang, Y., Huang, H., Miao, P., Liu, Y., Chen, A., Nan, C.W., and Gao, C. 2014. Bipolar loop-like non-volatile strain in the (001)-oriented $Pb(Mg_{1/3}Nb_{2/3})O_3$-$PbTiO_3$ single crystals. *Scientific Reports,* 4, 4591, 2014.

51. Nagarajan, V., A. Roytburd, A. Stanishevsky, S. Prasertchoung, T. Zhao, L. Chen, J. Melngailis, O. Auciello, and R. Ramesh. 2003. Dynamics of ferroelastic domains in ferroelectric thin films. *Nature Materials* 2:43–47.

52. Zhang, S., Y. G. Zhao, P. S. Li, J. J. Yang, S. Rizwan, J. X. Zhang, J. Seidel, T. L. Qu, Y. J. Yang, Z. L. Luo, Q. He, T. Zou, Q. P. Chen, J. W. Wang, L. F. Yang, Y. Sun, Y. Z. Wu, X. Xiao, X. F. Jin, J. Huang, C. Gao, X. F. Han, and R. Ramesh. 2012. Electric-field control of nonvolatile magnetization in $Co_{40}Fe_{40}B_{20}/Pb(Mg_{1/3}Nb_{2/3})_{0.7}Ti_{0.3}O_3$ structure at room temperature. *Physical Review Letters* 108:137203.

53. Cruz, M. P., Y. H. Chu, J. X. Zhang, P. L. Yang, F. Zavaliche, Q. He, P. Shafer, L. Q. Chen, and R. Ramesh. 2007. Strain control of domain-wall stability in epitaxial $BiFeO_3$ (110) films. *Physical Review Letters* 99:217601.

54. Baek, S. H., H. W. Jang, C. M. Folkman, Y. L. Li, B. Winchester, J. X. Zhang, Q. He, Y. H. Chu, C. T. Nelson, M. S. Rzchowski, X. Q. Pan, R. Ramesh, L. Q. Chen, and C. B. Eom. 2010. Ferroelastic switching for nanoscale nonvolatile magnetoelectric devices. *Nature Materials* 9:309–314.

55. Nogues, J., J. Sort, V. Langlais, V. Skumryev, S. Surinach, J. S. Munoz, and M. D. Baro. 2005. Exchange bias in nanostructures. *Physics Reports-Review Section of Physics Letters* 422:65–117.

56. Chu, Y.-H., L. W. Martin, M. B. Holcomb, M. Gajek, S.-J. Han, Q. He, N. Balke, C.-H. Yang, D. Lee, W. Hu, Q. Zhan, P.-L. Yang, A. Fraile-Rodriguez, A. Scholl, S. X. Wang, and R. Ramesh. 2008. Electric-field control of local ferromagnetism using a magnetoelectric multiferroic. *Nature Materials* 7:478–482.

57. Heron, J. T., M. Trassin, K. Ashraf, M. Gajek, Q. He, S. Y. Yang, D. E. Nikonov, Y. H. Chu, S. Salahuddin, and R. Ramesh. 2011. Electric-field-induced magnetization reversal in a ferromagnet-multiferroic heterostructure. *Physical Review Letters* 107:217202.

58. Bea, H., B. Dupe, S. Fusil, R. Mattana, E. Jacquet, B. Warot-Fonrose, F. Wilhelm, A. Rogalev, S. Petit, V. Cros, A. Anane, F. Petroff, K. Bouzehouane, G. Geneste, B. Dkhil, S. Lisenkov, I. Ponomareva, L. Bellaiche, M. Bibes, and A. Barthelemy. 2009. Evidence for room-temperature multiferroicity in a compound with a giant axial ratio. *Physical Review Letters* 102:217603.

59. He, Q., Y. H. Chu, J. T. Heron, S. Y. Yang, W. I. Liang, C. Y. Kuo, H. J. Lin, P. Yu, C. W. Liang, R. J. Zeches, W. C. Kuo, J. Y. Juang, C. T. Chen, E. Arenholz, A. Scholl, and R. Ramesh. 2011. Electrically controllable spontaneous magnetism in nanoscale mixed phase multiferroics. *Nature Communications* 2:225.

60. Zheng, H., F. Straub, Q. Zhan, P.-L. Yang, W.-K. Hsieh, F. Zavaliche, Y.-H. Chu, U. Dahmen, and R. Ramesh. 2006. Self-assembled growth of $BiFeO_3$-$CoFe_2O_4$ nanostructures. *Advanced Materials* 18:2747–2752.

61. Zavaliche, F., T. Zhao, H. Zheng, F. Straub, M. P. Cruz, P. L. Yang, D. Hao, and R. Ramesh. 2007. Electrically assisted magnetic recording in multiferroic nanostructures. *Nano Letters* 7:1586–1590.

62. MaCmanus-Driscoll, J. L., P. Zerrer, H. Y. Wang, H. Yang, J. Yoon, A. Fouchet, R. Yu, M. G. Blamire, and Q. X. Jia. 2008. Strain control and spontaneous phase ordering in vertical nanocomposite heteroepitaxial thin films. *Nature Materials* 7:314–320.

63. Bhatnagar, A., A. R. Chaudhuri, Y. H. Kim, D. Hesse, and M. Alexe. 2013. Role of domain walls in the abnormal photovoltaic effect in $BiFeO_3$. *Nature Communications* 4:2835.

64. Lee, J. H., I. Fina, X. Marti, Y. H. Kim, D. Hesse, and M. Alexe. 2014. Spintronic functionality of $BiFeO_3$ domain walls. *Advanced Materials* 26:7078–7082.

65. Lei, N., T. Devolder, G. Agnus, P. Aubert, L. Daniel, J.-V. Kim, W. Zhao, T. Trypiniotis, R. P. Cowburn, C. Chappert, D. Ravelosona, and P. Lecoeur. 2013. Strain-controlled magnetic domain wall propagation in hybrid piezoelectric/ferromagnetic structures. *Nature Communications* 4:1378.

66. Liu, M., B. M. Howe, L. Grazulis, K. Mahalingam, T. Nan, N. X. Sun, and G. J. Brown. 2013. Voltage-impulse-induced non-volatile ferroelastic switching of FMR for reconfigurable magnetoelectric microwave devices. *Advanced Materials* 25:4886–4892.

67. Matsukura, F., Y. Tokura, and H. Ohno. 2015. Control of magnetism by electric fields. *Nature Nanotechnology* 10:209–220.

SECTION II

Multiferroic Composites

6

Bulk Magnetoelectric Composites
Direct and Converse Magnetoelectric Effects

Jia-Mian Hu and Ce-Wen Nan

Contents

6.1 Introduction

In bulk magnetoelectric (ME) composites, the linear ME effect is a product of piezomagnetic and piezoelectric effects of the constituting phases (while the nonlinear ME effect involves nonlinear magnetostriction and/or electrostriction), and can be expressed as [1]

$$\text{Direct ME effect} = \frac{\text{electric}}{\text{mechanical}} \times \frac{\text{mechanical}}{\text{magnetic}}$$

$$\text{Converse ME effect} = \frac{\text{magnetic}}{\text{mechanical}} \times \frac{\text{mechanical}}{\text{electric}}$$

(6.1)

In this case, magnetic field (H)–induced charge/voltage output (direct ME effect), or electric field (E)–controlled magnetism (converse ME effect) is achieved via strain transfer across the interface. The cross coupling leads to new functionalities that are not available in either of the constituting phases and opens tantalizing possibilities for various technologies. For example, the direct ME effect converts the magnetic field into electric signals at room temperature without any other power source. It is very sensitive and can potentially revolutionize the current superconducting quantum interference devices (SQUID)–based technology, which is bulky, expensive, and only operates at low temperatures [2]. The converse ME effect enables the modulation of magnetization by electric fields rather than by power-consuming electric currents, and can be used to design next generation nonvolatile memories and low-power electric-field tunable microwave devices [3]. A detailed discussion on various applications of bulk ME composites is presented in Chapter 8 of this book.

There are various types of bulk ME composites [4,5], including (1) 0–3 particulate composites, which contain magnetostrictive particles embedded in a piezoelectric matrix or vice versa (Figure 6.1a); (2) 2–2 laminate composites with alternating magnetostrictive and piezoelectric layers (Figure 6.1b); and (3) 1–3 rod/fiber composites with magnetic rods embedded in piezoelectric matrix or vice versa (Figure 6.1c). Among them, laminate ME composites are of particular interest as they are easy to prepare, and usually show much lower leakage currents compared to 0–3 and 1–3 type composites. Industrially mass-produced multilayer ceramic capacitors (MLCCs) serve as one sagaciously designed laminate ME composites [2] in which the base metal electrode Ni and doped $BaTiO_3$ are the magnetostrictive and piezoelectric layers, respectively. $Pb(Zr,Ti)O_3(PZT)$ ceramics, $Pb(MgNb)O_3–PbTiO_3(PMN–PT)$ single crystals, P(VDF–TrFE) polymer, and so on, are other frequently used piezoelectric materials.

The main body of this chapter is divided into two sections, covering direct (Section 6.2) and converse (Section 6.3) ME effects. We begin in each section with a short introduction to the characterization methods, from which the readers can get an intuitive understanding of the cross-coupling phenomena. This is followed by a brief historical background introduction where important advances are highlighted. More detailed information can be found in two comprehensive review articles [5,6]. We then focus on the most recent progress regarding the direct and the converse ME effects in bulk composites, and end with some concluding remarks (Section 6.4). In Figure 6.2, we summarize the pivotal moments during the development of bulk ME composites over the past 50 years.

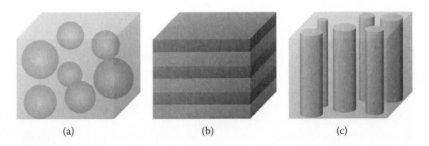

FIGURE 6.1 Schematics of (a) 0–3 particulate, (b) 2–2 laminate, and (c) 1–3 rod/fiber magnetoelectric (ME) composites.

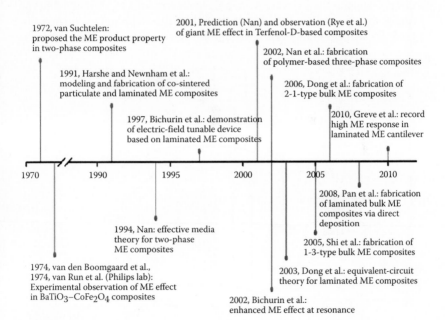

FIGURE 6.2 Pivotal moments in the development of bulk ME composites.

6.2 Direct ME Effect in Bulk Multiferroic Composites

6.2.1 Characterizing the Direct ME Effect

The time-dependent voltage output, $V_{ME}(t)$, of a multiferroic composite due to direct ME effect can be expressed as a Taylor series expansion of H [7]:

$$V_{ME}(t) = \alpha_0 + \alpha_1 H(t) + \alpha_2 H^2(t) + \cdots, \qquad (6.2)$$

where α_i ($i = 0, 1, 2, 3\ldots$) is the ith order Taylor expansion coefficient. For bulk ME composites, $V(t)$ is linearly proportional to the magnetomechanical deformation $\lambda(H)$, which can also be written as a Taylor series:

$$\lambda(H(t)) = \lambda(H_{dc}) + H(t) \frac{d\lambda}{dH}\bigg|_{H=H_{dc}} + \frac{1}{2} H^2(t) \frac{d^2\lambda}{dH^2}\bigg|_{H=H_{dc}} + \cdots. \qquad (6.3)$$

The linear direct ME coefficient α_E, dV_{ME}/dH (V/Oe) or dE/dH (V/cm Oe), is then proportional to the linear (first-order) piezomagnetic coefficient $d\lambda/dH$ under a magnetic field H_{dc}. Up to now, most theoretical and experimental studies have been devoted to the linear ME effect.

When it comes to measuring the linear ME coefficient, the most commonly used experimental setup is schematically shown in Figure 6.3. Here, a trilayer (piezomagnetic-piezoelectric-piezomagnetic) 2–2 ME composite is used as an example. During the measurement, a dc magnetic field H_{dc} is applied to the sample, with a small ac field H_{ac} generated by the Helmholtz coil superimposed. The ac field induces an ac voltage due to the direct ME effect, which is

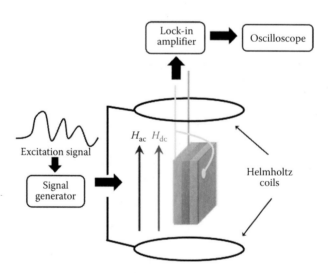

FIGURE 6.3 Sketch of the experimental setup for measuring the linear direct ME effect.

measured using a lock-in amplifier for high sensitivity. The linear ME coefficient is simply the ac voltage divided by H_{ac}. By sweeping the dc magnetic field, ME coefficient of the composite under various external fields can be obtained. Very recently, Shi et al. [8] developed a new high-speed dynamic method that can reduce the data-acquiring time from several seconds down to 300 ms with tolerable loss of accuracy, enabling high-speed, automatic recording of the ME loop (α_E vs. H_{dc}).

6.2.2 Historical Background of Direct ME Effect

Though the particulate ceramic composite of $BaTiO_3$–$CoFe_2O_4$ was fabricated at Philips Laboratory in the year 1974 using sophisticated unidirectional solidification method [9], the development of bulk ME composites gained momentum only during the past two decades. The $BaTiO_3$–$CoFe_2O_4$ composite had a direct ME voltage coefficient of $\alpha_E = 0.13$ V/cm Oe under a bias field of around 560 Oe at room temperature [10]. Figure 6.4a shows the mosaic-like grain structure of such composites, and the ME hysteresis loop obtained using an ac excitation field of 10 Oe superimposed on the dc magnetic field.

After the first report, efforts have been devoted to developing ME composites that have larger ME coefficient and are easier to fabricate. This goal had not been achieved until the early 2000s, when giant direct ME effect ($\alpha_E > 1$ V/cm Oe) was predicted [11,12] (Figure 6.5a) and later observed [13,14] (Figure 6.5b) in the laminate composites made of Terfenol-D ($Tb_{1-x}Dy_xFe_2$) alloys and PZT ceramic pellets (or P(VDF–TrFE) polymer) bonded by silver epoxy or epoxy resin. The Terfenol-D-based laminate composites have quickly become one of the most investigated bulk ME composite systems since then [15–18].

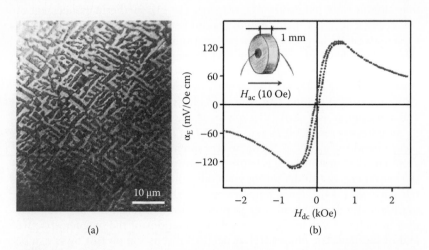

(a) (b)

FIGURE 6.4 (a) Microstructure and (b) ME hysteresis loop of the $BaTiO_3$–$CoFe_2O_4$ particulate composite prepared by unidirectional solidification. (Adapted from Van den Boomgaard, J. et al., *Ferroelectrics*, 10, 295–298, 1976.)

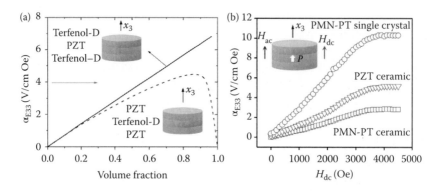

FIGURE 6.5 (a) Theoretical prediction (Adapted from Nan, C.W. et al., *Physical Review B*, 63, 144415, 2001.) and (b) experimental observation of giant ME effect in Terfenol-based laminated composites. (Adapted from Ryu, J. et al., *J. Electroceram.*, 8, 107–119, 2002.)

By replacing the Terfenol-D with a soft magnetic alloy, for example, amorphous iron-based Metglas (FeSiBC or FeSiCo), that has larger magnetic permeability and higher piezomagnetic coefficient, the ME voltage coefficient α_E can be further enhanced to 7.2 V/cm Oe at 1 Hz and up to 310 V/cm Oe at resonance frequency (~50 kHz) under a dc magnetic bias field as low as 8 Oe [19], making these Metglas-based laminate ME composites very promising as magnetic-field sensors. A push-pull design (with a piezofiber layer laminated between two Metglas layers) can further increase the α_E to 22 V/cm Oe at 1 Hz and close to 500 V/cm Oe at resonance (~22 kHz), both under a small bias magnetic field of 5 Oe [20]. Given that imperfect interfacial bonding may significantly reduce the ME response [12,21], further enhancement of ME coefficient in laminate composites can be achieved by directly growing a piezomagnetic layer (thickness ranges from μm to mm) onto the piezoelectric layer by electrodeposition [22] or magnetron sputtering [23]. Such direct deposition can improve the quality of interface by eliminating the epoxy bonding layer. For example, in a cantilever (20 mm × 2 mm × 140 μm) structure consisting of a laminate piezoelectric AlN (1.8 μm in thickness) and amorphous FeCoSiB (1.75 μm in thickness) composite sputtered on Si, an off-resonance (100 Hz) α_E of 3.1 V/cm Oe has been achieved, with a record-breaking α_E of 737 V/cm Oe at resonance [23]. Another approach to eliminate the bonding layer is to use insulating polymer (e.g., PVDF) as both the matrix and the binder for the embedded magnetostrictive and piezoelectric particles [24–27]. Such polymer-based ME composites are light-weight, mechanically flexible, and easy to process [6]. However, the ME response may be reduced due to the lower volume fractions of the magnetostrictive and/or piezoelectric phases [11]. A detailed discussion on this topic can be found in a recent review article [28].

6.2.3 Current Status of Direct ME Effect Study

6.2.3.1 Self-Biased Direct ME Effect

Most piezomagnetic materials show very small piezomagnetic coefficient $d\lambda/dH_{dc}$ (i.e., the slope of the λ–H_{dc} curve) at zero magnetic field, which leads to negligible linear ME voltage coefficient α_E of the composite. A bias dc magnetic field (ranging from several to hundreds of Oe, see References [10,20]) is usually required to achieve the peak value of $d\lambda/dH_{dc}$ in a composite. However, adding permanent hard magnet to the device not only increases the size but also causes cross-talk among neighboring units in an array, and provides a potential noise source, which is undesirable for many applications, for example, high-spatial-resolution magnetic-field sensors.

To overcome this problem, recent studies have explored the possibility of achieving finite α_E at zero magnetic field through a self-bias effect. This is made possible by the hysteric behavior of α_E when sweeping the dc field, H_{dc}, between negative and positive values [29,30]. One way to induce the hysteresis is illustrated by the three-phase PZT/Ni/Ni$_{1-x}$ZnFe$_2$O (NZFO) laminate composite, in which Ni and NZFO possess very different piezomagnetic and magnetostrictive responses [30]. Another approach is to exploit the intrinsic hysteric behavior of the linear piezomagnetic coefficient $d\lambda/dH_{dc}$ (that is proportional to $dM^2/\underline{d}H_{dc}$, where M is magnetization) in single-phase magnetic materials that show hysteresis in their magnetization versus field loop (M–H) [31–33]. This has already been observed in earlier works (see Figure 6.4b). Another example is shown in Figure 6.6. Clear nonhysteric and hysteric ME loops are observed in bilayer composites of Metglas ribbon (0.15 mm in thickness) and Ni foil (0.15 mm in thickness) bonded with PMT ($0.8[Pb(Zr_{0.52}Ti_{0.48})O_3] - 0.2[Pb(Zn_{1/3}Nb_{2/3})O_3] + 2$ mol% MnO$_2$) ceramic pellet (0.3 mm in thickness) through epoxy resin, respectively. The corresponding magnetization loops are shown in Figure 6.6b [31].

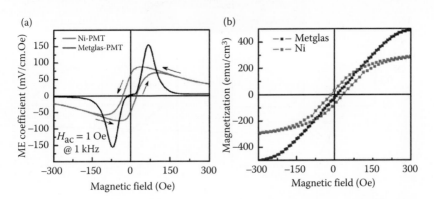

FIGURE 6.6 (a) Hysteric and nonhysteric ME loops of Ni–PMT and Metglas–PMT bilayers, respectively. (b) Magnetization versus magnetic field loops of Ni and Metglas. (From Zhou, Y. et al. *Appl. Phys. Lett.*, 101, 232905, 2012.)

Multiferroic Materials

As seen in Figure 6.6a, the α_E is about 63 mV/cm Oe at low frequency under zero dc bias magnetic field for the Ni-PMT composite, which is clearly related to the remnant magnetization shown in Figure 6.6b. More recently, in a low-temperature co-fired trilayer composite of textured PMN-PT layer sandwiched by two $(Ni_{0.6}Cu_{0.2}Zn_{0.2})Fe_2O_4$ (NCZF) ferrite layers that exhibit good interface quality, the off-resonance α_E at zero bias field is enhanced to 1.2 V/cm Oe [33]. Such low-temperature co-firing technique is also compatible with that used for industrially manufactured MLCCs, demonstrating the possibility of commercialization.

Nevertheless, one drawback of the hysteresis in ME loop is the strong dependence on magnetizing history [34]. A more promising approach toward self-biased direct ME effect is to add an antiferromagnetic layer on top of the ferromagnetic layer, whereby the exchange bias field, appearing as a unidirectional shift in the magnetization loop [35], provides a built-in dc bias field. Exchange bias is widely used in tunneling magnetoresistance (TMR)–based magnetic read-heads, and has recently been employed in multilayer ME nanocomposites [34], where at least one of the phases has a dimension at the nanometer scale. Though promising, achieving large self-biased α_E in these thin-film nanocomposites requires delicate control of the interface among phases during preparation, and thus may raise the cost substantially. Alternatively, exchange bias shift of magnetic hysteresis loops has been observed in partially crystallized bulk amorphous alloys (e.g., Metglas) through annealing in air [36]. This can also provide a built-in dc bias for magnetostrictive soft magnets in ME composites.

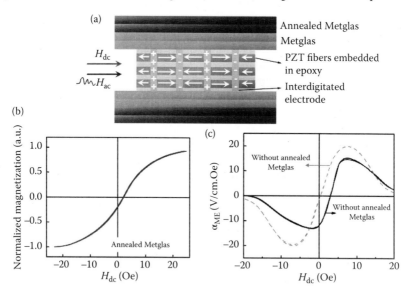

FIGURE 6.7 (a) Schematics of the multiple push–pull mode ME laminate composite with the annealed Metglas layers providing a unidirectional bias field. (b) Magnetization curve under low magnetic field. (c) ME coefficients of samples with and without the annealed Metglas. (From Li, M. et al. *Appl. Phys. Lett.*, 102, 082404, 2013.)

This has been demonstrated in a Metglas–PZT–Metglas laminate composite as shown in Figure 6.7. The in-plane oriented PZT piezofibers embedded in epoxy is sandwiched between two sets of annealed and amorphous Metglas layers. Shifts in both the magnetization curve (Figure 6.7b) and the ME curve (Figure 6.7c) are clearly observed, leading to a giant self-biased α_E of 12 V/cm Oe at low frequency under zero external field [37]. The ME curve shows negligible hysteresis, and hence is less history-dependent.

6.2.3.2 Nonlinear Direct ME Effect

Given that the linear direct ME voltage coefficient α_E under low-frequency excitation field (i.e., the magnetic field to be detected if it is used as a sensor) is much lower than that under resonance drive, and that the effective noise level of the lock-in amplifier-based ME sensor detection unit (see Figure 6.3) is much higher at low frequencies [38], the sensitivity of ME-based sensors is likely low at low frequencies. This may limit the possible application of ME sensors for biomagnetic detection such as magnetoencephalography (MEG) and magnetocardiography (MCG), where the target magnetic fields (produced by brain and heart) are usually at low frequencies, for example, 0.1–100 Hz. By using the record-high low-frequency (1 Hz) α_E of 52 V/cm Oe in a multiple push–pull mode laminate composites of Metglas and PMN–PT piezofibers and reducing every possible internal noise source, an ultra-high sensitivity of about 10 pT (pico-Tesla)/Hz$^{1/2}$ has been demonstrated [39]. However, it is still three to four orders of magnitude lower than that of SQUID (~ 0.001 pT/Hz$^{1/2}$) [40].

Recently, it was reported that the detection of target magnetic field ($H_{ac}\cos(2\pi f_0)$) of a low-frequency f_0 can be achieved at or close to the electromechanical resonance frequency [38,41], by superimposing a modulating magnetic field ($H_{mod}\cos(2\pi f_1)$) of a higher-frequency f_1, in addition to the dc magnetic field H_{dc}. The approach can significantly reduce the noise level and allow accurate measurements at low frequencies [41]. Building on Equation 6.2, the ME voltage output $V_{ME}(t)$ induced by the combined magnetic field $\mathbf{H}_{tot} = H_{dc} + H_{ac}\cos(2\pi f_0) + H_{mod}\cos(2\pi f_1)$ can be expressed as

$$
\begin{aligned}
v_{ME}(t) &= \alpha_0 + \alpha_A^{(1)}\cos\left(2\pi f_0 t\right) + \alpha_B^{(1)}\cos\left(2\pi f_1 t\right) + \alpha_A^{(2)}\cos\left(4\pi f_0 t\right) \\
&\quad + \alpha_B^{(2)}\cos\left(4\pi f_1 t\right) + \alpha_C^{(2)}\cos\left[2\pi\left(f_0 + f_1\right)t\right] + \alpha_C^{(2)}\cos\left[2\pi\left(f_0 - f_1\right)t\right] + \cdots, \\
\alpha_0 &\propto \left\{\lambda(H_{dc}) + qH_{dc} + \frac{1}{2}p\left[H_{dc}^2 + \frac{1}{2}\left(H_{ac}^2 + H_{mod}^2\right)\right]\right\}, \\
\alpha_A^{(1)} &\propto \left(q + pH_{dc}\right)H_{ac}, \quad \alpha_B^{(1)} \propto \left(q + pH_{dc}\right)H_{mod}, \\
\alpha_A^{(2)} &\propto \frac{p}{4}H_{ac}^2, \quad \alpha_B^{(2)} \propto \frac{p}{4}H_{mod}^2, \quad \alpha_C^{(2)} \propto pH_{ac}H_{mod}, \\
q &= \left.\frac{d\lambda}{dH}\right|_{H=H_{dc}}, \quad p = \left.\frac{d^2\lambda}{dH^2}\right|_{H=H_{dc}}
\end{aligned}
$$

$$(6.4)$$

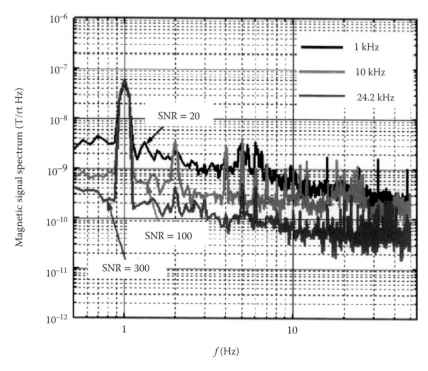

FIGURE 6.8 Equivalent magnetic noise spectrum for a magnetic signal (H_{ac}) of $f_0 = 1$ Hz, after applying a modulation field H_{mod} of $f_1 = 1$, 10, and 24.2 kHz [44].

where α_0 is the dc component, which only changes the absolute value without affecting the shape of the waveform, and can be screened using the ac coupling mode of an oscilloscope. $\alpha_A^{(2)}$ and $\alpha_B^{(2)}$ are related to the frequency doubling that has been observed in PZT–FeBSiC laminate composites [42], while $\alpha_C^{(2)}$ describes the efficiency of the frequency shifting, which is proportional to the nonlinear piezomagnetic coefficient $p = [d^2\lambda/dH^2]$, thus linearly dependent on $d\alpha_E/dH$, where α_E is proportional to the linear piezomagnetic coefficient [43,44]. Experimentally, $\alpha_C^{(2)}$ is obtained as $\alpha_C^{(2)} = V_{ME}/(H_{ac}H_{mod}d)$ with the unit of V/(cm Oe²), where V_{ME} is the measured ME voltage, d is the thickness of the insulating composite, and H_{ac} and H_{mod} are the magnitudes of the two excitation magnetic fields.

Superimposing the high-frequency modulating magnetic field, H_{mod} can effectively reduce the noise level when detecting magnetic fields. As shown in Figure 6.8, the noise level of a Metglas/PMN–PT fiber laminate is about 0.2 $nT/Hz^{1/2}$ at $f_1 = 24.2$ kHz (at resonance), which is much lower than the 3 $nT/Hz^{1/2}$ value at $f_1 = 1$ kHz, resulting in a 15 times larger signal-to-noise ratio (SNR) [44].

6.3 Converse ME Effect

6.3.1 Characterizing the Converse ME Effect

The converse ME effect describes the electric-field-induced change in magnetization. Accordingly, most characterization methods for magnetic properties are applicable to investigate the converse ME effect as long as the experimental setup allows the application of external electric fields onto the sample.

A simple approach is to use pick-up coils [45–53] wound about the composite under investigation as shown in Figure 6.9a. An ac electric field is applied across the piezoelectric layer during the measurement, causing time-dependent variations of magnetization and magnetic induction B (i.e., $\partial \mathbf{B}/\partial t \neq 0$) due to the strain-mediated converse ME effect. The varying B leads to ac voltage in the pick-up coil (i.e., Faraday's law of induction), which is measured by a lock-in amplifier. The converse ME coefficient can thus be obtained as $\alpha_M = \partial \mathbf{B}/\partial V$ [in the unit of G(Gauss)/V] or $\alpha_M = \partial \mathbf{B}/\partial E$ [in the unit of s (second)/m]. Similar to the case of direct ME voltage coefficient α_E, α_M also reaches its peak at the electromechanical resonance [45,47,49]. A moderate bias dc magnetic field H_{dc} is normally required to optimize the converse ME coupling, possibly by

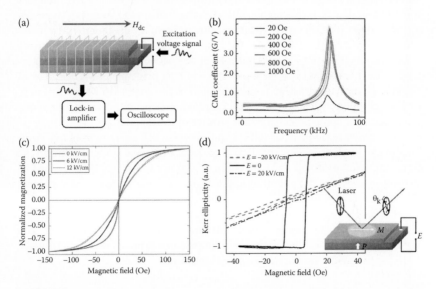

FIGURE 6.9 (a) Sketch of the pick-up coil-based experimental setup for measuring the converse ME coefficient. (b) Frequency-dependent converse ME coefficient of Terfenol-D/PMN–PT laminate. (From Wang, Y. et al., *Appl. Phys. Lett.*, 93,113503, 2008.) Electric field–modulated magnetization versus field loops of (c) Metglas/PZT/Metglas laminate measured using VSM (From Lou, J. et al., *Appl. Phys. Lett.*, 94,112508, 2009.) and (d) FeBSiC–PZT bilayer measured using MOKE magnetometer (From Ma, J. et al., *J. Phys. D Appl. Phys.* 43, 012001, 2010.) whose principle is briefly illustrated in the inset. The circle at the surface of the magnet indicates the laser spot.

maximizing the contribution of non-180° magnetic domain wall motion to the piezomagnetic coefficient [45].

As an example, the converse ME coefficient α_M of a Terfenol–D/PMN–PT laminate composite shows a maximum value of 4.4 G/V at the electromechanical resonance frequency of 74 kHz and an off-resonance value of 0.5 G/V under the optimum H_{dc} of 400 Oe (Figure 6.9b) [49]. In addition, the converse ME coupling in bulk ME composites also depends on a dc electric bias in some cases [53]. If the piezoelectric phase is also ferroelectric, applying a dc electric field allows us to measure the linear converse ME coefficient or susceptibility without the contribution of ferroelectric domain switching [54]. Furthermore, the experimental setup shown in Figure 6.9a for converse ME effect is quite similar to that for direct ME effect (cf. Figure 6.3). It is, therefore, possible to measure the converse or direct ME effect using one common set of experimental apparatus [51,54,55].

Alternatively, the converse ME effects can be characterized by recording voltage-induced changes in magnetization curves [e.g., M–H loop or temperature-dependent magnetization (see Reference [56])]. This can be achieved using the vibrating sample magnetometer (VSM) that is essentially based on pick-up coil as well [57], SQUID, or the magneto-optical Kerr effect (MOKE) magnetometer that detects local magnetization change within the laser spot (typically at the scale of μm). Figure 6.9c and d shows examples of static electric-field-modified M–H loops of a Metglas/PZT/Metglas laminate [52] measured using VSM and a FeBSiC–PZT bilayer sample [58] by MOKE magnetometer, respectively. The inset of Figure 6.9d shows the working principle of MOKE; that is, the change in magnetization is proportional to the change in Kerr rotation θ_k or Kerr ellipticity when the linearly polarized laser beam becomes elliptically polarized after being reflected by the surface of the magnet.

At microwave frequencies (GHz), the converse ME effect in bulk ME composites can be characterized by detecting the electric field–induced shift in the ferromagnetic resonance (FMR) field or resonance frequency [59] Such information is essential for potential applications in tunable microwave devices [3]. More recently, magnetic force microscope (MFM) has also been used to observe the magnetic domain structure evolution upon applying static electric fields in bulk ME composites [60,61], further improving our understanding of the converse ME effect by relating the macroscopic properties (magnetization, FMR) to the magnetic microstructures (domains).

6.3.2 Historical Background of Converse ME Effect

The study of converse ME effect is motivated by the possibility of achieving dual-field (magnetic and electric fields) tunable microwave devices (see a recent review [62] for a detailed discussion), which was first demonstrated by Bichurin et al. in 1997 [59]. Compared with traditional microwave devices that can only be tuned by a magnetic field, the dual field–tunable devices are potentially faster, less noisy, and consume much less power during operation [63]. Figure 6.10 shows the electric field–induced shift of FMR frequency of a prototypical ME resonator based on a bilayer composite consisting of low-loss ferrimagnet yttrium iron garnet

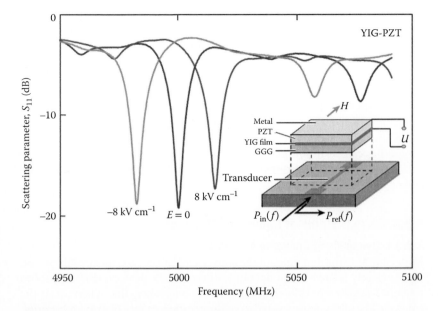

FIGURE 6.10 A prototypical ME resonator where the frequency-dependent re-flected power is measured. The PZT/YIG laminate prepared on a gadolinium gallium garnet (GGG) substrate is placed on a microstripline transducer and excited with microwaves. The dips in the absorption spectrums are due to FMR in YIG. (From Srinivasan, G., *Annu. Rev. Mater. Res.* 40,153–178, 2010.)

(YIG, 15 μm in thickness) and PZT (0.5 mm in thickness) [62]. Depending on the polarity of the applied electric field, the resonance frequency shifts to either higher or lower frequency, with an average electrically induced frequency change Δf_E of 24 MHz under 8 kV/cm electric field.

However, one major problem of these ferrite-ferroelectric–based bulk microwave devices is the limited tunable range. For instance, electric field–induced changes in magnetic resonance (Δf_E) and resonance field (ΔH_r) are below 150 MHz and 50 Oe (see References [63–65] for a detailed discussion), respectively, much smaller than the changes that can be induced by magnetic field [65]. This is mainly attributed to the large loss of the ferrite at microwave frequencies. On the other hand, giant Δf_E of about 5820 MHz has been demonstrated in a layered thin-film composite with a nanometer-thick (100 nm) FeGaB film grown on lead zinc niobate-lead titanate (PZN–PT, 0.5 mm in thickness) substrate [66]. A detailed discussion on these thin-film ME composites-based tunable microwave devices can be found in a recent review [67].

In addition to the converse ME effect at microwave frequencies, a low-frequency effect has also been subjected to increasing research efforts since it was reported by Wan et al. [45], which mainly focuses on the electric field–induced change in magnetization as briefly discussed in the previous section (see Figure 6.9b through d). Such functionality has great potential for ultralow power spintronic devices [68] such as magnetoresistive random access memories (MRAM)

[69,70] and ME logic devices [71]. Research in these areas is stimulated largely by the milestone work on $BaTiO_3$–$CoFe_2O_4$ ME nanocomposite thin films by Zheng et al. [72], or perhaps an earlier work that reports the modulation of magnetization in $La_{0.67}Sr_{0.33}MnO_3$ thin film (50 nm in thickness) under strains generated by cooling the $BaTiO_3$ single crystal substrate [73].

In general, research activities on the converse ME effect have gradually shifted from bulk materials to thin films, driven by the pressing need for device miniaturization and facilitated by the advances in thin-film growth techniques [74]. Nevertheless, studying the converse ME effect in bulk ME composites can be, in some cases, a simple and more economical approach for understanding such strain-mediated coupling effect. Two examples are discussed in the following section.

6.3.3 Current Status of Converse ME Effect

6.3.3.1 Dynamics of Converse ME Effect

Understanding the dynamics of the converse ME effect is critically important, as it is closely related to the operation speed of ME devices. However, it has remained largely unexplored [75]. In bulk composites, the dynamics of the converse ME effect mainly consists of time responses for (1) strain generation in the piezoelectric or ferroelectric component (maybe accompanied by ferroelectric polarization switching); (2) transfer of strain across the interface in the form of mechanical waves at the speed of sound [69]; and (3) achieving equilibrium magnetic state under strain.

For the first part, polarization switching time ranging from 10^{-8} to 10^2 s has been reported for bulk ferroelectric single crystals and ceramics depending on the magnitude of applied electric field [76]. If the frequency of the applied electric field is too high to allow a complete polarization switching, the generated strain would become smaller, leading to a weaker converse ME effect. For the second part, with the following assumptions: (1) a well-bonded interface without significant elastic loss; (2) an average strain transfer speed of 3000 m/s (the sound speed in solids is high, for example, about 5120 m/s in iron at 293 K); and (3) an average length of 1 mm along the thickness direction of the sample, the time span can be roughly estimated as about 0.33 μs. For the third part, the magnetization evolution in bulk samples generally occurs by domain wall motion, whose velocity depends on various factors such as the magnitude of strain, materials quality, device geometries, and so on. The details require a case-by-case study.

Thus, the dynamics of converse ME effect in bulk composites involve complex mechanisms and may vary from sample to sample. Recently, Chen et al. [77] studied the time-dependent converse ME effect of a laminate ME composite of Metglas (25 μm in thickness) and PMN–PT single crystal under low-frequency (0.5–8 Hz) excitation electric field. As shown in Figure 6.11, upon applying a static electric field of 8 kV/cm, the magnetic moment M shows a relaxation time t_s of about 0.6 s before achieving the new equilibrium.

Apparently, electric fields with frequencies higher than 1 Hz lead to a considerably smaller converse ME effect. Such time delay of 0.6 s is mainly

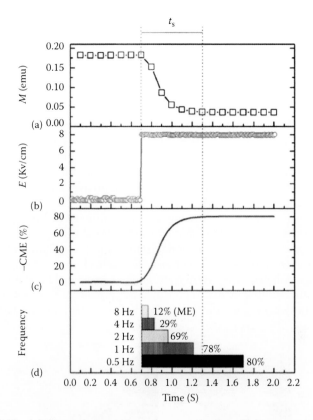

FIGURE 6.11 (a) Dynamic evolution of magnetization M in response to (b) a static electric field E of 8 kV/cm applied across the Metglas/PMN-PT. (c) Time-dependent converse ME effect, defined as $(M(E)-M(0))/M(0)$, under static E. (d) The maximum converse ME effect under ac electric fields of different frequencies. (Reproduced from Chen, Y. et al., *Appl. Phys. A*, 100, 1149–1155, 2010.)

contributed to the ferroelectric switching in PMN–PT single crystals, given the very fast magnetic relaxation (on the order of 10^{-6} s) in the amorphous Metglas ribbon [77]. This work sets up a good example of studying the dynamics of converse ME effect in bulk ME composites.

6.3.3.2 Converse ME Effect at the Mesoscale

Mesoscale, or "in-between" scale, corresponds to an intermediate region between macroscopic and atomic scales. In materials science, it normally refers to the scale of various microstructures (grains, precipitates, ferroic domains, etc.), ranging from nano- to micron meters. Studying the converse ME effect at the mesoscale is to understand how the ferroelectric and magnetic domains (domain walls) evolve under the external fields, which allows a better understanding of structure-property correlation in bulk ME composites.

Multiferroic Materials

Remarkably, the modulation of materials properties at mesoscale can, in some cases, exhibit features that cannot appear in the whole sample, which may lead to promising routes to reproduce these new features in nanocomposites.

For example, it is generally accepted that electric field is a time invariant and cannot be used to change the polarity of the magnetization vector (i.e., magnetization reversal), that is, time-variant, though up to 90° magnetization rotation by electric field–induced strain has recently been demonstrated both theoretically [78] and experimentally [58] in laminated ME composites (a detailed discussion can be found in recent reviews [79,80]). However, electric field–induced (180°) magnetization reversal is highly desirable for low-power spintronic and magnetic devices [81]. The seemingly impossible goal has recently been achieved in a commercial multilayer capacitor with Ni as the internal electrode and $BaTiO_3$ as the dielectric layer [60]. As seen from the MFM images (Figure 6.12a through d), nonvolatile (i.e., all images are captured after the electric field is shut off) and repeatable magnetization reversal is observed in the central region (around 14 μm, see Figure 6.12e) of the scanned area in Ni. This unconventional electric field–induced magnetization reversal is attributed to the transverse static magnetic field assumed to arise from the surrounding media and the dynamic process of reversible ferroelectric domain switching within 1–10 ns (see Reference [60] for details).

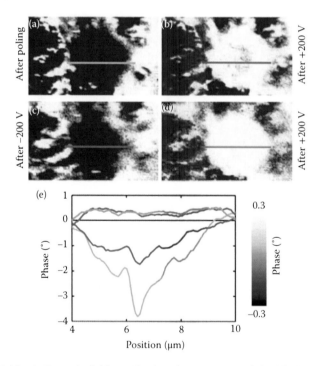

FIGURE 6.12 (a through d) Magnetic domain structures of the Ni electrode after successive electric poling. (e) Phase contrast profile along the central of the MFM images. Reproduced from Ghidini, M. et al., *Nat. Commun.*, 4, 1453, 2013.

6.4 Conclusion and Outlook

Motivated by the pressing need for device miniaturization and facilitated by advances in the growth of high-quality thin films, research interests in ME composites have gradually shifted from bulk to thin films in recent years [6]. For direct ME effect, thin-film-based laminate ME composites allow the design of highly sensitive magnetic-field sensors [34]. For the converse ME effect, composite thin films are not only more suitable for on-chip integration but also exhibit novel coupling mechanisms due to enhanced contributions from the interface such as electrically driven changes in spin-polarized charge densities, and interfacial exchange interaction (see recent review papers [6,75,80,82,83]). Of particular interest are thin-film composites with ferro-electric/piezoelectric films (nanometer-thick) grown on thick magnetic layer (micrometer or thicker) for the direct ME effect [34,84–87] and vice versa for the converse ME effect [56,66,88–93]. These composite thin films are easier to prepare than all-film nanocomposites and generally have higher-quality interfaces than those in bulk ME composites.

Nevertheless, studies on bulk ME composites are equally important not only for developing device applications at larger scales but also for providing clues for understanding and designing ME nanocomposites. And there are many issues about bulk ME composites that deserve further investigations, including the following.

1. *Further enhancement of direct ME coupling under low or zero bias field.* The record of off-resonance direct ME voltage coefficient is 52 V/cm Oe at 1 Hz with H_{dc} = 5 Oe [39], which, however, is largely based on the enhanced voltage output in the Metglas/piezofiber/ Metglas laminate. Recently, by enhancing the piezoelectric and electromechanical coefficients of PVDF, a record-high intrinsic ME voltage coefficient up to 20 V/cm Oe at 20 Hz at $H_{dc} \approx 4$ Oe has been observed in a Metglas–PVDF bilayer laminate [94,95]. It is also possible to enhance the direct ME effect by controlling magnetic domain structures through internal residual stress engineering [96] or by tuning the thickness of magnetic layer [97].

2. *Practical applications of ME devices.* Continuous academic interests have led to tremendous progress in the development of bulk ME composites. Now it is time to apply these composite materials in practical devices, including novel magnetic field sensors, for example, in biomagnetic imaging [34,97], novel microwave devices for aircraft, satellite, and portable communication systems [98,99].

Acknowledgments

This work was supported by the NSF of China under Grants 11234005 and 51221291 and Beijing Education Committee under Grant 20121000301.

References

1. Nan, C.-W. 1994. Magnetoelectric effect in composites of piezoelectric and piezomagnetic phases. *Physical Review B* 50:6082.
2. Israel, C., N. Mathur, and J. Scott. 2008. A one-cent room-temperature magnetoelectric sensor. *Nature Materials* 7:93–94.
3. Hu, J.-M., T. Nan, N. X. Sun, and L.-Q. Chen. 2015. Multiferroic magnetoelectric nanostructures for novel device applications. *MRS Bulletin* 40:728–735.
4. Newnham, R., D. Skinner, and L. Cross. 1978. Connectivity and piezoelectric-pyroelectric composites. *Materials Research Bulletin* 13:525–536.
5. Nan, C.-W., M. Bichurin, S. Dong, D. Viehland, and G. Srinivasan. 2008. Multiferroic magnetoelectric composites: Historical perspective, status, and future directions. *Journal of Applied Physics* 103:031101.
6. Ma, J., J. Hu, Z. Li, and C. W. Nan. 2011. Recent progress in multiferroic magnetoelectric composites: From bulk to thin films. *Advanced Materials* 23:1062–1087.
7. Zhuang, X., M. L. C. Sing, C. Cordier, S. Saez, C. Dolabdjian, L. Shen, J. F. Li, M. Li, and D. Viehland. 2011. Evaluation of applied axial field modulation technique on ME sensor input equivalent magnetic noise rejection. *Sensors Journal, IEEE* 11:2266–2272.
8. Shi, Z., L. Chen, Y. Tong, H. Xue, S. Yang, Y. Lu, C. Wang, and X. Liu. 2013. High speed characterization of the magnetoelectric hysteresis loop. *Magnetics, IEEE Transactions on* 49:5671–5674.
9. Van den Boomgaard, J., D. Terrell, R. Born, and H. Giller. 1974. An in situ grown eutectic magnetoelectric composite material. *Journal of Materials Science* 9:1705–1709.
10. Van den Boomgaard, J., A. Van Run, and J. V. Suchtelen. 1976. Magnetoelectricity in piezoelectric-magnetostrictive composites. *Ferroelectrics* 10:295–298.
11. Nan, C. W., M. Li, and J. H. Huang. 2001. Calculations of giant magnetoelectric effects in ferroic composites of rare-earth–iron alloys and ferroelectric polymers. *Physical Review B* 63:144415.
12. Nan, C.-W., G. Liu, and Y. Lin. 2003. Influence of interfacial bonding on giant magnetoelectric response of multiferroic laminated composites of $Tb_{1-x}Dy_xFe_2$ and $PbZrxTi_{1-x}O_3$. *Applied Physics Letters* 83:4366–4368.
13. Mori, K., and M. Wuttig. 2002. Magnetoelectric coupling in terfenol-D/polyvinylidenedifluoride composites. *Applied Physics Letters* 81:100–101.
14. Ryu, J., S. Priya, K. Uchino, and H.-E. Kim. 2002. Magnetoelectric effect in composites of magnetostrictive and piezoelectric materials. *Journal of Electroceramics* 8:107–119.
15. Dong, S., J.-F. Li, and D. Viehland. 2003. Longitudinal and transverse magnetoelectric voltage coefficients of magnetostrictive/piezoelectric laminate composite: Theory. *Ultrasonics, Ferroelectrics, and Frequency Control, IEEE Transactions on* 50:1253–1261.
16. Dong, S. X., J. F. Li, and D. Viehland. 2004. A longitudinal-longitudinal mode TERFENOL-D/$Pb(Mg_{1/3}Nb_{2/3})O_3PbTiO_3$ laminate composite. *Applied Physics Letters* 85:5305–5306.
17. Dong, S., J.-F. Li, and D. Viehland. 2004. Characterization of magnetoelectric laminate composites operated in longitudinal-transverse and transverse-transverse modes. *Journal of Applied Physics* 95:2625–2630.
18. Jia, Y., H. Luo, X. Zhao, and F. Wang. 2008. Giant magnetoelectric response from a piezoelectric/magnetostrictive laminated composite combined with a piezoelectric transformer. *Advanced Materials* 20:4776–4779.

19. Zhai, J. Y., S. X. Dong, Z. P. Xing, J. F. Li, and D. Viehland. 2006. Giant magnetoelectric effect in Metglas/polyvinylidene-fluoride laminates. *Applied Physics Letters* 89:083507.

20. Dong, S., J. Zhai, J. Li, and D. Viehland. 2006. Near-ideal magnetoelectricity in high-permeability magnetostrictive/piezofiber laminates with a (2-1) connectivity. *Applied Physics Letters* 89:252904.

21. Wang, X., and E. Pan. 2007. Magnetoelectric effects in multiferroic fibrous composite with imperfect interface. *Physical Review B* 76:214107.

22. Pan, D., Y. Bai, W. Chu, and L. Qiao. 2008. Ni–PZT–Ni trilayered magnetoelectric composites synthesized by electro-deposition. *Journal of Physics: Condensed Matter* 20:025203.

23. Greve, H., E. Woltermann, H.-J. Quenzer, B. Wagner, and E. Quandt. 2010. Giant magnetoelectric coefficients in $(Fe_{90}Co_{10})_{78}Si_{12}B_{10}$-AlN thin film composites. *Applied Physics Letters* 96:182501.

24. Nan, C.-W., L. Liu, N. Cai, J. Zhai, Y. Ye, Y. Lin, L. Dong, and C. Xiong. 2002. A three-phase magnetoelectric composite of piezoelectric ceramics, rare-earth iron alloys, and polymer. *Applied Physics Letters* 81:3831–3833.

25. Nan, C.-W., N. Cai, L. Liu, J. Zhai, Y. Ye, and Y. Lin. 2003. Coupled magnetic–electric properties and critical behavior in multiferroic particulate composites. *Journal of Applied Physics* 94:5930–5936.

26. Cai, N., C.-W. Nan, J. Zhai, and Y. Lin. 2004. Large high-frequency magnetoelectric response in laminated composites of piezoelectric ceramics, rare-earth iron alloys and polymer. *Applied Physics Letters* 84:3516–3518.

27. Nan, C.-W., N. Cai, Z. Shi, J. Zhai, G. Liu, and Y. Lin. 2005. Large magnetoelectric response in multiferroic polymer-based composites. *Physical Review B* 71:014102.

28. Martins, P., and S. Lanceros-Méndez. 2013. Polymer-based magnetoelectric materials. *Advanced Functional Materials* 23:3371–3385.

29. Yang, S.-C., C.-S. Park, K.-H. Cho, and S. Priya. 2010. Self-biased magnetoelectric response in three-phase laminates. *Journal of Applied Physics* 108:093706.

30. Laletin, U., G. Sreenivasulu, V. Petrov, T. Garg, A. Kulkarni, N. Venkataramani, and G. Srinivasan. 2012. Hysteresis and remanence in magnetoelectric effects in functionally graded magnetostrictive-piezoelectric layered composites. *Physical Review B* 85:104404.

31. Zhou, Y., S. C. Yang, D. J. Apo, D. Maurya, and S. Priya. 2012. Tunable self-biased magnetoelectric response in homogenous laminates. *Applied Physics Letters* 101:232905.

32. Zhang, J., P. Li, Y. Wen, W. He, A. Yang, and C. Lu. 2013. Giant self-biased magnetoelectric response with obvious hysteresis in layered homogeneous composites of negative magnetostrictive material Samfenol and piezoelectric ceramics. *Applied Physics Letters* 103:202902.

33. Yan, Y., Y. Zhou, and S. Priya. 2013. Giant self-biased magnetoelectric coupling in co-fired textured layered composites. *Applied Physics Letters* 102:052907.

34. Lage, E., C. Kirchhof, V. Hrkac, L. Kienle, R. Jahns, R. Knöchel, E. Quandt, and D. Meyners. 2012. Exchange biasing of magnetoelectric composites. *Nature Materials* 11:523–529.

35. Meiklejohn, W. H., and C. P. Bean. 1956. New magnetic anisotropy. *Physical Review* 102:1413.

36. Zhou, L., J. He, X. Li, B. Li, D. Zhao, and X. Wang. 2009. Exchange bias behaviour of amorphous CoFeNiSiB ribbons. *Journal of Physics D: Applied Physics* 42:195001.

37. Li, M., Z. Wang, Y. Wang, J. Li, and D. Viehland. 2013. Giant magnetoelectric effect in self-biased laminates under zero magnetic field. *Applied Physics Letters* 102:082404.

38. Shen, L., M. Li, J. Gao, Y. Shen, J. Li, D. Viehland, X. Zhuang, M. L. C. Sing, C. Cordier, and S. Saez. 2011. Magnetoelectric nonlinearity in magnetoelectric laminate sensors. *Journal of Applied Physics* 110:114510.

39. Wang, Y., D. Gray, D. Berry, J. Gao, M. Li, J. Li, and D. Viehland. 2011. An extremely low equivalent magnetic noise magnetoelectric sensor. *Advanced Materials* 23:4111–4114.

40. Zhai, J. Y., Z. P. Xing, S. X. Dong, J. F. Li, and D. Viehland. 2006. Detection of pico-Tesla magnetic fields using magneto-electric sensors at room temperature. *Applied Physics Letters* 88:062510.

41. Jahns, R., H. Greve, E. Woltermann, E. Quandt, and R. Knöchel. 2012. Sensitivity enhancement of magnetoelectric sensors through frequency-conversion. *Sensors and Actuators A: Physical* 183:16–21.

42. Ma, J., Z. Li, Y. Lin, and C. Nan. 2011. A novel frequency multiplier based on magnetoelectric laminate. *Journal of Magnetism and Magnetic Materials* 323:101–103.

43. Shen, Y., J. Gao, Y. Wang, P. Finkel, J. Li, and D. Viehland. 2013. Piezomagnetic strain-dependent non-linear magnetoelectric response enhancement by flux concentration effect. *Applied Physics Letters* 102:172904.

44. Shen, Y., J. Gao, Y. Wang, J. Li, and D. Viehland. 2014. High non-linear magnetoelectric coefficient in Metglas/PMN-PT laminate composites under zero direct current magnetic bias. *Journal of Applied Physics* 115:094102.

45. Wan, J., J.-M. Liu, G. Wang, and C. Nan. 2006. Electric-field-induced magnetization in Pb(Zr, Ti)O$_3$/Terfenol-D composite structures. *Applied Physics Letters* 88:2502.

46. Jia, Y., D. S.-W. Or, H. L. Chan, X. Zhao, and H. Luo. 2006. Converse magnetoelectric effect in laminated composites of PMN-PT single crystal and Terfenol-D alloy. *Applied Physics Letters* 88:242902.

47. Fetisov, Y., V. Petrov, and G. Srinivasan. 2007. Inverse magnetoelectric effects in a ferromagnetic–piezoelectric layered structure. *Journal of Materials Research* 22:2074–2080.

48. Zhou, J.-P., Y.-Y. Guo, Z. Xi, P. Liu, S. Lin, G. Liu, and H.-W. Zhang. 2008. Giant electric-field-induced magnetization in a magnetoelectric composite at high frequency. *Applied Physics Letters* 93:152501.

49. Wang, Y., F. Wang, S. W. Or, H. L. W. Chan, X. Zhao, and H. Luo. 2008. Giant sharp converse magnetoelectric effect from the combination of a piezoelectric transformer with a piezoelectric/magnetostrictive laminated composite. *Applied Physics Letters* 93:113503.

50. Chen, S., D. Wang, Z. Han, C. Zhang, Y. Du, and Z. Huang. 2009. Converse magnetoelectric effect in ferromagnetic shape memory alloy/piezoelectric laminate. *Applied Physics Letters* 95:022501.

51. Fetisov, Y., K. Kamentsev, D. Chashin, L. Fetisov, and G. Srinivasan. 2009. Converse magnetoelectric effects in a galfenol and lead zirconate titanate bilayer. *Journal of Applied Physics* 105:123918.

52. Lou, J., D. Reed, M. Liu, and N. Sun. 2009. Electrostatically tunable magnetoelectric inductors with large inductance tunability. *Applied Physics Letters* 94:112508.

53. Wu, T., T.-K. Chung, C.-M. Chang, S. Keller, and G. P. Carman. 2009. Influence of electric voltage bias on converse magnetoelectric coefficient in piezofiber/Metglas bilayer laminate composites. *Journal of Applied Physics* 106:054114.

54. Lou, J., G. N. Pellegrini, M. Liu, N. D. Mathur, and N. X. Sun. 2012. Equivalence of direct and converse magnetoelectric coefficients in strain-coupled two-phase systems. *Applied Physics Letters* 100:102907.

55. Timopheev, A., J. Vidal, A. Kholkin, and N. Sobolev. 2013. Direct and converse magnetoelectric effects in Metglas/LiNbO$_3$/Metglas trilayers. *Journal of Applied Physics* 114:044102.

56. Cherifi, R., V. Ivanovskaya, L. Phillips, A. Zobelli, I. Infante, E. Jacquet, V. Garcia, S. Fusil, P. Briddon, and N. Guiblin. 2014. Electric-field control of magnetic order above room temperature. *Nature Materials* 13:345–351.

57. Smith, D. 1956. Development of a vibrating-coil magnetometer. *Review of Scientific Instruments* 27:261–268.

58. Ma, J., Y. Lin, and C. Nan. 2010. Anomalous electric field-induced switching of local magnetization vector in a simple FeBSiC-on-Pb (Zr,Ti)O$_3$ multiferroic bilayer. *Journal of Physics D: Applied Physics* 43:012001.

59. Bichurin, M., R. Petrov, and Y. V. Kiliba. 1997. Magnetoelectric microwave phase shifters. *Ferroelectrics* 204:311–319.

60. Ghidini, M., R. Pellicelli, J. Prieto, X. Moya, J. Soussi, J. Briscoe, S. Dunn, and N. Mathur. 2013. Non-volatile electrically-driven repeatable magnetization reversal with no applied magnetic field. *Nature Communications* 4:1453.

61. Ghidini, M., R. Pellicelli, and N. Mathur. 2014. Non-volatile magnetoelectric edge effects observed using magnetic force microscopy. *Applied Physics Letters* 104:142401.

62. Srinivasan, G. 2010. Magnetoelectric composites. *Annual Review of Materials Research* 40:153–178.

63. Semenov, A., S. Karmanenko, V. Demidov, B. Kalinikos, G. Srinivasan, A. Slavin, and J. Mantese. 2006. Ferrite-ferroelectric layered structures for electrically and magnetically tunable microwave resonators. *Applied Physics Letters* 88:033503.

64. Tatarenko, A., G. Srinivasan, and M. Bichurin. 2006. Magnetoelectric microwave phase shifter. *Applied Physics Letters* 88:183507.

65. Ustinov, A., G. Srinivasan, and B. Kalinikos. 2007. Ferrite-ferroelectric hybrid wave phase shifters. *Applied Physics Letters* 90:031913.

66. Lou, J., M. Liu, D. Reed, Y. Ren, and N. X. Sun. 2009. Giant electric field tuning of magnetism in novel multiferroic FeGaB/lead zinc niobate–lead titanate (PZN-PT) heterostructures. *Advanced Materials* 21:4711–4715.

67. Liu, M., and N. X. Sun. 2014. Voltage control of magnetism in multiferroic heterostructures. *Philosophical Transactions of the Royal Society of London A: Mathematical, Physical and Engineering Sciences* 372:20120439.

68. Eerenstein, W., N. Mathur, and J. F. Scott. 2006. Multiferroic and magnetoelectric materials. *Nature* 442:759–765.

69. Hu, J.-M., Z. Li, L.-Q. Chen, and C.-W. Nan. 2011. High-density magnetoresistive random access memory operating at ultralow voltage at room temperature. *Nature Communications* 2:553.

70. Hu, J. M., Z. Li, L. Q. Chen, and C. W. Nan. 2012. Design of a voltage-controlled magnetic random access memory based on anisotropic magnetoresistance in a single magnetic layer. *Advanced Materials* 24:2869–2873.

71. Hu, J. M., Z. Li, Y. Lin, and C. Nan. 2010. A magnetoelectric logic gate. *Physica Status Solidi (RRL)-Rapid Research Letters* 4:106–108.

72. Zheng, H., J. Wang, S. Lofland, Z. Ma, L. Mohaddes-Ardabili, T. Zhao, L. Salamanca-Riba, S. Shinde, S. Ogale, and F. Bai. 2004. Multiferroic BaTiO$_3$-CoFe$_2$O$_4$ nanostructures. *Science* 303:661–663.

73. Lee, M., T. K. Nath, C.-B. Eom, M. C. Smoak, and F. Tsui. 2000. Strain modification of epitaxial perovskite oxide thin films using structural transitions of ferroelectric BaTiO$_3$ substrate. *Applied Physics Letters* 77:3547–3549.

74. Schlom, D. G., L.-Q. Chen, C.-B. Eom, K. M. Rabe, S. K. Streiffer, and J.-M. Triscone. 2007. Strain tuning of ferroelectric thin films. *Annual Review of Materials. Research* 37:589–626.

75. Vaz, C. A. 2012. Electric field control of magnetism in multiferroic heterostructures. *Journal of Physics: Condensed Matter* 24:333201.

76. Jullian, C., J. F. Li, and D. Viehland. 2003. Polarization dynamics over broad time and field domains in modified ferroelectrics. *Applied Physics Letters* 83:1196–1198.

77. Chen, Y., T. Fitchorov, A. L. Geiler, J. Gao, C. Vittoria, and V. G. Harris. 2010. Dynamic response of converse magnetoelectric effect in a PMN-PT-based multiferroic heterostructure. *Applied Physics A* 100:1149–1155.

78. Hu, J.-M., and C. Nan. 2009. Electric-field-induced magnetic easy-axis reorientation in ferromagnetic/ferroelectric layered heterostructures. *Physical Review B* 80:224416.

79. Hu, J.-M., L. Shu, Z. Li, Y. Gao, Y. Shen, Y. Lin, L. Chen, and C. Nan. 2014. Film size-dependent voltage-modulated magnetism in multiferroic heterostructures. *Philosophical Transactions of the Royal Society of London A: Mathematical, Physical and Engineering Sciences* 372:20120444.

80. Hu, J.-M., L.-Q. Chen, and C.-W. Nan. 2016. Multiferroic heterostructures integrating ferroelectric and magnetic materials. *Advanced Materials* 28:15–39.

81. Fechner, M., P. Zahn, S. Ostanin, M. Bibes, and I. Mertig. 2012. Switching magnetization by 180° with an electric field. *Physical Review Letters* 108:197206.

82. Vaz, C. A. F., and U. Staub. 2013. Artificial multiferroic heterostructures. *Journal of Materials Chemistry C* 1:6731–6742.

83. Hu, J.-M., J. Ma, J. Wang, Z. Li, Y.-H. Lin, and C. Nan. 2011. Magnetoelectric responses in multiferroic composite thin films. *Journal of Advanced Dielectrics* 1:1–16.

84. Wu, T., M. Zurbuchen, S. Saha, R.-V. Wang, S. Streiffer, and J. Mitchell. 2006. Observation of magnetoelectric effect in epitaxial ferroelectric film/manganite crystal heterostructures. *Physical Review B* 73:134416.

85. Wang, J., L. Wang, G. Liu, Z. Shen, Y. Lin, and C. Nan. 2009. Substrate effect on the magnetoelectric behavior of Pb(Zr$_{0.52}$Ti$_{0.48}$)O$_3$ film-on-CoFe$_2$O$_4$ bulk ceramic composites prepared by direct solution spin coating. *Journal of the American Ceramic Society* 92:2654–2660.

86. Wang, J., Z. Li, Y. Shen, Y. Lin, and C. Nan. 2013. Enhanced magnetoelectric coupling in Pb(Zr$_{0.52}$Ti$_{0.48}$)O$_3$ film-on-CoFe$_2$O$_4$ bulk ceramic composite with LaNiO$_3$ bottom electrode. *Journal of Materials Science* 48:1021–1026.

87. Liang, W., Z. Li, Z. Bi, T. Nan, H. Du, C. Nan, C. Chen, Q. Jia, and Y. Lin. 2014. Role of the interface on the magnetoelectric properties of BaTiO$_3$ thin films deposited on polycrystalline Ni foils. *Journal of Materials Chemistry C* 2:708–714.

88. Eerenstein, W., M. Wiora, J. Prieto, J. Scott, and N. Mathur. 2007. Giant sharp and persistent converse magnetoelectric effects in multiferroic epitaxial heterostructures. *Nature Materials* 6:348–351.

89. Thiele, C., K. Dörr, O. Bilani, J. Rödel, and L. Schultz. 2007. Influence of strain on the magnetization and magnetoelectric effect in La$_{0.7}$A$_{0.3}$MnO$_3$/PMN–PT (001)(A = Sr, Ca). *Physical Review B* 75:054408.

90. Wang, J., J. Ma, Z. Li, Y. Shen, Y. Lin, and C. Nan. 2011. Switchable voltage control of the magnetic coercive field via magnetoelectric effect. *Journal of Applied Physics* 110:043919.

91. Wu, T., A. Bur, P. Zhao, K. P. Mohanchandra, K. Wong, K. L. Wang, C. S. Lynch, and G. P. Carman. 2011. Giant electric-field-induced reversible and permanent magnetization reorientation on magnetoelectric Ni/(011)[Pb(Mg$_{1/3}$Nb$_{2/3}$)O$_3$]$_{1-x}$-[PbTiO3]$_x$ heterostructure. *Applied Physics Letters* 98:2504.

92. Lahtinen, T. H. E., K. J. A. Franke, and S. van Dijken. 2012. Electric-field control of magnetic domain wall motion and local magnetization reversal. *Scientific Reports* 2:258.

93. Zhang, S., Y. Zhao, P. Li, J. Yang, S. Rizwan, J. Zhang, J. Seidel, T. Qu, Y. Yang, and Z. Luo. 2012. Electric-field control of nonvolatile magnetization in Co$_{40}$Fe$_{40}$B$_{20}$/Pb(Mg$_{1/3}$Nb $_{2/3}$)$_{0.7}$Ti$_{0.3}$O$_3$ structure at room temperature. *Physical Review Letters* 108:137203.

94. Jin, J., S. G. Lu, C. Chanthad, Q. Zhang, M. Haque, and Q. Wang. 2011. Multiferroic polymer composites with greatly enhanced magnetoelectric effect under a low magnetic bias. *Advanced Materials* 23:3853–3858.

95. Jin, J., F. Zhao, K. Han, M. Haque, L. Dong, and Q. Wang. 2014. Multiferroic polymer laminate composites exhibiting high magnetoelectric response induced by hydrogen-bonding interactions. *Advanced Functional Materials* 24:1067–1073.

96. Ma, F. D., Y. M. Jin, Y. U. Wang, S. Kampe, and S. Dong. 2014. Effect of magnetic domain structure on longitudinal and transverse magnetoelectric response of particulate magnetostrictive-piezoelectric composites. *Applied Physics Letters* 104:112903.

97. Lage, E., N. Urs, V. Röbisch, I. Teliban, R. Knöchel, D. Meyners, J. McCord, and E. Quandt. 2014. Magnetic domain control and voltage response of exchange biased magnetoelectric composites. *Applied Physics Letters* 104:132405.

98. Liu, M., B. M. Howe, L. Grazulis, K. Mahalingam, T. Nan, N. X. Sun, and G. J. Brown. 2013. Voltage-impulse-induced non-volatile ferroelastic switching of ferromagnetic resonance for reconfigurable magnetoelectric microwave devices. *Advanced Materials* 25:4886–4892.

99. Sun, N. X., and G. Srinivasan. 2012. Voltage control of magnetism in multiferroic heterostructures and devices. *Spin* 02:1240004.

7

Multiferroic Nanocomposite Thin Films

Aiping Chen and Quanxi Jia

Contents

Multiferroic composites allow us to choose components with large piezoelectric and magnetostrictive coefficients, so the resulting materials exhibit superior multiferroic properties and magnetoelectric (ME) couplings. In Chapter 6, bulk composites have been discussed. Here we focus on multiferroic nanocomposites, in three different thin-film forms, that is, nanoparticles in matrix (0-3 type), bilayers or multilayers (2-2 type), and vertically aligned nanocomposites (1-3 type).

7.1 Heteroepitaxial Nanocomposite Thin Films

7.1.1 Introduction of Heteroepitaxial Nanocomposite Thin Films

Although tremendous progress has been made for bulk multiferroic composites and very large ME coefficients have been achieved, they have inherent limitions, such as imperfect interfaces and being bulky in size. There are also issues related to specific composite designs and materials selection. For example, bulk composites with ferrites in piezoelectric ceramic matrix structure suffer from large leakage current as ferrites are semiconductive [1]. Previous work has been summarized in several comprehensive review articles [1–4].

Recent advances in thin-film fabrication techniques have made it possible to prepare different kinds of nanostructures, and this provides great opportunities to control interface quality, microstructures and strain effect, and ultimately leads to improved functionalities [5]. Figure 7.1a through c shows schematically three types of thin-film nanocomposites. Among them, vertically aligned heteroepitaxial nanocomposite thin films have attracted much attention because of their unique advantages in controlling functional properties by tuning the microstructure, vertical interfacial area and vertical strain.

The term "epitaxy" originates from the Greek root "epi," meaning "above" and "taxis," meaning "in ordered manner." In thin-film preparation, epitaxy refers to the growth of a single-crystal film on a single-crystal substrate with matching structure and lattice parameters. Epitaxy can be classified into two types: homoepitaxy and heteroepitaxy. Homoepitaxy means that the film and substrate are the same material. For instance, the epitaxial growth of Si films on Si substrates and $SrTiO_3$ films on $SrTiO_3$ substrates are homoepitaxy. Heteroepitaxy indicates that the film and substrate are different materials such as the growth of VO_2 films on $SrTiO_3$ substrates [6].

When it comes to epitaxial films, strain is usually inevitable and it plays a significant role in tuning physical properties. In case of heteroepitaxy, there is

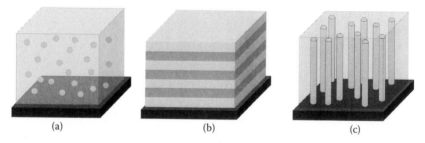

FIGURE 7.1 Schematic diagrams of the three types of nanocomposite thin films. (a) Nanoparticles in matrix-type nanocomposites, (b) lamellar multilayer nanocomposites, and (c) vertically alligned nanocomposites. (Reprinted with permission from Chen, A.P. et al., *Acta Mater.*, 61, 2783, 2013, Copyright 2013, Elsevier.)

always a strain due to lattice mismatch between film and substrate. The unrelaxed misfit strain f is given by

$$f = 2 \times \left(a_f - a_s\right) \big/ \left(a_f + a_s\right) \tag{7.1}$$

where a_f and a_s are the unstrained lattice parameters of film and substrate, respectively.

When the misfit strain f is less than ~7%, strained lattice-matching heteroepitaxy can occur. The epitaxial layer grows pseudomorphically up to a "critical thickness," where it becomes energetically favorable for dislocations to form. Interestingly, domain-matching epitaxy is possible for $f > $ ~7%. In this case, the initially very large misfit strain could be relaxed by matching m planes of the film with n planes of the substrate, leaving a much smaller residual strain (f_r):

$$f_r = \left(ma_f - na_s\right) \big/ na_s \tag{7.2}$$

$$n = m + 1 \tag{7.3}$$

where m and n are integers. Narayan et al. [7,8] have pioneered the domain-matching epitaxial growth of TiN/Si (100) with 3/4 matching, AlN/Si (100) with 4/5 matching, and ZnO/α-Al$_2$O$_3$ (0001) with 6/7 matching of major planes across the film/substrate interface [9]. Considering the TiN/Si (100) case, the original misfit strain f is as large as ~24.6% based on the strained lattice-matching epitaxy. However, the domain-matching epitaxy could produce a much smaller residual strain f_r of ~4.4%. In the nanocomposite films, domain matching also exists in the vertical interface. For example, along the vertical interface of the LSMO:CeO$_2$ system, 2/3 and 3/4 matching exist and the residual strain is significantly reduced [10].

7.1.2 Introduction of Vertically Aligned Nanocomposite Thin Films

Heteroepitaxial vertically aligned-nanocomposite thin films are self-assembled composites (two phases) grown on single-crystal substrates. In the out-of-plane direction, both phases grow epitaxially on the substrate. The nanostructured thin films could form different in-plane morphologies such as nanocheckerboard, nanopillar in matrix, and nanomaze [11,12]. Vertically aligned nanocomposites exhibit a wide range of desirable physical properties because of their unique microstructure, controllable vertical strain, and variable materials selections. The growth of such nanocomposite films is a complex process. The substrate symmetry and lattice parameters, lattice matching between the two components, composition and surface energy are critical factors when designing new vertically aligned nanocomposites. As shown in Figure 7.2a, both microstructure and vertical lattice strain need to be taken into account to achieve the desired functionalities.

During the last few years, spontaneous nanostructure formation has been extensively studied in various systems [10,13–36]. Perovskite oxides are the most widely investigated materials for nanocomposites owing to their rich physical properties such as ferromagnetism, ferroelectricity, and multiferroicity. In nanocomposites, coexisting of various components leads to interesting properties that are different from or even absent in the constitute materials.

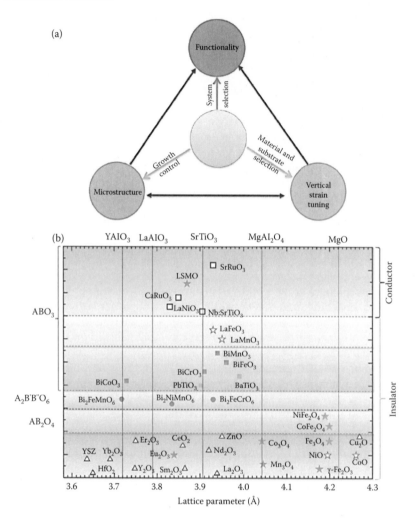

FIGURE 7.2 (a) Relationship between the microstructure, strain, and functionality of vertically aligned nanocomposite thin films. (b) Commonly used oxides for vertically aligned nanocomposites. □, Conducting perovskite oxides, which could be used as bottom electrodes. ■, Perovskite multiferroic materials. ▨, Perovskite FE materials. ●, Double perovskite multiferroic materials. △, Common insulating metal oxides. ★, Ferromagnetic (FM) materials. ☆, Antiferromagnetic materials. (Reprinted from *Acta Mater.*, 61, Chen, A.P. et al., Microstructure, vertical strain control and tunable functionalities in self-assembled, vertically aligned nanocomposite thin films, 2783, Copyright 2013, with permission from Elsevier.)

For example, vertically aligned nanocomposite films of $La_{0.67}Ca_{0.33}MnO_3$:MgO (LCMO:MgO) exhibit tunable magnetotransport properties [22,29]. In the thin-film composite of $CoFe_2O_4$:$BaTiO_3$, a sudden change in the magnetization at the $BaTiO_3$ phase transition temperature indicates a possible ME coupling [37]. In $CoFe_2O_4$:$BiFeO_3$ nanocomposites, the magnetic anisotropy is controlled by the composite structure [26,32,38]. A light-induced magnetization change in $CoFe_2O_4$ has been demonstrated in $CoFe_2O_4$:$SrRuO_3$ vertically aligned nanocomposites [39]. Incorporation of a second phase (nanoparticles) into YBCO films proved to be an effective approach to control the strain effect and flux pinning in superconducting materials [40,41]. During the past decade, many two-phase nanocomposite systems have been developed and investigated. Several comprehensive reviews about the design, growth, and functionality control of vertically aligned nanocomposites can be found in the literature [5,42–45]. The most commonly studied materials for vertically aligned nanocomposite films are summarized in Figure 7.2b. The crystal structures, elastic constants, and lattice parameters of the two phases and substrates have to be carefully selected to design new functional nanocomposites.

7.2 Characterizing ME Coupling in Thin Films

In ME composites, tuning the polarization (magnetization) by magnetic (electric) field is realized through strain coupling at the heterointerface, which is schematically shown in Figure 7.3. Direct ME effect refers to the process of strain generated in the magnetostrictive phase under a magnetic field that induces surface charges in the piezoelectric phase through the piezoelectric effect. The converse ME effect refers to the process of strain in the piezoelectric phase under an electric field inducing magnetization change in the FM

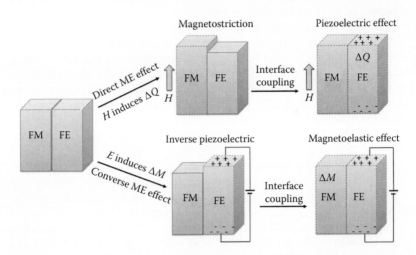

FIGURE 7.3 Strain-mediated couplings in composites with an FM layer and a ferroelectric (FE) layer. The top panel depicts the direct ME effect. A magnetic field-induced deformation in the FM layer (magnetostrictive effect) is transferred to the FE layer and it induces surface charges through the piezoelectric effect. The bottom panel depicts the converse ME effect. An electric field deforms the FE layer (converse piezoelectric effect), and induces a magnetization change or domain reorientation in the FM layer.

phases through the piezomagnetic effect. The strain mediated ME coupling is well established in FE/FM bilayers and heterostructures. However, the coupling in vertical nanocomposite could be more complicated. The contribution from the strain mediated interface coupling effect seems always opposite to the volume contribution. Composites using components with large piezoelectric and magnetostrictive coefficients usually exhibit large ME coupling coefficients. Other factors such as the resistivity of the magnetic phase, chemical stability, and mechanical strength of the interface are important for ME composite designs.

The ME coupling strength can be characterized by measuring the direct or converse ME coefficients. In the case of direct ME effect, a static magnetic field, superimposed with a small ac magnetic field (ΔH_{ac}), is applied to the ME composites and surface charge or voltage signal is recorded. The direct ME voltage coefficient is given by

$$\alpha_E = \Delta E/\Delta H_{ac} = \Delta V/\left(h\Delta H_{ac}\right) = \Delta Q/\left(\varepsilon_0 \varepsilon S \Delta H_{ac}\right) \tag{7.4}$$

where h is the thickness of ME composites along the electric field direction, ΔV is the ME voltage, ΔQ is the generated charge, S is the area, ΔH_{ac} is the ac magnetic field, and ε_0 is equal to 8.85×10^{-12} F/m. The output voltage signal is related to the composite thickness. When the film has a thickness of less than 1 µm, the ME voltage is usually less than 1 µV [46]. Therefore, in bulk composites (hundreds of micrometers to millimeters in thickness), one can directly measure the output voltage (open circuit condition), while in thin-film composites (a few hundred nanometers in thickness), we usually measure the charge generated (short-circuit condition).

Besides direct measurements, the magnetocapacitance effect can also be used to evaluate ME coupling strength. Typically, the capacitance change induced by dc magnetic fields is measured. It is noted that the magnetocapacitance effect can stem from a combination of magnetoresistance and the Maxwell–Wagner effect [47]. This possibility can be limited by conducting experiments at low temperatures to reduce the leakage. Using this method, an ME coefficient of 0.9 µC/cm Oe was estimated for $CoFe_2O_4$:$BiFeO_3$ nanocomposites at 10 K [48].

In the case of the converse ME effect, electric fields are applied to ME composites and the change of magnetization is recorded. Thus, various techniques that can detect magnetic properties have been employed, including ferromagnetic resonance (FMR), vibrating sample magnetometer (VSM) or superconducting quantum interference device (SQUID), magneto-optic Kerr effect (MOKE), and x-ray absorption near edge spectroscopy (XANES). It is noted that FMR and VSM methods are suitable for both bulk and thin-film ME composites, but MOKE and XANES are surface-sensitive techniques that are usually used for thin-film samples.

The resonant frequency of a magnetic material under an external field B is given by $f = \dfrac{\gamma}{2\pi}\sqrt{B\left(B + \mu_0 M\right)}$, where M is the magnetization and γ is the gyromagnetic ratio. Apparently, when an electric field changes the magnetization of an ME composite, the FMR field shifts and the converse ME coefficient is then given by $\Delta f/\Delta E$, where ΔE is the applied electric field (kV/cm) change and Δf is the FMR field shift (Oe). The unit of converse ME coefficient is Oe cm/V. This method has been used in bulk particulate composites [49] and heterostructures [50,51].

Measuring the converse ME effect using VSM is straightforward and electric fields are directly applied to composites during measurements. This method has been used in thin-film heterostructures [52,53]. MOKE measures the changes to light reflected from a magnetized surface, which can be used to quantitatively estimate the converse ME effect. In-charge/strain-mediated ME heterostructures, Vaz et al. reported a converse ME effect in $La_{0.8}Sr_{0.2}MnO_3/$ $PbZr_{0.2}Ti_{0.8}O_3$ on $SrTiO_3$ substrates by MOKE. $\Delta M/\Delta E = 0.8$ Oe cm/kV at 100 K has been reported [54]. In $FeBSiC/Pb(ZrTi)O_3$ heterostructures, a significant change in Kerr hysteresis loops was observed when the applied electric field (20 kV/cm) is larger than the coercive field of PZT substrates [55].

XANES has been used to probe the absorption edge of Mn or Fe and to characterize the magnetization change in FM materials such as magnetic metals, $La_xSr_{1-x}MnO_3$, $CoFe_2O_4$, $NiFe_2O_4$, and $BiFeO_3$. These FM (or ferrimagnetic) films usually deposited on FE layers or substrates include $PbZr_{0.2}Ti_{0.8}O_3$, PMN–PT, and $BaTiO_3$. For example, XANES was used to determine the converse ME effect and the origin of the ME coupling in $La_{0.8}Sr_{0.2}MnO_3/$ $PbZr_{0.2}Ti_{0.8}O_3$ heterostructures [56]. It should be noted that ME couplings can be also characterized by using piezoresponse force microscopy or magnetic force microscopy. However, research along this direction is rare.

7.3 Multiferroic Properties and ME Coupling in Nanocomposites

As shown in Figure 7.1, there are three types of ME nanocomposite films: (1) 0-3 type with FM nanoparticles embedded in the FE matrix, (2) 2-2 type heterostructures with alternating FE and FM layers, and (3) 1-3 type vertically aligned nanocomposite films with FM nanopillars embedded into a FE matrix. In the following subsections, we will cover the recent progress in these areas.

7.3.1 Nanoparticles in Matrix (0-3 Type) Nanocomposite Thin Films

For 0-3 type ME nanocomposite films, most research has focused on the effects of size and concentration of FM nanoparticles. In Table 7.1, we summarize the recent studies on ME coupling effects in 0-3 nanocomposite films. It shows ME coefficients ranging from 16 to 549 mV/cm Oe, similar to that of bulk particulate composites but one to two orders of magnitude lower than that of bulk laminate composites.

Wan et al. investigated $CoFe_2O_4$:$PbZr_{0.52}Ti_{0.48}O_3$ nanocomposite films with $CoFe_2O_4$ nanoparticles embedded in the $PbZr_{0.52}Ti_{0.48}O_3$ matrix [57]. As shown in Figure 7.4a, the nanoparticles have an average diameter of ~150 nm and they

TABLE 7.1 Properties of 0-3 Nanocomposite Films in Recent Studies

Material System	Method	α_E (mV/ cm Oe)	H_{dc} (Oe) [f (kHz)]	Reference
31% CFO:PZT	Sol–gel	220	2000 [1]	[57]
31% CFO:PZT	Sol–gel	317	2000 [50]	[57]
35% NFO:PZT	PLD	16	2500 [0.194]	[20]
0.065% CFO:PZT	Sol–gel	549	450 [1]	[58]
400 nm(Co:BTO)/100 nm BTO	RF sputtering	170	500 [50]	[59]
50% $La_{0.6}Ca_{0.4}MnO_3$: $Bi_{0.6}Nd_{0.4}TiO_3$	CSD	29.8	5000 [20]	[60]

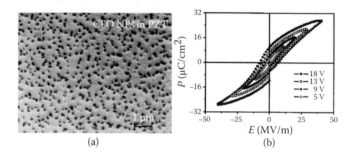

(a)   (b)

FIGURE 7.4 (a) A top-view SEM image of the $CoFe_2O_4:PbZr_{0.52}Ti_{0.48}O_3$ nanocomposite thin films. (b) P–E hysteresis loops of $CoFe_2O_4:PbZr_{0.52}Ti_{0.48}O_3$ composite thin films. (Reprinted with permission from Wan, J.G. et al., *Appl. Phys. Lett.*, 86, 122501. Copyright 2005, American Institute of Physics.)

are randomly distributed. The volume fraction of nanoparticles is ~31%. Electric measurement shows that the nanocomposite has a high resistivity of ~5×10^9 Ω cm at zero bias. Figure 7.4b shows well-defined P-E loops, and the maximum saturation polarization (P_s) and remnant polarization (P_r) reach 28 µC/cm² and 11.0 µC/cm², respectively. An ME coefficient α_E of 220 mV/cm Oe was estimated in these 0-3 nanocomposites. The α_E increases with increasing magnetic field frequency f_r, and the maximum α_E value is 317 mV/cm Oe at f_r = 50 kHz.

Later, another 0-3 type nanocomposite films with $NiFe_2O_4$ (35%) nanoparticles embedded in $PbZr_{0.52}Ti_{0.48}O_3$ matrix was prepared on Nb:STO substrates, where the transverse and longitudinal ME voltage coefficients were measured [20]. After poling, the α_E values are deduced based on charges generated under an *ac* magnetic field at a frequency of 194 Hz. The maximum value of α_{E31} was 4 mV/cm Oe, and the value of α_{E33} was 16 mV/cm Oe. The smaller coefficients in nanocomposites compared with that of the bulk particulate composites (80 mV/cm Oe) might be due to substrate clamping effects [61].

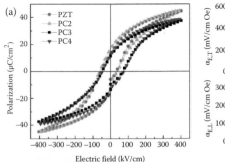 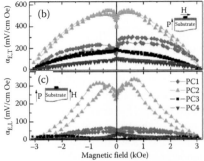

FIGURE 7.5 (a) Room temperature *P–E* hysteresis loops of pure $PbZr_{0.52}Ti_{0.48}O_3$ and $CoFe_2O_4$:$PbZr_{0.52}Ti_{0.48}O_3$ nanocomposite films. (b) The transverse and (c) longitudinal ME coupling coefficients of the nanocomposite films. The insets depict the field orientations relative to the sample for the respective measurements. (Reprinted with permission from McDannald, A. et al., *Appl. Phys. Lett.*, 102, 122905. Copyright 2013, American Institute of Physics.)

In 2013, McDannald et al. [58] reported ME properties of 0-3 type nanocomposite films with various concentrations of much smaller $CoFe_2O_4$ nanoparticles (5–8 nm) dispersed in $PbZr_{0.52}Ti_{0.48}O_3$ matrix. The solution containing $CoFe_2O_4$ nanoparticles and $PbZr_{0.52}Ti_{0.48}O_3$ precursors was spin-coated onto Pt-coated Si substrates and annealed at 700°C. The nanocomposites exhibit good *P-E* hysteresis loops, as shown in Figure 7.5a. An *ac* magnetic field (1 Oe) superimposed onto a *dc* magnetic field generated by an electromagnet was used to measure the ME coupling coefficients. Nanocomposite films with 0.065% $CoFe_2O_4$ nanoparticles (molar ratio) exhibited the highest transverse and longitudinal α_E of 549 mV/cm Oe and 338 mV/cm Oe, respectively (Figure 7.5b). With further increase of the nanoparticles concentration, the ME coefficient decreases, probably due to the agglomeration of nanoparticles. Recently, Li et al. [62] reported a hybrid structure of $CoFe_2O_4$:$BiFeO_3$ (0-3 type)nanocomposite by alternately growing 2-2 and 1-3 type composite layers using pulsed laser deposition. It was proposed that such a design reduces the clamping effect from substrate and significantly suppresses the overall leakage current. The ME coupling was characterized by monitoring the piezoelectric response under magnetic field and the lateral ME coefficient was about 338 mV/cm Oe.

The converse ME coupling of Fe_3O_4/$CoFe_2O_4$ nanoparticulate thin films on $PbZr_{0.53}Ti_{0.47}O_3$ substrates has also been investigated [63]. Clear magnetization change under an electric field of 10 kV/cm has been observed and a large ME coupling coefficient up to 10.1 V/Oe cm was reported.

7.3.2 Bilayer and Multilayer (2-2 Type) Nanocomposite Thin Films

In bulk form, ME laminate composites usually exhibit superior properties than particulate composites. This is mainly due to the insulating FE layer blocking the leakage current in laminates. However, the ME coupling coefficient for 2-2 type ME composite films is usually comparable to that of the 0-3 type nano-composite films.

Ortega et al. [64] investigated ME couplings in $CoFe_2O_4/PbZr_{0.53}Ti_{0.47}O_3$ multilayers prepared by pulsed laser deposition. The multilayer films show well-defined FE loops with P_r and coercive field (E_c) of ~25 $\mu C/cm^2$ and 68 kV/cm, respectively. The P_r and E_c of pure $PbZr_{0.53}Ti_{0.47}O_3$ film are 33 $\mu C/cm^2$ and 37 kV/cm, respectively. The coupling between the FE and FM layers was demonstrated by measuring the change in FE hysteresis loops under magnetic fields. But the ME coupling coefficient was not reported.

In Table 7.2, we summarize the recent studies on the ME coupling in 2-2 type nanocomposite films. The ME voltage coefficients of most bilayer and multilayer nanocomposites are comparable to those of the 0-3 type nanocomposite films.

Nan et al. [67] have investigated a bilayer ($CoFe_2O_4/BaTiO_3$) ME compos-ite as shown in Figure 7.6a, which exhibits an ME coefficient of 104 mV/cm Oe under a *dc* magnetic field of 100 Oe. The FE and FM hysteresis loops are shown in Figure 7.6b. The ME measurement was performed under an open circuit condition. When an in-plane altering magnetic field was applied, no ME output was observed from the single-phase $BaTiO_3$ and $CoFe_2O_4$ films, as shown in Figure 7.6c. However, the bilayer composite shows an ME voltage output following the on/off of the magnetic excitation field ΔH_{ac}. The output ΔV increases linearly with the *ac* magnetic field (Figure 7.6d) and the ME volt-age coefficients can be obtained from the slopes (see Equation 7.4). The out-of-plane and in-plane ME voltage coefficients for the $CoFe_2O_4/BaTiO_3/SrTiO_3$ heterostructures are ~104 and 66 mV/cm Oe, respectively. It has been reported that the magnetostriction of epitaxial $CoFe_2O_4$ thin films, $\lambda(001) = -188 \times 10^{-6}$ [77], is larger than that of $NiFe_2O_4$, which is why the ME voltage coefficients in $CoFe_2O_4/BaTiO_3$ are larger than that of $NiFe_2O_4/BaTiO_3$ bilayers [69,70]. For 2-2 type ME composites, the substrate clamping effect could reduce the ME coupling significantly. By choosing the right substrates, much larger ME coupling coefficients have been achieved. For example, Feng et al. synthesized $Pb(Zr_{0.52}Ti_{0.48})O_3/Ni$ films on Ni foils and a large α_{E31} of 772 mV/cm Oe was reported [67]. A strong ME coupling was also observed in $La_{0.7}Sr_{0.3}MnO_3/Ba_{0.7}Sr_{0.3}TiO_3$ bilayers and multilayers, in which the longitudinal ME coeffi-cient can be as large as 300 mV/cm Oe [68].

However, it has to be noted that a magnetic field–induced change in FE hys-teresis loop (*P–E* loop) may not always be related to a real ME coupling effect. Dussan et al. [78] have investigated the FE properties of $PbZr_{0.52}Ti_{0.48}O_3$ in $La_{0.67}Sr_{0.33}MnO_3/PbZr_{0.52}Ti_{0.48}O_3$ heterostructures and reported clear changes in the *P–E* loop under different magnetic fields, as shown in Figure 7.7a. When the magnetic field is increased, the *P–E* loop first broadens and then disap-pears at 0.34 T (300 K) for the sample with 0.55 μm $PbZr_{0.52}Ti_{0.48}O_3$ layer. Figure 7.7b shows the recovery of *P–E* loop after the removal of the external magnetic field. It was proposed that the negative magnetoresistance of $La_{0.67}Sr_{0.33}MnO_3$ under low magnetic field rather than magnetocapacitance is responsible for the observed results. The dramatic increase in leakage current through the device due to magnetoresistance dominates this process. No significant magnetic field dependence is observed in the sample with a 1.4 μm $PbZr_{0.52}Ti_{0.48}O_3$ layer.

TABLE 7.2 ME Coupling Coefficients of 2-2 Type Nanocomposite Films in Recent Studies

Material System	Method	α_E (mV/ cm Oe)	H_{dc} (Oe) [f (kHz)]	References
$Pb(Zr_{0.4}Ti_{0.6})O_3/$ $Ni_{0.8}Zn_{0.2}Fe_2O_4$	PLD	15	2000 [1]	[65]
$Pb(Zr_{0.4}Ti_{0.6})O_3/$ $Ni_{0.8}Zn_{0.2}Fe_2O_4$	PLD	30	2000 [20]	[65]
$Pb(Zr_{0.52}Ti_{0.48})O_3/$ $La_{0.7}Sr_{0.3}MnO_3$	PLD	4	4000 [1]	[66]
400 nm $Pb(Zr_{0.52}Ti_{0.48})O_3/$ 30 nmPt/Ni foil	Sol–gel	772	86[–]	[67]
300 nm $(La_{0.7}Sr_{0.3}MnO_3/$ $Ba_{0.7}Sr_{0.3}TiO_3)_m$	PLD	35–300	50~100 [1]	[68]
CFO/BTO	PLD	104	100 [1]	[69]
NFO/BTO	PLD	12.1	100 [1]	[70]
NFO/BTO	PLD	79/44 (OOP/IP)	100 [1]	[71]
BTO/NFO	PLD	37/31 (OOP/IP)	100 [1]	[71]
PZT film/ CFO ceramic	Sol–gel	60	~600 [1]	[72]
PZT film/LaNiO$_3$/CFO ceramic	Sol–gel	115	~600 [1]	[72]
15 μm NFO/15 μm BTO multilayers	Tape casting	18	2000 [0.194]	[73]
3.5 μm Co/PZT ceramic/3.5 μm Co	Sputtering	9	50 [0.1]	[74]
300 μm $La_{0.7}Sr_{0.3}MnO_3$/PZT	Doctor blade	60 (120K)	35 [0.1]	[75]
2 μm Py/80 μm PZT/2 μm Py multilayers	Sputtering	240	50 [0.1]	[76]

In another study, Chen et al. [79] proposed that the magnetocapacitance effect in $La_{0.7}Sr_{0.3}MnO_3/Pb_{0.7}Sr_{0.3}TiO_3$ can be understood by combining the Maxwell–Wagner capacitor model and the magnetoresistance and magnetostriction of $La_{0.7}Sr_{0.3}MnO_3$. Nevertheless, the FE hysteresis loops show no change under different magnetic fields. More systematic investigations are needed in this area.

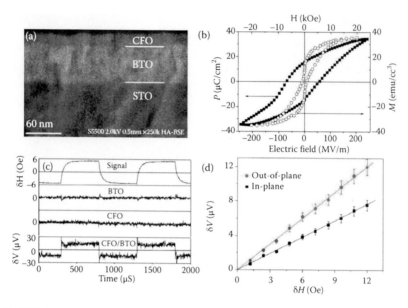

FIGURE 7.6 (a) A typical Cross-sectional SEM image of the $CoFe_2O_4$–$BaTiO_3$ bilayer films. The solid lines indicate the interfaces. (b) FE hysteresis loop and in-plane magnetic hysteresis loop of the $CoFe_2O_4$–$BaTiO_3$ bilayer films. (c) The ME responses of $BaTiO_3$, $CoFe_2O_4$ and the composite films when an in-plane magnetic field is applied. (d) ME voltage output ΔV as a function of ac magnetic field (1 kHz) with a dc magnetic field of 100 Oe. The in-plane and out-of-plane represent the measurement modes with the magnetic fields parallel and perpendicular to the surface of the films, respectively. (Reprinted with permission from Zhang, Y. et al., *Appl. Phys. Lett.*, 92, 062911. Copyright 2008, American Institute of Physics.)

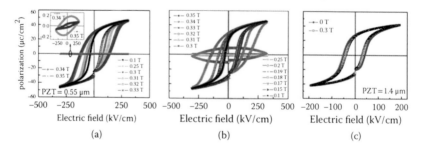

FIGURE 7.7 (a) P–E loops of a $PbZr_{0.52}Ti_{0.48}O_3$ film (0.55 μm) with $La_{0.67}Sr_{0.33}MnO_3$ electrode on $LaAlO_3$ substrate under different magnetic fields. A significant magnetic field dependence is observed near the critical field of $H = 0.34$ T, which is shown more clearly in the inset. (b) The recovery of FE hysteresis loop after the removal of the external magnetic field. (c) P–E loops of a thicker $PbZr_{0.52}Ti_{0.48}O_3$ films under 0 T and 0.3 T magnetic fields. No significant magnetic field dependence is observed in this case. (Reprinted with permission from Dussan, S. et al., *J. Phys. Condens. Matt.* 23, 202203, 2011, Copyright 2011, Institute of Physics.)

7.3.3 Vertically Aligned-Nanocomposite (1-3 Type) Films

Due to the large vertical interface and reduced substrate clamping effect in vertically aligned nanocomposite films, they are more favorable for ME couplings. By using Green's function technique, Nan et al. [80] have predicted that vertically aligned nanocomposites exhibit much larger ME coupling coefficients than multilayer structures. Recently, Wu et al. [81] simulated the misfit strain and FM composition effects on the ME coupling in vertically aligned nanocomposites. It was found that the ME voltage coefficients of vertical $CoFe_2O_4$:$BaTiO_3$ and $CoFe_2O_4$:$PaTiO_3$ nanocomposite films strongly depend on the in-plane and out-of-plane misfit strains. At the optimized FM phase content of ~60%, a large ME voltage coefficient of up to 2 V/cm Oe is predicted for $CoFe_2O_4$:$BaTiO_3$ nanocomposites.

Zheng et al. [37] first reported the multiferroic $CoFe_2O_4$:$BaTiO_3$ (35:65 molar ratio) vertically aligned nanocomposite in 2004. As shown in Figure 7.8a, the nanocomposite showed clear P–E loops with P_s = 23 μC/cm². The plan-view transmission electron microscope (TEM) image revealed $CoFe_2O_4$ nanopillars of ~25 nm in diameter embedded in the $BaTiO_3$ matrix (Figure 7.8b). Large uniaxial magnetic anisotropy was observed due to the vertical compressive strain in the $CoFe_2O_4$ phase induced by the $BaTiO_3$ matrix. Since $CoFe_2O_4$ has negative magnetostriction, compressive vertical strain leads to the out-of-plane easy axis (Figure 7.8c). A magnetization change was observed close to the $BaTiO_3$ phase transition temperature, which was absent in the $CoFe_2O_4$/$BaTiO_3$ multilayer structures (Figure 7.8d), indicating that the ME coupling is larger in vertically aligned nanocomposites than that in multilayers. The $CoFe_2O_4$:$BaTiO_3$ nanocomposites have recently been revisited by Schmitz-Antoniak et al. [82]. Rectangular $CoFe_2O_4$ nanopillars with 100–200 nm in size have been synthesized by pulsed laser deposition at 950°C, and the couplings studied using soft x-ray absorption spectroscopy and its associated linear dichroism. Besides the $CoFe_2O_4$:$BaTiO_3$ systems, many other similar ME nanocomposites have been developed over the years [83–86].

The $CoFe_2O_4$:$BiFeO_3$ system is well-studied vertically aligned nanocomposites with large ME coupling effect. Zavaliche et al. [87,88] investigated the electric field-controlled magnetization using magnetic force microscopy (MFM). The composite film was magnetized out-of-plane in a 20 kOe magnetic field, which resulted in a predominantly upward magnetization in the $CoFe_2O_4$ columnar structures, as shown in Figure 7.9a. The film was sequentially poled under –12 and 12 V electric fields, and magnetization switching was confirmed by MFM, as shown in Figure 7.9b. To quantify the strength of the ME coupling, capacitors were fabricated. The sample was magnetized in a downward 20 kOe magnetic field and M–H loops were recorded before and after electrical poling as shown in Figure 7.9c and d. The static perpendicular ME coupling coefficient, α_{E33} = $\Delta M/\Delta E$ was determined to be 0.01 Oe cm/V. The ΔM and ΔE are the electric field–induced magnetization change and the poling field, respectively. To reduce the leakage, Wang et al. [53] grew $CoFe_2O_4$:$BiFeO_3$ nanocomposite films on insulating PMN–PT (001) substrates, and the electric field–induced magnetization change was investigated.

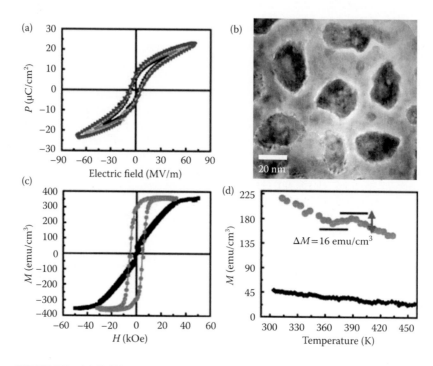

FIGURE 7.8 (a) *P–E* hysteresis loops showing that the $CoFe_2O_4$:$BaTiO_3$ nanocomposite film is FE with a saturation polarization $P_s = 23$ µC/cm². (b) TEM planar view image reveals the $CoFe_2O_4$ nanopillars in the $BaTiO_3$ matrix. (c) Out-of-plane (gray) and in-plane (black) *M–H* hysteresis loops show a large uniaxial anisotropy. (d) *M–T* curve measured at $H = 100$ Oe shows a distinct drop in magnetization at the FE Curie temperature of $BaTiO_3$ for the nanocomposite, while the multi-layer film (black curve) shows negligible change. (Reprinted with permission from Zheng, H. et al., *Science*, 303, 661–663, 2004, Copyright 2004, Science.)

The electric field poling dramatically reduced the out-of-plane remnant magnetization and enhanced the in-plane magnetization. The static ME coupling coefficient, $\alpha_{33} = \Delta M/\Delta E$ was estimated to be ~0.5 Oe cm/V.

Direct ME coupling effect in $CoFe_2O_4$:$BiFeO_3$ vertically aligned nanocomposites have also been investigated. Yan et al. [46] grew $CoFe_2O_4$:$BiFeO_3$ nanocomposite films (150–2400 nm) on $SrRuO_3$-buffered $SrTiO_3$ (001) substrates at 700°C. A magnetic cantilever method was used to measure the ME coupling. The maximum ME coefficient was determined to be ~20 mV/cm Oe. The reduced piezoelectric and effective piezomagnetic coefficients of the FE and FM phases, due to substrate clamping effect, were suggested to be responsible for the relatively small coefficient. Yan et al. [89] further studied the orientation effect on the ME coupling by growing composite films on $SrTiO_3$ substrates with different orientations. The maximum ME coefficient for the *L–L* mode (see chapter 6) followed the trend of ME$_{(001)}$ > ME$_{(110)}$ > ME$_{(111)}$ (16, 15, and 8 mV/cm Oe, respectively), which was correlated with the

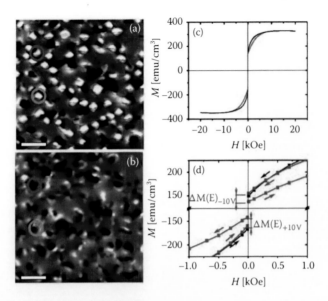

FIGURE 7.9 Magnetic force microscopy images taken (a) after magnetization in an upward 20 kOe field, and (b) after electrical poling at 12 V clearly reveal changes in the magnetization of CoFe$_2$O$_4$:BiFeO$_3$ (35:65) nanocomposites. The scale bars are 1 μm. (c) M–H loops taken before (black) and after (red) electrical poling of ~10% of the total film area. (d) An enlarged view of the central part of (c). (Reprinted with permission from Zavaliche, F. *et al.*, *Nano Letters*, 5, 1793, 2005, Zavaliche, F. et al., Electric field-induced magnetization switching in epitaxial columnar nanostructures. *Nano Letters* 5:1793–1796. Copyright 2005, American Chemical Society.)

longitudinal piezoelectric coefficient d_{33} of the BiFeO$_3$ phase following $d_{33(001)}$ >$d_{33(110)}$ >$d_{33(111)}$. In addition, the strains experienced by films with different orientations, following the trend of $\sigma_{(111)} > \sigma_{(110)} > \sigma_{(001)}$, were also suggested to contribute to the larger ME coupling in (001) nanocomposite films.

Oh et al. [90] investigated the ME coupling of a 300 nm CoFe$_2$O$_4$:BiFeO$_3$ nanocomposite film with 1:1 volume ratio (CoFe$_2$O$_4$ pillars in BiFeO$_3$ matrix). A small *ac* magnetic field (H_{ac} = 4 Oe) superimposed on a *dc* magnetic field was used to excite the sample and the induced charges (instead of voltages) were measured. The transverse ME coefficient α_{E31} was ~60 mV/cm Oe, which is five times larger than the longitudinal ME coefficient α_{E33}. The larger transverse ME coefficient was attributed to enhanced transverse magnetostriction of the CoFe$_2$O$_4$ nanopillars.

Another factor that affects the ME coupling in vertically aligned nanocomposite is the volume ratio between the two constituting phases. Wan et al. observed that, for the CoFe$_2$O$_4$:Pb(ZrTi)O$_3$ nanocomposites, 25% CoFe$_2$O$_4$ leads to an α_E as large as 390 mV/cm Oe [92]. With higher CoFe$_2$O$_4$ contents, the α_E decreases significantly. The increased leakage and variation of microstructure were suggested to be responsible for the reduced ME coupling effect. In Table 7.3, we summarize the recent studies on ME coupling in 1-3 type

TABLE 7.3   Converse and Direct ME Coupling Coefficients of 1-3 Type Vertically Aligned Nanocomposite Films in Recent Studies

Material System	Method	α	H_{dc} (Oe) [f (kHz)]	References
35% CFO:BFO/ SRO/ STO(001)	PLD	20 mV/cm Oe	100 [–]	[46]
35% CFO:BFO/ SRO/ STO(001)	PLD	16 mV/cm Oe	100 [–]	[89]
35% CFO:BFO/ SRO/ STO(011)	PLD	15 mV/cm Oe	100 [–]	[89]
35% CFO:BFO/ SRO/ STO(111)	PLD	8 mV/cm Oe	100 [–]	[89]
50% CFO:BFO/ SRO/ STO(001)	PLD	60 mV/cm Oe	700 [1]	[90]
Polycrystalline 65% CFO:BFO/Pt/ TiO_2/SiO_2/Si	PLD	102 mV/cm Oe	PFM	[91]
35% CFO:BFO/ PMN-PT (001)	PLD	0.5 Oe cm/V	–	[53]
35% CFO:BFO/ Nb:STO(001)	PLD	0.01 Oe cm/V	–	[87]
25% CFO:PZT/ Pt/Ti/SiO2/Si	PLD	390 mV/cm Oe	0 [10]	[92]

nanocomposite films. The reported direct ME voltage coefficients of vertically aligned nanocomposites is close to 10–100 mV/cm Oe.

7.4 Conclusions and Outlook

Significant progress has been achieved in multiferroic research during the past decade. In Figure 7.10, we have summarized the ME coefficients obtained for various single-phase and composite multiferroic materials. In single-phase multiferroic materials, such as Cr_2O_3 and $BiFeO_3$, relatively small ME coupling coefficients of less than 20 mV/cm Oe have been reported. Among the bulk composites, particulate composites exhibit relative small ME coefficients

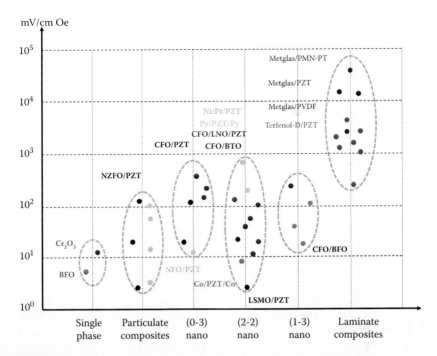

FIGURE 7.10 Off-resonance ME coupling coefficients of different types of multiferroic materials.

ranging from a few mV/cm Oe to ~100 mV/cm Oe. Laminates exhibit the largest ME coefficients up to 50 V/cm Oe for Metglas/PMN-PT. The ME coefficients of nanocomposite films are typically below 1 V/cm Oe. Most of the earlier studies focus on 2-2 type heterostructures, where the ME coefficients range from 4 to 772 mV/cm Oe. One of the advantages of 2-2 type nanocomposite films is the reduced leakage current, which is a serious issue in 0-3 or 1-3 type nanocomposites due to the low resistivity magnetic component across the sample. On the other hand, substrate clamping effect is the most serious for 2-2 type heterostructures, but much less so for 1-3 type nanocomposite films. Besides strain-mediated ME composites, exchange bias and charge-mediated ME couplings may provide alternative approaches to manipulate magnetization via electric field [93,94].

It is worth noting that the ME coupling coefficients we have discussed are values obtained at low frequencies. At resonance frequency, it usually increases by one to two orders of magnitude. For example, Zhai et al. [95] observed an ME coefficient of 7.2 V/cm Oe at 1 kHz and of 310 V/cm Oe at resonance (50 kHz) in Metglas/PVDF laminate composites. In Metglas/PZT-fiber composites, Dong et al. [96] reported a high ME voltage coefficient of 22 V/cm Oe at 1 Hz and ~500 V/cm Oe at resonance (22 kHz). $(Fe_{90}Co_{10})_{78}Si_{12}B_{10}$/AlN-based thick films fabricated by MEMS technique exhibited 3.1 V/cm Oe at low frequencies but an extremely high ME coefficient of 737 V/cm Oe at resonance frequency [97]. The laminate composites with exceptionally high ME coefficient are well suited for sensors [98].

Although the ME coupling coefficients of nanocomposite films are far smaller than those of laminates, and most current studies are focusing on the fascinating physics involved, they are indeed promising for low-power spintronic-based memory devices [99–109]. Detailed discussions on this topic can be found in Chapter 11 of this book and in a number of comprehensive reviews [1–4,110–116].

References

1. Ma, J., J. M. Hu, Z. Li, and C. W. Nan. 2011. Recent progress in multiferroic magnetoelectric composites: from bulk to thin films. *Advanced Materials* 23:1062–1087.
2. Vaz, C. A. F., J. Hoffman, C. H. Ahn, and R. Ramesh. 2010. Magnetoelectric coupling effects in multiferroic complex oxide composite structures. *Advanced Materials* 22:2900–2918.
3. Martin, L., S. P. Crane, Y. H. Chu, M. B. Holcomb, M. Gajek, M. Huijben, C. H. Yang, N. Balke, and R. Ramesh. 2008. Multiferroics and magnetoelectrics: thin films and nanostructures. *Journal of Physics Condensed Matter* 20:434220.
4. Nan, C. W., M. I. Bichurin, S. X. Dong, D. Viehland, and G. Srinivasan. 2008. Multiferroic magnetoelectric composites: Historical perspective, status, and future directions. *Journal of Applied Physics* 103:031101.
5. Chen, A. P., Z. X. Bi, Q. X. Jia, J. L. MacManus-Driscoll, and H. Y. Wang. 2013. Microstructure, vertical strain control and tunable functionalities in self-assembled, vertically aligned nanocomposite thin films. *Acta Materialia* 61:2783.
6. Chen, A. P., Z. Bi, W. Zhang, J. Jian, Q. X. Jia, and H. Y. Wang. 2014. Textured metastable VO_2 (B) thin films on $SrTiO_3$ substrates with significantly enhanced conductivity. *Applied Physics Letters* 104:071909.
7. Narayan, J., and B. C. Larson. 2003. Domain epitaxy: A unified paradigm for thin film growth. *Journal of Applied Physics* 93:278–285.
8. Wang, H., A. Sharma, A. Kvit, Q. Wei, X. Zhang, C. C. Koch, and J. Narayan. 2001. Mechanical properties of nanocrystalline and epitaxial TiN films on (100) silicon. *Journal of Materials Research* 16:2733–2738.
9. Wang, H., S. R. Foltyn, P. N. Arendt, Q. X. Jia, J. L. MacManus-Driscoll, L. Stan, Y. Li, X. Zhang, and P. C. Dowden. 2004. Microstructure of $SrTiO_3$ buffer layers and its effects on superconducting properties of $YBa_2Cu_3O_7$-(delta) coated conductors. *Journal of Materials Research* 19:1869–1875.
10. Chen, A. P., Z. X. Bi, H. Hazariwala, X. H. Zhang, Q. Su, L. Chen, Q. X. Jia, J. L. MacManus-Driscoll, and H. Y. Wang. 2011. Microstructure, magnetic, and low-field magnetotransport properties of self-assembled $(La_{0.7}Sr_{0.3}MnO_3)_{0.5}$:$(CeO_2)_{0.5}$ vertically aligned nanocomposite thin films. *Nanotechnology* 22:315712.
11. Chen, A. P., M. Weigand, Z. Bi, W. Zhang, X. Lu, P. Dowden, J. L. MacManus-Driscoll, H. Wang, and Q. X. Jia. 2014. Evolution of microstructure, strain and physical properties in oxide nanocomposite films. *Scientific Reports* 4:5426.
12. Chen, A. P., W. Zhang, F. Khatkhatay, Q. Su, C.-F. Tsai, L. Chen, Q. X. Jia, J. L. MacManus Driscoll, and H. Wang. 2013. Magnetotransport properties of quasi-one-dimensionally channeled vertically aligned heteroepitaxial nanomazes. *Applied Physics Letters* 102:093114.

13. Zheng, H. M., F. Straub, Q. Zhan, P. L. Yang, W. K. Hsieh, F. Zavaliche, Y. H. Chu, U. Dahmen, and R. Ramesh. 2006. Self-assembled growth of $BiFeO_3$-$CoFe_2O_4$ nanostructures. *Advanced Materials* 18:2747–2752.

14. Zheng, H., Q. Zhan, F. Zavaliche, M. Sherburne, F. Straub, M. P. Cruz, L. Q. Chen, U. Dahmen, and R. Ramesh. 2006. Controlling self-assembled perovskite-spinel nanostructures. *Nano Letters* 6:1401–1407.

15. Zhan, Q., R. Yu, S. P. Crane, H. Zheng, C. Kisielowski, and R. Ramesh. 2006. Structure and interface chemistry of perovskite-spinel nanocomposite thin films. *Applied Physics Letters* 89:172902.

16. Yoon, J., S. Cho, J. H. Kim, J. Lee, Z. X. Bi, A. Serquis, X. H. Zhang, A. Manthiram, and H. Y. Wang. 2009. Vertically aligned nanocomposite thin films as a cathode/electrolyte interface layer for thin-film solid-oxide fuel cells. *Advanced Functional Materials* 19:3868–3873.

17. Yang, H., H. Y. Wang, J. Yoon, Y. Q. Wang, M. Jain, D. M. Feldmann, P. C. Dowden, J. L. MacManus-Driscoll, and Q. X. Jia. 2009. Vertical interface effect on the physical properties of self-assembled nanocomposite epitaxial films. *Advanced Materials* 21:3969–3969.

18. Yan, L., Y. D. Yang, Z. G. Wang, Z. P. Xing, J. F. Li, and D. Viehland. 2009. Review of magnetoelectric perovskite-spinel self-assembled nano-composite thin films. *Journal of Materials Science* 44:5080–5094.

19. Stern, I., J. B. He, X. L. Zhou, P. Silwal, L. D. Miao, J. M. Vargas, L. Spinu, and D. H. Kim. 2011. Role of spinel substrate in the morphology of $BiFeO_3$-$CoFe_2O_4$ epitaxial nanocomposite films. *Applied Physics Letters* 99:082908.

20. Ryu, H., P. Murugavel, J. H. Lee, S. C. Chae, T. W. Noh, Y. S. Oh, H. J. Kim, K. H. Kim, J. H. Jang, M. Kim, C. Bae, and J. G. Park. 2006. Magnetoelectric effects of nanoparticulate $Pb(Zr_{0.52}Ti_{0.48})O_3$-$NiFe_2O_4$ composite films. *Applied Physics Letters* 89:102907.

21. Park, S., Y. Horibe, T. Asada, L. S. Wielunski, N. Lee, P. L. Bonanno, S. M. O'Malley, A. A. Sirenko, A. Kazimirov, M. Tanimura, T. Gustafsson, and S. W. Cheong. 2008. Highly aligned epitaxial nanorods with a checkerboard pattern in oxide films. *Nano Letters* 8:720–724.

22. Moshnyaga, V., B. Damaschke, O. Shapoval, A. Belenchuk, J. Faupel, O. I. Lebedev, J. Verbeeck, G. Van Tendeloo, M. Mucksch, V. Tsurkan, R. Tidecks, and K. Samwer. 2005. Structural phase transition at the percolation threshold in epitaxial $(La_{0.7}Ca_{0.3}MnO_3)_{1-x}$:$MgO_x$ nanocomposite films. *Nature Materials* 4:247–252.

23. Mohaddes-Ardabili, L., H. Zheng, S. B. Ogale, B. Hannoyer, W. Tian, J. Wang, S. E. Lofland, S. R. Shinde, T. Zhao, Y. Jia, L. Salamanca-Riba, D. G. Schlom, M. Wuttig, and R. Ramesh. 2004. Self-assembled single-crystal ferromagnetic iron nanowires formed by decomposition. *Nature Materials* 3:533–538.

24. MaCmanus-Driscoll, J. L., P. Zerrer, H. Y. Wang, H. Yang, J. Yoon, A. Fouchet, R. Yu, M. G. Blamire, and Q. X. Jia. 2008. Strain control and spontaneous phase ordering in vertical nanocomposite heteroepitaxial thin films. *Nature Materials* 7:314–320.

25. Luo, H. M., H. Yang, S. A. Bally, O. Ugurlu, M. Jain, M. E. Hawley, T. M. McCleskey, A. K. Burrell, E. Bauer, L. Civale, T. G. Holesinger, and Q. X. Jia. 2007. Self-assembled epitaxial nanocomposite $BaTiO_3$-$NiFe_2O_4$ films prepared by polymer-assisted deposition. *Journal of the American Ceramic Society* 129:14132.

26. Liao, S. C., P. Y. Tsai, C. W. Liang, H. J. Liu, J. C. Yang, S. J. Lin, C. H. Lai, and Y. H. Chu. 2011. Misorientation control and functionality design of nanopillars in self-assembled perovskite-spinel heteroepitaxial nanostructures. *ACS Nano* 5:4118–4122.

27. Li, J. H., I. Levin, J. Slutsker, V. Provenzano, P. K. Schenck, R. Ramesh, J. Ouyang, and A. L. Roytburd. 2005. Self-assembled multiferroic nanostructures in the $CoFe_2O_4$-$PbTiO_3$ system. *Applied Physics Letters* 87:072909.

28. Levin, I., J. H. Li, J. Slutsker, and A. L. Roytburd. 2006. Design of self-assembled multiferroic nanostructures in epitaxial films. *Advanced Materials* 18:2044.

29. Lebedev, O. I., J. Verbeeck, G. Van Tendeloo, O. Shapoval, A. Belenchuk, V. Moshnyaga, B. Damashcke, and K. Samwer. 2002. Structural phase transitions and stress accommodation in $(La_{0.67}Ca_{0.33}MnO_3)_{1-x}$:$(MgO)_x$ composite films. *Physical Review B* 66:104421.

30. Hsieh, Y. H., J. M. Liou, B. C. Huang, C. W. Liang, Q. He, Q. Zhan, Y. P. Chiu, Y. C. Chen, and Y. H. Chu. 2012. Local conduction at the $BiFeO_3$–$CoFe_2O_4$ tubular oxide interface. *Advanced Materials* 24:4564–4568.

31. Harrington, S. A., J. Y. Zhai, S. Denev, V. Gopalan, H. Y. Wang, Z. X. Bi, S. A. T. Redfern, S. H. Baek, C. W. Bark, C. B. Eom, Q. X. Jia, M. E. Vickers, and J. L. MacManus-Driscoll. 2011. Thick lead-free ferroelectric films with high Curie temperatures through nanocomposite-induced strain. *Nature Nanotechnology* 6:491–495.

32. Dix, N., R. Muralidharan, J. M. Rebled, S. Estrade, F. Peiro, M. Varela, J. Fontcuberta, and F. Sanchez. 2010. Selectable spontaneous polarization direction and magnetic anisotropy in $BiFeO_3$-$CoFe_2O_4$ epitaxial nanostructures. *ACS Nano* 4:4955–4961.

33. Crane, S. P., C. Bihler, M. S. Brandt, S. T. B. Goennenwein, M. Gajek, and R. Ramesh. 2009. Tuning magnetic properties of magnetoelectric $BiFeO_3$-$NiFe_2O_4$ nanostructures. *Journal of Magnetism and Magnetic Materials* 321:L5–L9.

34. Chen, A. P., Z. X. Bi, C. F. Tsai, J. Lee, Q. Su, X. H. Zhang, Q. X. Jia, J. L. MacManus-Driscoll, and H. Y. Wang. 2011. Tunable low-field magnetoresistance in $(La_{0.7}Sr_{0.3}MnO_3)_{0.5}$:$(ZnO)_{0.5}$ self-assembled vertically aligned nanocomposite thin films. *Advanced Functional Materials* 21:2423–2429.

35. Chaix-Pluchery, O., C. Cochard, P. Jadhav, J. Kreisel, N. Dix, F. Sanchez, and J. Fontcuberta. 2011. Strain analysis of multiferroic $BiFeO_3$-$CoFe_2O_4$ nanostructures by Raman scattering. *Applied Physics Letters* 99:072901.

36. Bi, Z. X., J. H. Lee, H. Yang, Q. X. Jia, J. L. MacManus-Driscoll, and H. Y. Wang. 2009. Tunable lattice strain in vertically aligned nanocomposite $(BiFeO_3)_x$:$(Sm2O3)_{1-x}$ thin films. *Journal of Applied Physics* 106:094309.

37. Zheng, H., J. Wang, S. E. Lofland, Z. Ma, L. Mohaddes-Ardabili, T. Zhao, L. Salamanca-Riba, S. R. Shinde, S. B. Ogale, F. Bai, D. Viehland, Y. Jia, D. G. Schlom, M. Wuttig, A. Roytburd, and R. Ramesh. 2004. Multiferroic $BaTiO_3$-$CoFe_2O_4$ nanostructures. *Science* 303:661–663.

38. Aimon, N. M., D. H. Kim, H. K. Choi, and C. A. Ross. 2012. Deposition of epitaxial $BiFeO_3$/$CoFe_2O_4$ nanocomposites on (001) $SrTiO_3$ by combinatorial pulsed laser deposition. *Applied Physics Letters* 100:092901.

39. Liu, H. J., L. Y. Chen, Q. He, C. W. Liang, Y. Z. Chen, Y. S. Chien, Y. H. Hsieh, S. J. Lin, E. Arenholz, C. W. Luo, Y. L. Chueh, Y. C. Chen, and Y. H. Chu. 2012. Epitaxial photostriction-magnetostriction coupled self-assembled nanostructures. *ACS Nano* 6:6952.

40. Tsai, C. F., J. H. Lee, and H. Y. Wang. 2012. Microstructure and superconducting properties of $YBa_2Cu_3O_7$-delta thin films incorporated with a self-assembled magnetic vertically aligned nanocomposite. *Superconductor Science and Technology* 25:075016.

41. Zhu, Y. Y., C. F. Tsai, J. Wang, J. H. Kwon, H. Y. Wang, C. V. Varanasi, J. Burke, L. Brunke, and P. N. Barnes. 2012. Interfacial defects distribution and strain coupling in the vertically aligned nanocomposite $YBa_2Cu_3O_{7-x}$/ $BaSnO_3$ thin films. *Journal of Materials Research* 27:1763.

42. Zhang, W., A. P. Chen, Z. X. Bi, Q. X. Jia, J. L. MacManus-Driscoll, and H. Wang. 2014. Interfacial coupling in heteroepitaxial vertically aligned nanocomposite thin films: From lateral to vertical control. *Current Opinion in Solid State and Materials Science* 18:6–18.

43. MacManus-Driscoll, J. L. 2010. Self-assembled heteroepitaxial oxide nanocomposite thin film structures: designing interface-induced functionality in electronic materials. *Advanced Functional Materials* 20:2035–2045.

44. Liu, H. J., W. I. Liang, Y. H. Chu, H. M. Zheng, and R. Ramesh. 2014. Self-assembled vertical heteroepitaxial nanostructures: from growth to functionalities. *MRS Communications* 4:31–44.

45. Tra, V. T., J. C. Yang, Y. H. Hsieh, J. Y. Lin, Y. C. Chen, and Y. H. Chu. 2014. Controllable electrical conduction at complex oxide interfaces. *Physica Status Solidi (RRL)—Rapid Research Letters* 8:478–500.

46. Yan, L., Z. P. Xing, Z. G. Wang, T. Wang, G. Y. Lei, J. F. Li, and D. Viehland. 2009. Direct measurement of magnetoelectric exchange in self-assembled epitaxial $BiFeO_3$-$CoFe_2O_4$ nanocomposite thin films. *Applied Physics Letters* 94:192902.

47. Catalan, G. 2006. Magnetocapacitance without magnetoelectric coupling. *Applied Physics Letters* 88:102902.

48. Stratulat, S. M., X. L. Lu, A. Morelli, D. Hesse, W. Erfurth, and M. Alexe. 2013. Nucleation-induced self-assembly of multiferroic $BiFeO_3$-$CoFe_2O_4$ nanocomposites. *Nano Letters* 13:3884–3889.

49. Babu, S. N., and L. Malkinski. 2012. Large converse magnetoelectric effect in $Na_{0.5}Bi_{0.5}TiO_3$-$CoFe_2O_4$ lead-free multiferroic composites. *Journal of Applied Physics* 111:07D919.

50. Liu, M., O. Obi, J. Lou, Y. J. Chen, Z. H. Cai, S. Stoute, M. Espanol, M. Lew, X. Situ, K. S. Ziemer, V. G. Harris, and N. X. Sun. 2009. Giant electric field tuning of magnetic properties in multiferroic ferrite/ferroelectric heterostructures. *Advanced Functional Materials* 19:1826–1831.

51. Lou, J., M. Liu, D. Reed, Y. H. Ren, and N. X. Sun. 2009. Giant electric field tuning of magnetism in novel multiferroic FeGaB/lead zinc niobate-lead titanate (PZN-PT) heterostructures. *Advanced Materials* 21:4711.

52. Eerenstein, W., M. Wiora, J. L. Prieto, J. F. Scott, and N. D. Mathur. 2007. Giant sharp and persistent converse magnetoelectric effects in multiferroic epitaxial heterostructures. *Nature Materials* 6:348–351.

53. Wang, Z. G., Y. D. Yang, R. Viswan, J. F. Li, and D. Viehland. 2011. Giant electric field controlled magnetic anisotropy in epitaxial $BiFeO_3$-$CoFe_2O_4$ thin film heterostructures on single crystal $Pb(Mg_{1/3}Nb_{2/3})_{0.7}Ti_{0.3}O_3$ substrate. *Applied Physics Letters* 99:043110.

54. Molegraaf, H. J. A., J. Hoffman, C. A. F. Vaz, S. Gariglio, D. van der Marel, C. H. Ahn, and J. M. Triscone. 2009. Magnetoelectric effects in complex oxides with competing ground states. *Advanced Materials* 21:3470.

55. Ma, J., Y. H. Lin, and C. W. Nan. 2010. Anomalous electric field-induced switching of local magnetization vector in a simple FeBSiC-on-Pb(Zr,Ti)O_3 multiferroic bilayer. *Journal of Physics D: Applied Physics* 43:012001.

56. Vaz, C. A. F., J. Hoffman, Y. Segal, J. W. Reiner, R. D. Grober, Z. Zhang, C. H. Ahn, and F. J. Walker. 2010. Origin of the magnetoelectric coupling effect in $Pb(Zr_{0.2}Ti_{0.8})O_3/La_{0.8}Sr_{0.2}MnO_3$ multiferroic heterostructures. *Physical Review Letters* 104:127202.

57. Wan, J. G., X. W. Wang, Y. J. Wu, M. Zeng, Y. Wang, H. Jiang, W. Q. Zhou, G. H. Wang, and J. M. Liu. 2005. Magnetoelectric $CoFe_2O_4$-$Pb(Zr,Ti)O_3$ composite thin films derived by a sol-gel process. *Applied Physics Letters* 86:122501.

58. McDannald, A., M. Staruch, G. Sreenivasulu, C. Cantoni, G. Srinivasan, and M. Jain. 2013. Magnetoelectric coupling in solution derived 3-0 type $PbZr_{0.52}Ti_{0.48}O_3$:$xCoFe_2O_4$ nanocomposite films. *Applied Physics Letters* 102:122905.

59. Park, J. H., H. M. Jang, H. S. Kim, C. G. Park, and S. G. Lee. 2008. Strain-mediated magnetoelectric coupling in $BaTiO_3$-Co nanocomposite thin films. *Applied Physics Letters* 92:062908.

60. Cheng, C. P., M. H. Tang, X. S. Lv, Z. H. Tang, and Y. G. Xiao. 2012. Magnetoelectric coupling in $La_{0.6}Ca_{0.4}MnO_3$-$Bi_{0.6}Nd_{0.4}TiO_3$ composite thin films derived by a chemical solution deposition method. *Applied Physics Letters* 101:212902.

61. Zhai, J. Y., N. Cai, Z. Shi, Y. H. Lin, and C. W. Nan. 2004. Magnetic-dielectric properties of $NiFe_2O_4$/PZT particulate composites. *Journal of Physics D: Applied Physics* 37:823–827.

62. Li, Y. X., Z. C. Wang, J. J. Yao, T. N. Yang, Z. G. Wang, J. M. Hu, C. L. Chen, R. Sun, Z. P. Tian, J. F. Li, L. Q. Chen, and D. Viehland. 2015. Magnetoelectric quasi-(0-3) nanocomposite heterostructures. *Nature Communications* 6:6680.

63. Ren, S. Q., and M. Wuttig. 2008. Magnetoelectric nano-Fe_3O_4/$CoFe_2O_4$ parallel to $PbZr_{0.53}Ti_{0.47}O_3$ composite. *Applied Physics Letters* 92:083502.

64. Ortega, N., P. Bhattacharya, R. S. Katiyar, P. Dutta, A. Manivannan, M. S. Seehra, I. Takeuchi, and S. B. Majumder. 2006. Multiferroic properties of $Pb(Zr,Ti)O_3$/$CoFe_2O_4$ composite thin films. *Journal of Applied Physics* 100:126105.

65. Ryu, S., J. H. Park, and H. M. Jang. 2007. Magnetoelectric coupling of [00l]-oriented $Pb(Zr_{0.4}Ti_{0.6})O_3$-$Ni_{0.8}Zn_{0.2}Fe_2O_4$ multilayered thin films. *Applied Physics Letters* 91:142910.

66. Ma, Y. G., W. N. Cheng, M. Ning, and C. K. Ong. 2007. Magnetoelectric effect in epitaxial $Pb(Zr_{0.52}Ti_{0.48})O_3/La_{0.7}Sr_{0.3}MnO_3$ composite thin film. *Applied Physics Letters* 90:152911.

67. Feng, M., J. Wang, J. Hu, J. Wang, J. Ma, H. Li, Y. Shen, Y. Lin, L. Q. Chen, and C. W. Nan. 2015. Optimizing direct magnetoelectric coupling in $Pb(Zr,Ti)O_3$/Ni multiferroic film heterostructures. *Applied Physics Letters* 106:072901.

68. Martinez, R., A. Kumar, R. Palai, G. Srinivasan, and R. S. Katiyar. 2012. Observation of strong magnetoelectric effects in $Ba_{0.7}Sr_{0.3}TiO_3/La_{0.7}Sr_{0.3}MnO_3$ thin film heterostructures. *Journal of Applied Physics* 111:104104.

69. Zhang, Y., C. Y. Deng, J. Ma, Y. H. Lin, and C. W. Nan. 2008. Enhancement in magnetoelectric response in $CoFe_2O_4$-$BaTiO_3$ heterostructure. *Applied Physics Letters* 92:062911.

70. Deng, C. Y., Y. Zhang, J. Ma, Y. H. Lin, and C. W. Nana. 2007. Magnetic-electric properties of epitaxial multiferroic $NiFe_2O_4$-$BaTiO_3$ heterostructure. *Journal of Applied Physics* 102:074114.

71. Deng, C. Y., Y. Zhang, J. Ma, Y. H. Lin, and C. W. Nan. 2008. Magnetoelectric effect in multiferroic heteroepitaxial $BaTiO_3$-$NiFe_2O_4$ composite thin films. *Acta Materialia* 56:405–412.

72. Wang, J., Z. Li, Y. Shen, Y. H. Lin, and C. W. Nan. 2013. Enhanced magneto-electric coupling in $Pb(Zr_{0.52}Ti_{0.48})O_3$ film-on-$CoFe_2O_4$ bulk ceramic composite with $LaNiO_3$ bottom electrode. *Journal of Materials Science* 48:1021–1026.

73. Patil, D., J. H. Kim, Y. S. Chai, J. H. Nam, J. H. Cho, B. I. Kim, and K. H. Kim. 2011. Large longitudinal magnetoelectric coupling in $NiFe_2O_4$-$BaTiO_3$ laminates. *Applied Physics Express* 4:073001.

74. Stognij, A. I., N. N. Novitskii, S. A. Sharko, A. V. Bespalov, O. L. Golikova, A. Sazanovich, V. Dyakonov, H. Szymczak, and V. A. Ketsko. 2014. Effect of interfaces on the magnetoelectric properties of Co/PZT/Co heterostructures. *Inorganic Materials* 50:280–284.

75. Srinivasan, G., E. T. Rasmussen, B. J. Levin, and R. Hayes. 2002. Magnetoelectric effects in bilayers and multilayers of magnetostrictive and piezoelectric perovskite oxides. *Physical Review B* 65:134402.

76. Stognij, A., N. Novitskii, A. Sazanovich, N. Poddubnaya, S. Sharko, V. Mikhailov, V. Nizhankovski, V. Dyakonov, and H. Szymczak. 2013. Ion-beam sputtering deposition and magnetoelectric properties of layered heterostructures (FM/PZT/FM)$_n$, where FM—Co or $Ni_{78}Fe_{22}$. *The European Physical Journal Applied Physics* 63:21301.

77. Chen, A. P., N. Poudyal, J. Xiong, J. P. Liu, and Q. X. Jia. 2015. Modification of structure and magnetic anisotropy of epitaxial $CoFe_2O_4$ films by hydrogen reduction. *Applied Physics Letters* 106:111907.

78. Dussan, S., A. Kumar, R. S. Katiyar, S. Priya, and J. F. Scott. 2011. Magnetic control of ferroelectric interfaces. *Journal of Physics Condensed Matter* 23:202203.

79. Chen, Y., G. S. Wang, S. A. Zhang, X. Y. Lei, J. Y. Zhu, X. D. Tang, Y. L. Wang, and X. L. Dong. 2011. Magnetocapacitance effects of $Pb_{0.7}Sr_{0.3}TiO_3$/$La_{0.7}Sr_{0.3}MnO_3$ thin film on Si substrate. *Applied Physics Letters* 98:052910.

80. Nan, C. W., G. Liu, Y. H. Lin, and H. D. Chen. 2005. Magnetic-field-induced electric polarization in multiferroic nanostructures. *Physical Review Letters* 94:197203.

81. Wu, H. P. 2014. Adjustable magnetoelectric effect of self-assembled vertical multiferroic nanocomposite films by the in-plane misfit strain and ferromagnetic volume fraction. *Journal of Applied Physics* 115:114105.

82. Schmitz-Antoniak, C., D. Schmitz, P. Borisov, F. M. F. de Groot, S. Stienen, A. Warland, B. Krumme, R. Feyerherm, E. Dudzik, W. Kleemann, and H. Wende. 2013. Electric in-plane polarization in multiferroic $CoFe_2O_4$/$BaTiO_3$ nanocomposite tuned by magnetic fields. *Nature Communications* 4:2051.

83. Zhang, J. X., J. Y. Dai, W. Lu, H. L. W. Chan, B. Wu, and D. X. Li. 2008. A novel nanostructure and multiferroic properties in $Pb(Zr_{0.52}Ti_{0.48})O_3$/$CoFe_2O_4$ nanocomposite films grown by pulsed-laser deposition. *Journal of Physics D Applied Physics* 41:235405.

84. Kim, K. S., S. H. Han, H. G. Kim, J. S. Kim, and C. I. Cheon. 2010. Phase separation and microstructure of $BaTiO_3$-$CoFe_2O_4$ epitaxial nanocomposite films deposited under low working pressure. *Journal of Vacuum Science and Technology B* 28:C5A14–C5A19.

85. Zhang, W., J. Jian, A. P. Chen, L. Jiao, F. Khatkhatay, L. Li, F. Chu, Q. X. Jia, and J. L. MacManus-Driscoll. 2014. Strain relaxation and enhanced perpendicular magnetic anisotropy in $BiFeO_3$:$CoFe_2O_4$ vertically aligned nanocomposite thin films. *Applied Physics Letters* 104:062402.

86. Li, Y. X., Y. D. Yang, J. J. Yao, R. Viswan, Z. G. Wang, J. F. Li, and D. Viehland. 2012. Controlled growth of epitaxial $BiFeO_3$ films using self-assembled $BiFeO_3$-$CoFe_2O_4$ multiferroic heterostructures as a template. *Applied Physics Letters* 101:022905.

87. Zavaliche, F., H. Zheng, L. Mohaddes-Ardabili, S. Y. Yang, Q. Zhan, P. Shafer, E. Reilly, R. Chopdekar, Y. Jia, P. Wright, D. G. Schlom, Y. Suzuki, and R. Ramesh. 2005. Electric field-induced magnetization switching in epitaxial columnar nanostructures. *Nano Letters* 5:1793–1796.

88. Zavaliche, F., T. Zhao, H. Zheng, F. Straub, M. P. Cruz, P. L. Yang, D. Hao, and R. Ramesh. 2007. Electrically assisted magnetic recording in multiferroic nanostructures. *Nano Letters* 7:1586–1590.

89. Yan, L., Z. G. Wang, Z. P. Xing, J. F. Li, and D. Viehland. 2010. Magnetoelectric and multiferroic properties of variously oriented epitaxial $BiFeO_3$-$CoFe_2O_4$ nanostructured thin films. *Journal of Applied Physics* 107:064106.

90. Oh, Y. S., S. Crane, H. Zheng, Y. H. Chu, R. Ramesh, and K. H. Kim. 2010. Quantitative determination of anisotropic magnetoelectric coupling in $BiFeO_3$-$CoFe_2O_4$ nanostructures. *Applied Physics Letters* 97:052902.

91. Yan, F., G. N. Chen, L. Lu, P. Finkel, and J. E. Spanier. 2013. Local probing of magnetoelectric coupling and magnetoelastic control of switching in $BiFeO_3$-$CoFe_2O_4$ thin-film nanocomposite. *Applied Physics Letters* 103:042906.

92. Wan, J. G., Y. Y. Weng, Y. J. Wu, Z. Y. Li, J. M. M. Liu, and G. H. Wang. 2007. Controllable phase connectivity and magnetoelectric coupling behavior in $CoFe_2O_4$-$Pb(Zr, Ti)O_3$ nanostructured films. *Nanotechnology* 18:465708.

93. Zhou, Z. Y., M. Trassin, Y. Gao, Y. Gao, D. N. Qiu, K. Ashraf, T. X. Nan, X. Yang, S. R. Bowden, D. T. Pierce, M. D. Stiles, J. Unguris, M. Liu, B. M. Howe, G. J. Brown, S. Salahuddin, R. Ramesh, and N. X. Sun. 2015. Probing electric field control of magnetism using ferromagnetic resonances. *Nature Communications* 6:6082.

94. Zhou, Z. Y., B. M. Howe, M. Liu, T. X. Nan, X. Chen, K. Mahalingam, N. X. Sun, and G. J. Brown. 2015. Interfacial charge-mediated non-volatile magnetoelectric coupling in $Co_{0.3}Fe_{0.7}/Ba_{0.6}Sr_{0.4}TiO_3/Nb$:$SrTiO_3$ multiferroic heterostructures. *Scientific Reports* 5:7740.

95. Zhai, J., S. Dong, Z. Xing, J. F. Li, and D. Viehland. 2006. Giant magnetoelectric effect in Metglas/polyvinylidene-fluoride laminates. *Applied Physics Letters* 89:083507.

96. Dong, S. X., J. Y. Zhai, J. F. Li, and D. Viehland. 2006. Near-ideal magnetoelectricity in high-permeability magnetostrictive/piezofiber laminates with a (2-1) connectivity. *Applied Physics Letters* 89:252904.

97. Greve, H., E. Woltermann, H. J. Quenzer, B. Wagner, and E. Quandt. 2010. Giant magnetoelectric coefficients in $(Fe_{90}Co_{10})_{78}Si_{12}B_{10}$-AlN thin film composites. *Applied Physics Letters* 96:182501.

98. Wang, Y. J., J. F. Li, and D. Viehland. 2014. Magnetoelectrics for magnetic sensor applications: status, challenges and perspectives. *Materials Today* 17:269–275.

99. Hur, N., S. Park, P. A. Sharma, J. S. Ahn, S. Guha, and S. W. Cheong. 2004. Electric polarization reversal and memory in a multiferroic material induced by magnetic fields. *Nature* 429:392–395.

100. Wu, S. M., S. A. Cybart, P. Yu, M. D. Rossell, J. X. Zhang, R. Ramesh, and R. C. Dynes. 2010. Reversible electric control of exchange bias in a multiferroic field-effect device. *Nature Materials* 9:756–761.

101. Bocher, L., A. Gloter, A. Crassous, V. Garcia, K. March, A. Zobelli, S. Valencia, S. Enouz-Vedrenne, X. Moya, N. D. Marthur, C. Deranlot, S. Fusil, K. Bouzehouane, M. Bibes, A. Barthelemy, C. Colliex, and O. Stephan. 2012. Atomic and electronic structure of the BaTiO$_3$/Fe interface in multiferroic tunnel junctions. *Nano Letters* 12:376–382.
102. Kim, Y. M., A. Morozovska, E. Eliseev, M. P. Oxley, R. Mishra, S. M. Selbach, T. Grande, S. T. Pantelides, S. V. Kalinin, and A. Y. Borisevich. 2014. Direct observation of ferroelectric field effect and vacancy-controlled screening at the BiFeO$_3$/La$_x$Sr$_{1-x}$MnO$_3$ interface. *Nature Materials* 13:1019–1025.
103. Sun, D. L., M. Fang, X. S. Xu, L. Jiang, H. W. Guo, Y. M. Wang, W. T. Yang, L. F. Yin, P. C. Snijders, T. Z. Ward, Z. Gai, X. G. Zhang, H. N. Lee, and J. Shen. 2014. Active control of magnetoresistance of organic spin valves using ferroelectricity. *Nature Communications* 5:4396.
104. Radaelli, G., D. Petti, E. Plekhanov, I. Fina, P. Torelli, B. R. Salles, M. Cantoni, C. Rinaldi, D. Gutierrez, G. Panaccione, M. Varela, S. Picozzi, J. Fontcuberta, and R. Bertacco. 2014. Electric control of magnetism at the Fe/BaTiO$_3$ interface. *Nature Communications* 5:3404.
105. Zhang, S., Y. G. Zhao, X. Xiao, Y. Z. Wu, S. Rizwan, L. F. Yang, P. S. Li, J. W. Wang, M. H. Zhu, H. Y. Zhang, X. F. Jin, and X. F. Han. 2014. Giant electrical modulation of magnetization in Co$_{40}$Fe$_{40}$B$_{20}$/Pb(Mg$_{1/3}$Nb$_{2/3}$)$_{0.7}$Ti$_{0.3}$O$_3$(011) heterostructure. *Scientific Reports* 4:3727.
106. Cherifi, R. O., V. Ivanovskaya, L. C. Phillips, A. Zobelli, I. C. Infante, E. Jacquet, V. Garcia, S. Fusil, P. R. Briddon, N. Guiblin, A. Mougin, A. A. Unal, F. Kronast, S. Valencia, B. Dkhil, A. Barthelemy, and M. Bibes. 2014. Electric-field control of magnetic order above room temperature. *Nature Materials* 13:345–351.
107. Nan, T. X. 2014. Non-volatile switching of magnetism in multiferroic heterostructures. *Master Thesis*, Northeastern University, Boston, MA.
108. Zhou, Z. Y. 2014. Voltage control of magnetism. Ph.D thesis, Northeastern University, Boston, MA.
109. Zhang, S. 2014. Electric-Field Control of Magnetization and Electronic Transport in Ferromagnetic/Ferroelectric Heterostructures. Ph.D thesis, Tsinghua University, Beijing, China.
110. Prellier, W., M. P. Singh, and P. Murugavel. 2005. The single-phase multiferroic oxides: from bulk to thin film. *Journal of Physics Condensed Matter* 17:R803–R832.
111. Wang, Y., J. M. Hu, Y. H. Lin, and C. W. Nan. 2010. Multiferroic magnetoelectric composite nanostructures. *NPG Asia Materials* 2:61–68.
112. Vaz, C. A. F. 2012. Electric field control of magnetism in multiferroic heterostructures. *Journal of Physics Condensed Matter* 24:333201.
113. Martin, L. W., and R. Ramesh. 2012. Multiferroic and magnetoelectric heterostructures. *Acta Materialia* 60:2449–2470.
114. Fusil, S., V. Garcia, A. Barthelemy, and M. Bibes. 2014. Magnetoelectric devices for spintronics. *Annual Review of Materials Research* 44:91–116.
115. Matsukura, F., Y. Tokura, and H. Ohno. 2015. Control of magnetism by electric fields. *Nature Nanotechnology* 10:209.
116. Vaz, C. A. F., F. J. Walker, C. H. Ahn, and S. Ismail-Beigi. 2015. Intrinsic interfacial phenomena in manganite heterostructures. *Journal of Physics Condensed Matter* 27:123001.

Applications of Multiferroic Magnetoelectric Composites

Yuan Zhou, Jong-Woo Kim, Shuxiang Dong,
Shashank Priya, Junling Wang, and Jungho Ryu

Contents

8.1 Introduction

Multiferroic magnetoelectric (ME) materials have attracted considerable interest due to their technological importance in various applications including magnetic field sensors, filters, transformers, information storage devices, and energy harvesters [1]. In general, ferroic materials, including ferroelectric, ferromagnetic, and ferroelastic materials exhibit hystersis behavior between physical parameters (magnetization M, polarization P, strain ε) and their conjugate external stimuli (magnetic field H, electric field E, external stress σ) [2]. The term "multiferroic" is used for materials that have more than one ferroic order, where the ME coupling allows tuning of two or more physical parameters under external stimuli [3]. One of the most appealing properties of multiferroics is the ME coupling, which allows for the modulation of electric polarization by an external magnetic field or the change of magnetization by an applied electric field.

The intrinsic ME effect was first theoretically predicted by Pierre Curie in 1894, after the observation of the movement of a magnetized dielectric in an electric field by Röntgen [4]. In 1926, Debye coined the term "magnetoelectric" after the first attempt of a static ME effect [5]. However, it was unsuccessful until the experimental confirmation of ME effect in single-phase chromium oxide (Cr_2O_3) [6,7], after which the development of ME materials has been pursued for several decades. Various other single-phase compounds that possess ME effect have been discovered, such as the well-known $BiFeO_3$, $YMnO_3$, $TbMnO_3$, and so on [8]. However, the magnitude of ME coupling in these single-phase materials is usually small at room temperature, which dramatically limits their potential for practical applications [4,9,10].

To obtain larger ME coupling at room temperature, various composites consisting of ferromagnetic and ferroelectric materials have been developed. The idea of a composite displaying ME effect was first proposed by van Suchtelen (Philips Research Lab) in 1972. van den Boomgard et al. [11–14] then synthesized the first ME composite consisting of $BaTiO_3$ and $CoFe_2O_4$ using a unidirectional solidification process, which demonstrated a much higher ME response of 0.13 V/cm Oe than that of single-phase multiferroics. In these composites, the ME response is generated based on the elastic coupling between constituent phases. Subsequently, a variety of two-phase particulate ME composite materials have been investigated [1,4,8]. However, these efforts were not able to extend the ME coupling coefficient beyond the level of ~0.1 V/cm Oe, although the corresponding theoretical calculation predicted a much higher response on the order of 1 V/cm Oe.

The situation remained unchanged until 2001, when Ryu et al. [15] reported the fabrication of ME composites of a PZT ceramic disc sandwiched between Terfenol-D discs using Ag epoxy in a laminate configuration. This configuration not only preserved the physical properties of each constituent but also provided the highest ME voltage coefficient of up to 4.68 V/cm Oe at that time (voltage value was peak-peak) [15]. This work provided a new path for the fabrication of ME composites through a simple epoxy-bonding technique. To date, many types of ME laminate composites have been experimentally and theoretically investigated, and thoroughly reviewed by Fiebig [4], Priya et al. [8], Ramesh and Spaldin [16], Nan et al. [1], and Srinivasan [17].

In view of the design of ME composites, many studies on the constituting materials (e.g., physical properties, preparation methods, and geometry), the constituent interface control (e.g., structural compatibility, bonding techniques), and the measurement conditions (e.g., operation modes, excitation field frequency, and bias field) have been conducted, as summarized in Table 8.1.

One of the most crucial parameters for the practical application of ME composites has been the magnitude of ME voltage coefficient, which determines the sensitivity/voltage/power of ME-based devices. The characteristic ME behavior, in turn, depends on the properties of the piezoelectric and magnetostrictive components of the composites, which makes selecting the appropriate components the first step toward ME composites with high performance. Second, the manner in which the constituting materials are combined should also be considered. These include the appropriate phase connectivity (i.e., 0–3,

TABLE 8.1 Various Parameters That Are Important for the Development of ME Composites

Design Factors	Features	Challenges	Solutions
Materials	Piezoelectric	Physical properties/ anisotropy	Ceramic/crystal/ polymer
	Magnetostrictive		Ceramic/metal/ alloy
Combination	Connectivity	Coupling	0–3, 1–3, 2–2
Synthesis	Epoxy bonding	Low coupling	Co-firing
	Thin-film deposition	Low power	Thick film printing
Characterization	Operation mode	Stress transfer	(L–L)>(L–T/ T–L)>(T–T)
	Frequency	F_r/bandwidth	Low f_r/broadband
	Magnetic field	H_{bias}/bandwidth	Self-bias/ wideband
	Configuration	Volume ratio/ size	Demagnetizaiton/ MEMS

1–3, or 2–2) as well as the synthesis techniques, which should be considered based on the selected materials and composite structures. In general, adhesive bonding is widely used for synthesizing ME laminates due to the ease of fabrication and low costs [18]. However, to optimize the ME interface coupling, direct bonding that completely eliminates the low mechanical strength epoxy in the composite is desirable. In this respect, the co-firing technique not only enhances the mechanical coupling at the interfaces, but may also decrease the production cost due to its compatibility with the industrial production process that is commonly used for multilayer capacitors.

Besides the constituents and interfaces, the coupling effect of an ME composite also depends on several other factors including: (1) the operation mode of the composite, which is related to how the polarization and magnetization are poled with respect to each other. These include L–T, T–L, T–T, and L–L modes, where the two letters represent magnetization and polarization orientations, respectively. L refers to longitudinal and T refers to transverse. Both theoretical calculations and experiments have demonstrated that the magnitude of the ME voltage coefficient for laminates follows the order (L–L) > (L–T/T–L) > (T–T) [18]. (2) The excitation field frequency. When the ac excitation field is at a frequency corresponding to the electromechanical resonance frequency of the piezoelectric phase or ferromagnetic resonance frequency of the magnetic phase in the composite, the ME voltage coefficient shows a peak with the magnitude increased by a factor of up to 100 [17,18]. (3) DC bias field. The ME coupling coefficient generally varies with the dc bias field, reaching a maximum at an optimized value. (4) The volume ratio of constituent layers and their size scaling effect. In summary, a comprehensive understanding of these factors provides guidance toward the design of high-performance ME composites.

A variety of ME devices have been demonstrated, including magnetic sensors, energy harvesters, electric field–tunable multiferroic devices, and transformers, and so on. In this chapter, the current state-of-the-art ME devices are summarized based on the ME effect used: direct ME effect (the change of electrical polarization under external magnetic field) or converse ME effect (the change of magnetization under external electric field). In the first category, the development of magnetic sensors, energy harvesters, and other functional devices (transformer, gyrator, and surgical tool) is covered. Subsequently, converse ME coupling–based multiferroic devices are presented with emphasis on the electric-field control of magnetization, permeability, and spin wave. Lastly, considering the limitations of conventional ME composites and devices, the recent development of self-biased, MEMS scalable, flexible ME devices, and their future perspectives are discussed.

8.2 Applications of Direct ME Effect

When a magnetic field is applied to the ME composite, strain will be generated from the magnetostrictive layer and subsequently transferred to the piezoelectric layer, producing electrical charges on the surface. This is referred to as the direct ME effect. Accordingly, by monitoring or harvesting the magnetically induced electric signal, one can use ME composites in various devices such as magnetic sensors, energy harvesters, transducers, gyrators, and antennas.

8.2.1 Magnetic Sensors

ME composites as sensors have attracted considerable attention due to their simple implementation mechanism and passive nature; that is, no external power source is required. In general, the operation of conventional ME composites requires both a dc (H_{dc}) and an ac (H_{ac}) magnetic fields, when using the direct ME coupling effect. The ME voltage output depends on these magnetic field inputs. Thus, by monitoring the change of ME voltage using appropriate electronic signal processing components, an ME composite can serve as an ac or dc magnetic field sensor [19,20]. Furthermore, since a current (I) passing through a wire generates magnetic field in the surrounding space, an ME composite can also act as a current sensor by monitoring the corresponding magnetic flux. Highly sensitive magnetic sensors can be obtained using ME composites with high ME coupling coefficients. However, for practical applications, understanding the sources of signal noises is equally important toward establishing high sensitivity.

8.2.1.1 Sensor Design and Characterization
8.2.1.1.1 Magnetic Field Sensor

As discussed previously, by selecting high-performance piezoelectric and magnetostrictive materials and combining them with a suitable structure, we can obtain high-performance ME composites capable of detecting small magnetic fields (ac or dc) in the room temperature. In general, for a given H_{ac}, the ME coupling coefficient first increases with increasing H_{dc}, reaching a maximum at an optimized dc bias (H_{bias}), and then decreases with further increasing H_{dc}. This behavior can be understood by taking into account the relationship given as follows [21]:

$$\alpha_{ME} \propto \frac{d\lambda}{dH} = q \qquad (8.1)$$

This indicates that the ME coefficient is directly related to the nature of the ferromagnetic phase (λ: strain, q: piezomagnetic coefficient) and the effectiveness of elastic coupling between the two phases. Since q is a function of H_{dc}, it is necessary to apply optimum bias magnetic field to the ME sensor to obtain maximum sensitivity. Depending on the materials utilized, ME composites will have different ME behaviors due to the difference in the coercive field, permeability, and magnetization. Thus, by optimizing an ME composite with small optimum dc bias, a high-sensitive ac magnetic field sensor can be obtained.

Over the last decade, various ME laminate–based sensors have been developed. Among them, bimorphs and multiple push–pull configurations have demonstrated considerable potential with sub-nanotesla sensitivity [1,17–20]. Terfenol-D and Metglas are the most commonly used magnetostrictive materials, whereas lead zirconate titanate ceramics and piezoelectric relaxor single crystals (such as $Pb(Zn_{1/3}Nb_{2/3})O_3$–$PbTiO_3$ (PZN–PT) and $Pb(Mg_{1/3}Nb_{2/3})O_3$–$PbTiO_3$ (PMN–PT)) [22,23] are usually selected as the piezoelectric layers. The typical structure of a push–pull mode ME laminate is illustrated in Figure 8.1a [24]. This laminate consists of a symmetric longitudinally poled piezoelectric PMN–PT crystal and two longitudinally magnetized magnetostrictive Terfenol-D layers. The symmetric nature allows for optimized elastic coupling between consecutive layers. As a result, a large $\alpha_{ME} \sim 30$ V/cm Oe under optimized $H_{dc} = 450$ Oe and resonance drive (~77.8 kHz) was obtained with a high magnetic field sensitivity of ~1.2×10^{-12} T, as shown in Figure 8.1b. The composition and geometry of the ferromagnetic phase can be further tuned to tailor the optimum dc bias [1,17,25]. Figure 8.1c shows the modified push–pull ME laminate consisting of piezofibers and high-permeability Metglas ($\mu_r > 40,000$) that exhibit a large piezomagnetic coefficient at low H_{dc} (Figure 8.1d) [25]. Very high ME coefficients of up to 22 V/cm Oe at 1 kHz, $H_{dc} = 4$ Oe and ~500 V/cm Oe at $f_r \sim 22$ kHz, $H_{dc} = 4$ Oe were achieved. These values are much higher than the original values and require much lower bias fields. So far, the detection of small ac magnetic field as

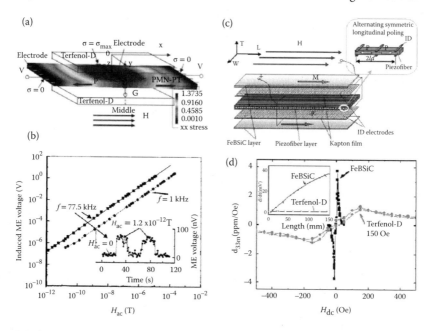

FIGURE 8.1 (a) Push–pull mode ME laminate having two longitudinal magnetization Terfenol-D layers sandwiching one symmetric longitudinally polarized PMN–PT layer. (b) Magnetic field sensitivity with a minute magnetic variation of 1.2×10^{-12} T taken under $H_{dc} = 450$ Oe, $f_{ac} = 1$ Hz and 77.5 kHz (resonance) (From Dong, S. et al., *Appl. Phys. Lett.*, 87, 062502, 2005.). (c) Optimized push–pull mode ME laminate with Metglas as magnetostrictive layer and piezoelectric fiber as piezoelectric layer, each piezofiber has numerous alternating symmetric longitudinally poled push–pull units. (d) Effective piezomagnetic coefficient as a function of dc magnetic bias H_{dc} for ferromagnetic Metglas and Terfenol-D. (From Dong, S. et al., *Appl. Phys. Lett.*, 89, 252904, 2006.)

low as 10 pT at room temperature at 1 Hz has been achieved with a push–pull Metglas/PMN–PT fiber laminate composite [19]. These high-sensitivity ME sensors show significant superiority in simple preparation and small size, low power consumption, and low costs compared with the sophisticated and very expensive superconducting quantum interference devices (SQUID).

As reflected in the example discussed earlier, another important feature of the ME composites is the change in ME coefficient as a function of the ac magnetic field H_{ac} frequency. When it matches the electromechanical resonance of the piezoelectric phase or ferromagnetic resonance of the magnetic phase in the ME composite, the ME voltage coefficient shows a peak with the magnitude increased by a factor of up to 100 [17,18]. Cho et al. [21] analyzed the direct and converse ME effects in laminate composites and showed that they are maximized at antiresonance frequency, f_a and resonance frequency, f_r, respectively. A detection sensitivity of 10 nT has been reported for small long-type Terfenol-D/PZT ME laminate under resonant drive (84 kHz), compared with 100 nT under a low-frequency drive (1 kHz) [20]. The detection limit was further enhanced by concentrating the magnetic flux through the magnetostrictive geometry effect. The corresponding dc magnetic field detection sensitivity of Metglas/PZT laminates was improved from 15 to 6 nT [$f = 1$ kHz and $H_{ac} = 0.1$Oe] [26]. By replacing the PZT with high-performance (high g_{33} and k_{33}) piezoelectric relaxor single crystals, PMN-PT, the maximum value of α_{ME} for these lamaintes (~45 V/cm Oe at 1 kHz) is about three times higher than that of the Metglas/PZT lamainte of similar geometry [27]. This enhancement results in a further increase in dc magnetic field sensitivity to 5 nT at 1 kHz and 1 nT at the resonant frequency of ~22.8 kHz, as shown in Figure 8.2. These high dc field–detection sensitivities show significant potential toward practical application in navigation systems.

In general, the fundamental resonance frequency can be tuned via the length of the laminate (l), where the first longitudinal resonance is as $f_L = 1/(2l\sqrt{\rho s_{11}})$ (ρ is the density and s_{11} is the elastic compliance of the laminate) [18]. However, for applications at low frequency, the corresponding

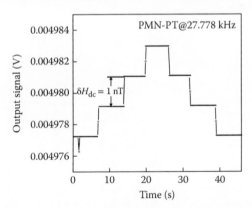

FIGURE 8.2 DC sensitivity of a Metglas/PMN–PT laminate under ac drive field of $H_{ac} = 0.1$Oe at resonance frequency. (From Gao, J. et al., *J. Appl. Phys.*, 109, 074507, 2011.)

composite dimension would be prohibitively large. In order to overcome this problem, an asymmetric unimorph configuration was proposed for trilayer ME composites (Terfenol-D/steel/PZT). It works at the bending mode that shows an enhanced coupling of ~40 V/cm Oe at a low frequency of ~5 kHz [28]. Alternative approaches that utilize unique laminate configurations (PZT-bimorph cantilever with attached permanent magnet) have been developed to achieve giant α_{ME} ~250 V/cm Oe at low resonance frequency (~60 Hz) with a reasonable size of the components [29]. Furthermore, a colossal low-frequency resonant magnetomechanical and ME effect has been observed in a three-phase composite made of PZT ceramic fibers/phosphor copper-sheet unimorph and NdFeB magnets, as shown in Figure 8.3 [30]. A very large ME voltage coefficient of ~16,000 V/cm Oe was obtained at the first-order bending

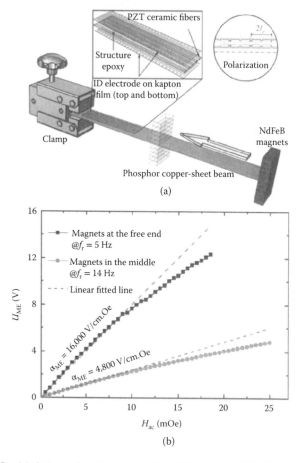

FIGURE 8.3 (a) Schematic diagram of the PZT ceramic fiber/phosphor copper sheet composite beam with NdFeB magnets. Each PZT fiber has numerous longitudinal push–pull units. (b) ME voltage outputs as a function of H_{ac} at resonant frequency for the composite beams with magnets at the free end and in the middle, respectively. (From Liu, G. et al., *Appl. Phys. Lett.*, 101, 142904, 2012.)

resonance frequency of ~5 Hz, which can be ascribed to the strong magnetic force effect and enhanced mechanical quality factor.

8.2.1.1.2 ME Current Sensor

For a given ac or dc current, the magnitude of the corresponding ac or dc magnetic field varies as a function of distance from the wire following:

$$H = \frac{I}{2\pi r} \tag{8.2}$$

where r is the distance from the wire, that is, the radius of the vortex magnetic field. The change in magnetic field can be detected by monitoring the variation of the ME voltage, which can then be used to derive the change in the corresponding I. Therefore, the current sensing mechanism is similar to that of magnetic field detection using ME composites. Figure 8.4a shows the conceptual design of an ME current sensor, where the rectangular-shaped ME laminate is mounted on the surface of an electrical cable. A pair of permanent magnets are placed at the two ends of the ME laminate to provide a dc bias field. This configuration is simple for sensor installation. However, the best sensitivity can be achieved only for current flowing in a certain direction, and the placement of the sensor may produce another level of difficulty for performance optimization. For stable and highly sensitive detection of current in electrical cables producing ac vortex magnetic fields, ring-type ME laminates were developed [31,32]. In this configuration, the cable can be placed in the center of the ring-type sensor, as shown in Figure 8.4b.

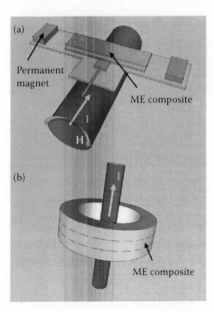

FIGURE 8.4 Schematic diagrams of (a) rectangular-shaped ME current sensor mounted on the surface of an electrical cable and (b) ring-shaped ME current sensor.

Dong et al. [33] demonstrated a ring-type ME laminate consisting of two Terfenol-D rings magnetized in the circumferential direction and one PZT ring made of segments also poled in the circumferential direction. A strain will be generated and transferred to the piezoelectric layer when a vortex magnetic field is presented. Thus, one can detect the electrical current by monitoring the induced ME voltage. Furthermore, by using single crystals with higher piezoelectric coefficients and optimizing the ring-type structure, Dong et al. [34] designed a quasi-ring-type ME sensor that is able to detect ac current as small as 10^{-7} A and/or a vortex magnetic field as small as 6 pT. However, the requirement and placement of a dc magnetic drive field is still a concern. To address this issue, Leung et al. [35] proposed a modified ring-type ME current sensor. In this design, instead of using the entire Terfenol-D ring, an epoxy-bonded Terfenol short-fiber ring with an embedded NdFeB magnetic plate was used as the magnetostrictive phase. The permanent magnetic plates not only provide an initial magnetic bias, but also eliminate the need for an external dc bias source for ME coupling optimization.

8.2.1.2 Noise Contribution and Reduction

While the detection limit of ME sensors depends on the ME coupling coefficient, it is also affected by noises appearing in the system, including extrinsic and intrinsic ones [19,36].

8.2.1.2.1 Extrinsic Noise

Extrinsic noises refer to any unwanted disturbances that originate from the environment. The main contributions include mechanical vibration, temperature fluctuation, and electromagnetic interference. When dealing with these extrinsic noises, direct rejection techniques can be used to reduce or eliminate the electromagnetic interference and thermal noise. This can be achieved by placing a metallic shield around the ME component as well as good thermal insulation. Alternatively, various composite designs can be used to further reduce the effects of thermal and vibration noises [37]. In a symmetrical ME laminate, the neutral line sits at the center of the bending piezoelectric layer, where the vibration-induced charges cancel out. In an asymmetrical structure, both bending and longitudinal modes will be excited due to the thermal expansion difference, where the thermal strain can be partially canceled. Xing et al. [37] have analyzed various ME laminate configurations and their capability to reject thermal or vibration noises. Accordingly, SS–AN (symmetrical signal, asymmetrical noise) modes can cancel vibration noise, whereas AS–SN (asymmetrical signal, symmetrical noise) modes can reject thermal noise. Accordingly, Zhai et al. [38] designed a Terfenol-D/PZT bimorph ME laminate that operates in a bending mode under an asymmetrical magnetic bias. Unlike a conventional bimorph, the magnetization directions are opposite to each other in the two Terfenol-D layers by using two U-shaped magnetic biases. When the temperature is changed, each layer will change shape longitudinally with induced thermal charges of the same sign, where the differential output of the thermal noise will be null. Following a similar strategy,

Israel et al. [39] eliminated the influence of thermal fluctuations on the output of an ME sensor comprising two BaTiO$_3$:Ni multilayer capacitors in-series. Therefore, it is crucial to identify the noise source and design the ME sensor accordingly for better sensitivity.

8.2.1.2.2 Intrinsic Noise

Unlike extrinsic noises, intrinsic noises, that is, Johnson noise and Flicker (1/f) noise, originated from the detection unit and circuit. Johnson noise is an electronic noise generated by the thermal agitation of electrons in resistors, and is random in nature. It can be represented either as a voltage or as a current noise source in series/parallel with a noiseless resistor. Flicker noise is present in all devices that have imperfect contact between two conductors. Figure 8.5a shows the equivalent noise circuit model for an ME-based sensor [36,40]. The ME composite is modeled as a magnetic field–induced charge source in parallel with a capacitor C and a resistor R. In this model, five noise sources are considered. Thus, the total noise charge density (N_T) can be given as

$$N_T = \sqrt{N_{\text{Loss}}^2 + N_R^2 + N_i^2 + N_v^2 + N_{R_f}^2} \qquad (8.3)$$

where N_{Loss} and N_R are the dielectric loss noise and resistance noise from the ME transducer, N_i and N_v are the current and voltage noise from the low-noise amplifier, and N_{R_f} is the thermal noise from the feedback resistor [41]. Details of the analysis can be found in recent publications [36,42]. Figure 8.5b shows the estimated contributions from various noise sources for a typical Metglas/piezofiber sensor [42]. From these results, we can see that dielectric loss noise, resistance noise, and current noise are the main contributing sources, whereas all the noise sources depend on frequency. To address the noise originated from the ME transducer, various strategies have been considered, including (1) using materials with ultra-high piezoelectric coefficients and low tangent losses, such as single-crystal PMN–PT [35]; (2) optimizing the poling of the piezoelectric phase for high piezoelectric coefficient and low dielectric loss [43,44]; and (3) optimizing the interfacial bonding in terms of strength, thickness, and uniformity of epoxy resin [45,46]. At the same time, to reduce the electronic noise from detection circuits, various charge amplifiers have been developed [36]. In summary, both the circuit and sensor configuration have to be considered for optimized performance.

8.2.2 Energy Harvesters

The deployment of wireless sensor networks and remote monitoring devices has made tremendous progress. This rapid proliferation is related to the significant progress being made in CMOS electronics that has reduced the power requirement considerably. At the same time, energy harvesters are also being developed to meet the power requirement of the wireless sensor networks and remote-monitoring devices in order to improve the device lifetime and combat the limitations of conventional batteries [22,47–49]. Various energy

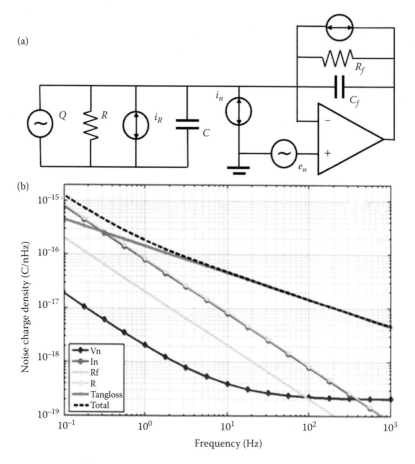

FIGURE 8.5 (a) Equivalent noise model for a magnetic sensor. (From Wang, Y.J. et al., *Philos. Trans. Ser. A Math. Phys. Eng. Sci.*, 372, 20120455, 2014.) (b) Contributions from various noise sources for a typical Metglas/piezofiber sensor. (From Wang, Y. et al., *Mater. Today*, 17, 269–275, 2014.)

sources are available for this purpose, including solar radiation, thermal gradient, mechanical vibration, magnetic field, ocean wave, and wind. Among them, energy associated with magnetic field and mechanical vibration can be converted into electricity through an ME device.

Conventional methods for mechanical vibration energy harvesting are based on electromagnetic, electrostatic, piezoelectric, and magnetostrictive mechanisms [48,50]. Among them, devices based on piezoelectric effect show relatively higher power density [51] with a small size. However, piezoelectric energy harvesters are restricted to a narrow bandwidth around the operating frequency. Piezoelectric unimorph and bimorph cantilever-based structures are widely used in vibration energy harvesters as they have low operating frequency that can be tuned by adding the tip mass [51,52]. However, the power density achievable is relatively small at the centimeter-scale [53].

Electromagnetic energy harvesters operating on the principle of Faraday's law can provide higher power density at larger dimensions [53–56]. Therefore, researchers have attempted to combine the cantilever structure with a moving magnetic tip mass that oscillates within an inductive core [53]. Multimode energy-harvesting devices with various configurations that combine electromagnetic and piezoelectric mechanisms have been proposed [56–58]. However, under small magnetic fields, the power generated from electromagnetic energy harvesters is disappointingly small [59].

An alternative method to harvest magnetic energy is to utilize an ME composite [58]. By selecting high-performance piezoelectric and magnetostrictive materials and by optimizing the composite structure, it is expected that a high-efficiency ME transducer with large voltage output under low magnetic field can be obtained [61]. The energy-harvesting mechanism can be described as follows: when the ME composite is placed in an ac magnetic field, the magnetostrictive layer responds by elongating or contracting, thereby straining the piezoelectric layer that results in an output voltage across the electrical load through the direct piezoelectric effect. Since the piezoelectric phase in the composite also responds to mechanical vibration directly, an ME-based energy harvester can harness energy from both mechanical vibrations and magnetic field at the same time. The combination is expected to enhance the power output and conversion efficiency.

8.2.2.1 ME-Based Magnetic Energy Harvester

In general, all ME transducers can be considered potential energy harvesters to scavenge energy from stray magnetic fields, which are widely present around power cables, subways, and industrial machines. For example, large electric motors commonly used in manufacturing plants emit a periodic stray magnetic field from the inductive windings. The oscillating magnetic field can induce an ac stress on the magnetostrictive layer in an ME composite, which is transferred to the piezoelectric layer and thereby creates electric charges through the direct piezoelectric effect. For practical applications, the size and weight of the harvester should be compatible with the millimeter/centimeter-scale electronics/sensors and the ME coupling characteristics must be tailored to match the low natural frequencies and magnitude inherent to most magnetic field sources. Li et al. [60] developed an ME composite consisting of a magnetostrictive Terfenol-D plate, multiple piezoelectric plates, and a copper ultrasonic horn, as shown in Figure 8.6a. This design with a high-Q ultrasonic horn can decrease the energy loss and gather the energy at resonance, producing a maximum power of 6.5 μW at f_{ac} ~26 kHz and H_{ac} = 1 Oe, as shown in Figure 8.6b. Alternatively, Gao et al. [61] designed an asymmetrical bilayered push–pull mode Metglas–Pb(Zr,Ti)O$_3$ laminate (Figure 8.6c), in which giant α_{ME} > 400V/cm Oe can be obtained with a tunable resonance frequency in the range of 60–220 Hz by tip mass loading. Based on this tunability, a 60 Hz magnetic field energy harvester was designed with a suitable tip mass capable of harvesting stray magnetic field from electronic instruments working on a 60 Hz ac power supply. Correspondingly, the maximum harvested power output was

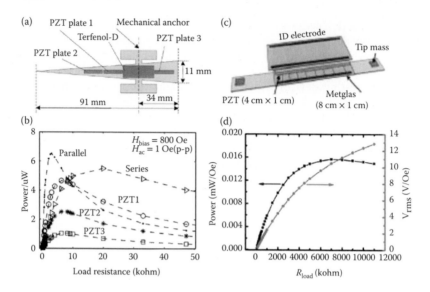

FIGURE 8.6 (a) Composite structure consisting of an ultrasonic horn, a Terfenol-D plate, and PZT plates. (b) ME power outputs of different configurations as functions of load resistance (From Li, P. et al., *Appl. Phys. Lett.*, 90, 022503, 2007.). (c) Metglas–PZT bending laminate with a tip mass. (d) Power output as a function of load resistance at the bending mode resonance frequency of 60 Hz. (From Gao, J.Q. et al., *J. Appl. Phys.*, 112, 104101, 2012.)

16 µW/Oe with a 6 MΩ resistance load (Figure 8.6d), with a power density of ≥200 µW/cm³. Furthermore, Dong et al. [62] achieved a power output of 420 µW/Oe across a 50 kΩ load under an ac magnetic field of 1Oe at ~21kHz using an ME cantilever based on the push–pull type Metglas–Pb(Zr,Ti)O₃ laminate.

Ryu et al. [63] recently demonstrated a novel energy-harvesting technique for wasted parasitic magnetic noise following the magneto-mechano-electric conversion mechanism. The 50/60 Hz parasitic magnetic noise surrounding us from electric power transmission infrastructure is captured by newly discovered ME composites. The device consists of an anisotropic piezoelectric single-crystal fiber composite and a Ni magnetostrictive metal plate in the form of a one-end clamped cantilever as shown in Figure 8.7. The device using anisotropic <011> fiber composite with d_{32} mode showed a giant output voltage of ~9.5 Vpp when it was operated under the condition of H_{ac} ~160 µT at 60 Hz without any dc magnetic bias field. Furthermore, when this device was tested under H_{ac}~500 µT field, it can turn on/off 35 high-intensity LEDs with frequency of ~1 Hz, clearly demonstrating the feasibility as power source for wireless sensor networks, portable electronics, and wireless charging systems. A wireless sensor network module composed of power management circuit, rechargeable batteries, microcontroller, and wireless transceiver was successfully driven by the generator working under H_{ac} ~700 µT at 60 Hz [63].

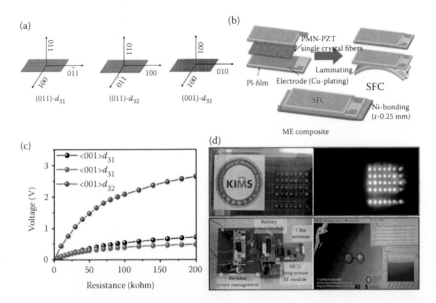

FIGURE 8.7 (a) Crystallographic orientations of piezoelectric single-crystal fibers used in the device, each piece is 0.35 (W) × 0.2 (T) × 28 (L) mm³. (b) Schematic diagrams showing the fabrication of ME laminate composites of Ni and single-crystal fibers for the energy harvester. (c) The voltage generated as a function of load resistance. (d) Combined with a capacitor, the generator was able to turn on/off 35 commercial high-intensity LEDs at a frequency of ~1 Hz. Also shown in the wireless sensor network module (power management circuit, rechargeable batteries, microcontroller, and RF communication circuit) driven by the generator and the control software screenshot. (From Ryu, J. et al., *Energy Environ. Sci.*, 8, 2402–2408, 2015.)

8.2.2.2 ME-Based Mechanical Vibration Energy Harvester

Conventional mechanical vibration energy-harvesting devices employ piezo-electric or electromagnetic transduction mechanisms [47,49,64]. However, the narrow bandwidth or the relatively low power density of these devices hinders their potential for practical usage. On the other hand, an ME composite can also harvest mechanical vibration energy due to the piezoelectric phase. Dai et al. and other research groups have demonstrated a series of devices based on this consideration [65–69], as shown in Figure 8.8a. In this design, the harvester uses four ME transducers to generate power in the air gap between multiple magnet pairs, where the magnets are arranged on the free end of a cantilever beam for ac magnetic fields. A maximum power of ~7.13 mW (1.1 mW/cm³) under an acceleration of 2.5 g at 35 Hz was achieved [66]. Moss et al. [70] reported a bi-axial oscillator using a permanent-magnet/ball bearing arrangement, which produces a peak power of 121 μW from a vibration of 61 mg at 9.8 Hz. Ju et al. [71] designed a vibration energy harvester using a freely movable spherical permanent magnet to create a time varying magnetic field for ME conversion.

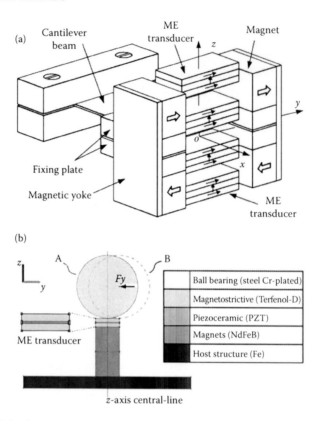

FIGURE 8.8 Schematic diagram of the ME vibration energy harvesters with (a) cantilever beam with attached magnets at the free end (From Dai, X. et al., *Sensors Actuat. A-Phys.*, 166, 94–101, 2011). (b) ME composite located between the bearing and the magnet. (From Moss, S.D. et al., *Sensors Actuat. A-Phys.*, 175, 165–168, 2012.)

Alternative approaches have been proposed in the literature to transform a mechanical rotation into magnetic field variation and thereby generate power through the ME transducer. When a permanent magnet rotates above an ME composite, the induced magnetostriction depends on the angular position and alternates between λ_{11} and λ_{12}. Thus, the piezoelectric layer is stressed and generates electrical charges [72]. In their experiment, Li et al. [67] attached a cantilever beam with a magnet at the free end on a rotating base, thereby creating a sinusoidal magnetic wave. The rotation-induced magnetic field variation was able to generate a power of 157 µW across a 3.3 MΩ load at 599 rpm.

8.2.2.3 Dual-Phase ME Energy Harvester

To further enhance the efficiency and power output of the ME energy harvester, researchers have made an attempt to generate power from both mechanical vibration and alternating magnetic field simultaneously, termed dual-phase mode energy harvester [62,73,74]. Dong et al. [62] first demonstrated a multimode system for harvesting magnetic and mechanical energy. The system consists of a cantilever beam with a tip mass and an ME laminate attached to the center of the beam (Figure 8.9a and b), which generates an open circuit voltage of 8 V peak to peak under vibration of 50 mg and ac magnetic field of 2 Oe. They also proposed an equivalent circuit model that predicts the summation effect of mechanical and magnetic energies, as shown in Figure 8.9c. Correspondingly, the induced voltage ($V_{induced}$) across the harvester under an open circuit condition can be given as [62]

$$V_{induced} = -\phi_p \left(\frac{Z_c}{Z_m} \right) (F + \phi_m H) \tag{8.4}$$

where ϕ_p is the electromechanical coupling factor, ϕ_m is the magnetoelastic coupling factor, Z_c is the capacitance impedance, and Z_m is the mechanical impedance. The induced voltage can clearly be seen to be a sum effect from two contributions: (1) an F-induced voltage via mechanical-to-electric conversion and (2) an H-induced voltage via magneto-mechano-electric conversion. By utilizing the magnetostrictive layer as both a magnetic field-strain conversion layer under magnetic field and a flexible substrate under mechanical vibration, the composite structure can provide an additive effect as shown in Figure 8.9d. In the dual mode operation, ME energy harvesters have demonstrated

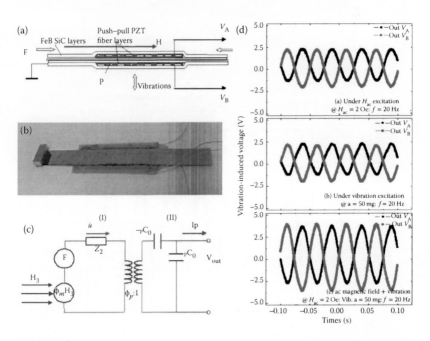

FIGURE 8.9 (a) Schematic of the ME energy harvester configuration, (b) image of the ME harvester prototype with tip mass, (c) equivalent circuit for the dual-phase mode energy harvester, (d) voltage output of the energy harvester under H_{ac}, vibration, and dual-phase mode. (From Dong, S. et al., *Appl. Phys. Lett.*, 93, 103511, 2008.)

high conversion efficiency. Furthermore, by utilizing a PMN–PZT single-crystal/Ni cantilever laminate, Kambale et al. [73] was able to obtain a maximum α_{ME} and power output of 7.28 V/cm Oe and 1.31 mW at resonance under 0.7g acceleration.

8.2.2.4 Self-Biased ME Energy Harvester

For the dual-phase energy harvesters, a solenoid and/or a permanent magnet are required as the dc bias magnetic field source, which dramatically increases the size and therefore limits the overall power density. To address this issue, Zhou et al. recently reported an energy-harvesting system utilizing self-biased ME composite with large ME coupling in the absence of a bias field [74,76]. The harvester combines an ME laminate (PZT-MFC/Ni) and piezoelectric unimorph structure in a cantilever configuration, as shown in Figure 8.10a through c. The Ni beam plays two important roles: (1) as magnetic-field-active cantilever for the piezoelectric bender and (2) as ferromagnetic phase with low field magnetic hysteresis and nonzero piezomagnetic coefficient that is essential for self-biased ME response [76]. Large ME coefficient of ~50 V/cm Oe and

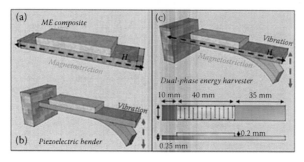

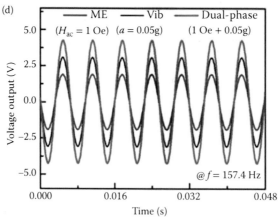

FIGURE 8.10 (a-c) Schematic diagram of self-biased dual-phase energy harvester consisting of magnetostrictive bender and ME composites. (d) Performance of the energy harvester working under H_{ac}, vibration, and the dual-phase mode at 157.4 Hz. (From Zhou, Y. et al., *Appl. Phys. Lett.*, 103, 192909, 2013.)

power density ~4.5 mW/cm³ (1g acceleration) were observed at the resonance frequency in the absence of a dc bias field. The ME voltage can be further increased with increasing H_{ac}. Under the dual-phase mode, an additive effect in voltage output was realized under zero-biased condition compared with that of the vibration or magnetic field only mode, as shown in Figure 8.10d. Alternatively, by using a single-crystal fiber composite, Patil et al. [75] were able to further enhance the ME coupling to ~82.9 V/cm Oe. This implies that there is enhancement in both power density and energy-harvesting efficiency. The simple structure of the self-biased ME energy harvester provides considerable potential toward a high-performance energy-harvesting system.

8.2.3 ME Transformers and Gyrators

ME transformers as voltage gain devices have been investigated due to the large gain at the resonance frequency [77,78]. It is also tunable with the application of a small dc magnetic bias. Over the past dacade, a variety of ME transformers have been developed [77,79–83].

The early design of ME transformer was proposed by Dong et al. [77], where a ring-type Terfenol-D/PZT laminate demonstrated a large voltage gain as shown in Figure 8.11a and b. In this design, an ac magnetic field is induced in the ME laminate through a solenoid wrapped around the laminate. When the ac field is at the resonance frequency of the laminate, a giant ME voltage

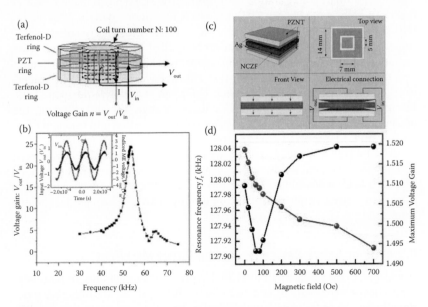

FIGURE 8.11 (a) Schematic of the ring-type coil-laminate ME transformer. (b) Voltage gain as a function of frequency for the ring-type ME transformer (From Dong, S. et al., *Appl. Phys. Lett.*, 84, 4188, 2004.). (c) Schematic of the co-fired ME transformer. (d) Frequency and voltage gain tunability as a function of magnetic field. (From Zhou, Y. et al., *Appl. Phys. Lett.*, 104, 232906, 2014.)

output can be obtained through the ME coupling effect, offering potential for high-voltage miniature transformer application. By optimizing the laminate structure, an extremely high voltage gain of ~300 was obtained at resonance and optimum H_{dc} [78]. Since the magnetic field is coupled to the piezoelectric layer, a change of magnetic field can lead to a large shift in the resonance and voltage output, providing the basis for a magnetic field–tunable device [81]. Furthermore, these ME laminates possess a unique current-to-voltage (I/V) gyration effect [84,85]. Thus, this design has the advantage of high voltage gain and wider working bandwidth [82] than that of conventional piezoelectric transformer. Compared with conventional electromagnetic transformer, it does not require the secondary coils with a high turn ratio.

However, the use of a coil around the ME composite poses a challenge for its implementation. To solve this issue, Islam et al. demonstrated a co-fired ME transformer made of a particulate $Pb(Zr_{0.52}Ti_{0.48})O_3–Ni_{0.8}Zn_{0.2}Fe_2O_4$ (PZT–NZF) composite with the ring-dot type electrode pattern of a conventional unipoled transformer in the step-up mode [79]. Since the NZF magnetostrictive particles are completely spread over the transformer disk through the mixed oxide sintering route, the transformer performance is sensitive with respect to the external magnetic field [80]. However, the weak ME coupling of the particulate composites hinders its tunability under dc magnetic field. Zhou et al. [86] designed and demonstrated a co-fired ME transformer consisting of two unipoled piezoelectric transformers and a magnetostrictive layer in a laminate configuration with Ag electrodes, as shown in Figure 8.11c. $0.2Pb(Zn_{1/3}Nb_{2/3})–0.8Pb(Zr_{0.5}Ti_{0.5})O_3$ (PZNT) and $Ni_{0.6}Cu_{0.2}Zn_{0.2}Fe_2O_4$ (NCZF) were selected as the piezoelectric and magnetostricitve components, respectively. When electrical excitation is applied to the outer square ring (input), contour extensional vibration is generated and transferred to the inner square ring (output). The mechanical strain is converted to electrical voltage through the direct piezoelectric effect. Under an external dc magnetic field, the magnetostrictive layer undergoes shape change, thereby straining the piezoelectric layer, resulting in an output voltage variation and resonance frequency shift. Figure 8.11d shows a large resonance frequency tunability [$(f_{r,o}–f_{r,H})/H = 1.4$ Hz/Oe, where $f_{r,o}$ and $f_{r,H}$ are the resonance frequencies under zero and nonzero magnetic fields, respectively], which was realized with a small dc magnetic field of 60–80 Oe. The performance of the ME transformer can be further improved by optimizing the input/output area ratio, material composition, and electrical connectivity [87,88].

8.2.4 Other Active Devices Based on Direct ME Effect

In addition to the applications discussed in previous subsections, the nonlinear properties of ferromagnetic and piezoelectric materials also enable other functionalities such as frequency multiplication by harmonic generation [89–92]. Demonstration of nonlinear ME effect was previously reported in a ceramic NZFO/PZT laminate, where frequency doubling and harmonic generation were explored at very low frequencies (1 mHz to 1 Hz) and high field (200–1000 Oe) [89]. Inspired by this work, Ma et al. [90] proposed a frequency multiplier based on an epoxy-bonded FeBSiC/PZT laminate and demonstrated its capability of doubling frequency in a broad frequency range (20 Hz to 2 kHz).

Furthermore, ME composites can also be used for in vivo medical and therapeutic applications. A microscale ME cantilever can be fixed to a catheter and inserted into the human body, acting as a scraping tool [93,94]. Utilizing the direct-ME coupling effect, the cantilever motion can be manipulated by a magnetic field, acting as a remotely controlled microblade. Meanwhile, the magnetic field–induced voltage output through the piezoelectric layer can be monitored as a feedback signal for the status of the device.

8.3 Applications of Converse ME Effect

In addition to the direct ME effect discussed previously, the converse ME effect, that is, the electric field control of magnetization, is also of great importance for functional devices. When an electric field is applied to the piezoelectric layer, mechanical deformation is generated and transferred to the magnetostrictive layer, producing a proportional magnetization modulation. This electric field control of magnetization can be further classified into three groups [95]: (1) Magnetization switching, which can be used in spintronics, including ME random access memory (MERAM); (2) Magnetic permeability variation, which can be used as voltage-tunable inductors, bandpass filters, and phase shifters; and (3) Spin wave, which can be used as voltage-tunable resonators, filters, and phase shifters. In this section, we will discuss the possibilities for novel potential ME devices such as electrically tunable microwave devices and memories.

8.3.1 Resonators

ME resonators are of significant importance for investigating the nature of electromagnetic coupling in the system and for devices such as filters, phase shifters, and antennas. In the microwave region of the electromagnetic spectrum, ME effect can be observed in the form of a shifted ferromagnetic resonance (FMR) peak by an external electric field. Studies so far have focused on two mechanisms: (1) Strain-mediated ME coupling, where a strong bonding between the ferrite and ferroelectric layers is needed for effective strain transfer. The control is gained through mechanical strains [96–99]. (2) Hybrid spin-electromagnetic mode, where bonding is not necessary due to the proximity of two materials having different dielectric and magnetic properties. Tuning is gained through changes in dielectric constant [100–102]. In general, low loss magnets including yttrium iron garnet (YIG), lithium ferrite, and hexagonal ferrites are used as the magnetic phase in microwave devices, whereas PZT, PMN-PT, and PZN-PT are used as the ferroelectric phase [95,103,104]. Figure 8.12a shows schematically a basic YIG–PZT resonator [105]. A PZT plate is bonded to a (111) YIG film epitaxially grown on a gallium gadolinium garnet (GGG) substrate. A magnetic field H is applied in order to excite the FMR in YIG, which is then tuned by an electric field applied to PZT. Since YIG resonators form the basis of band-pass and band-stop filters, YIG–PZT resonators would allow the electric field control of such devices.

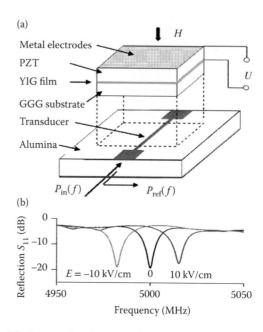

(a)

Metal electrodes

PZT

YIG film

GGG substrate

Transducer

Alumina

H

U

$P_{in}(f)$ $P_{ref}(f)$

(b)

Reflection S_{11} (dB)

$E = -10$ kV/cm 0 10 kV/cm

4950 5000 5050

Frequency (MHz)

FIGURE 8.12 (a) Schematic diagram of a YIG–PZT microstrip resonator. (b) Electric field tuning of FMR for H perpendicular to the sample plane. The frequency shift is ~22.5 MHz for $E = 10$ kV/cm and ~24.5 MHz for $E = -10$ kV/cm. (From Fetisov, Y.K. and Srinivasan, G., *Appl. Phys. Lett.*, 88, 143503, 2006.)

Figure 8.12b shows its tunable response (a shift of FMR) through ME interaction. A frequency shift of ~22.5 MHz was obtained under an electric field of $E = 10$ kV/cm. Upon reversal of E, the FMR shift is also reversed. With a proper choice of the piezoelectric phase, a tuning range of 0.5–1 GHz is quite possible [1,95]. Such tunability could form the basis for miniature microwave resonators and filters that are compatible with integrated circuit technology.

8.3.2 Filters

Band-pass and band-stop filters are widely used in RF/microwave communication and signal-processing systems. The development of tunable filters is of great technological importance. Microwave ME resonators possess the unique capability of being electrically tunable with large bandwidth, making them perfect candidates for electric field–tunable filters [1,95].

8.3.2.1 Band-Pass Filter

The design and analysis of an electric field–tunable microwave band-pass filter was first reported by Srinivasan et al. in 2005 [106]. The filter is based on the FMR of the ferrite component in (111) YIG/(001) PMN-PT, which can be tuned over a wide frequency range with a voltage applied to the piezoelectric component. Theoretical estimation shows that, under an electric field of 30 kV/cm, the mechanical deformation due to piezoelectric effect coupled to the ferrite layer can tune the FMR by as much as 420 MHz. The filter is predicted to

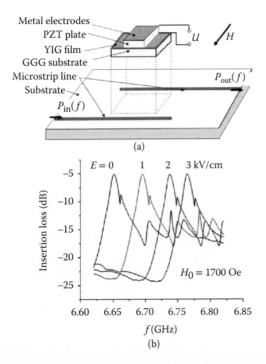

FIGURE 8.13 (a) Schematic of the YIG–PZT ME band-pass filter. (b) Corresponding transmission characteristic of the YIG–PZT filter. (From Tatarenko, A.S. et al., *Electron. Lett.*, 42, 540–541, 2006.)

have a bandwidth of 80 MHz. Subsequently, a YIG–PZT microstrip band-pass filter was designed and characterized, as shown in Figure 8.13 [107]. A magnetic field was applied parallel to the sample plane and perpendicular to the microstrips. When an input signal was applied to the filter, the transmitted power was tuned through the ME resonator under an applied electric field. Figure 8.13b shows the frequency dependence of the insertion loss L as a function of applied electric field, with an up shift in resonance frequency (up to 2% tunability under $E = 3$ kV/cm) [107]. This E-field resonant frequency tunability was further improved to 10% for a YIG/PZN-PT-based ME bandpass filter [95].

8.3.2.2 Band-Stop Filter

The FMR absorption of the magnetic field-tunable YIG-based resonator has also been used to achieve a tunable band-stop filter [108]. Accordingly, the YIG–PZT ME resonator combined with the broadband air-gap microstrip (Figure 8.14a) can act as an electric field–tunable band-stop filter [109]. Such a filter was demonstrated with a peak attenuation greater than 50 dB and a 40 dB rejection bandwidth of ~13 MHz, as shown in Figure 8.14b. The tunable range is ~30 MHz and can be further enhanced by optimizing the coupling coefficient of the ME heterostructure. These voltage-tunable band-stop filters would facilitate a more compact, lightweight, and power-efficient integrated circuits.

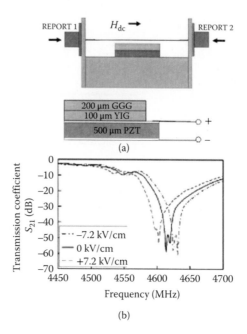

FIGURE 8.14 (a) Schematic diagram of the band-stop filter. (b) Transmission S_{21} of tuned microwave band-stop filter. (From Pettiford, C. et al., *IEEE Trans. Magn.*, 43, 3343–3345 © 2007 IEEE.)

8.3.3 Phase Shifter and Delay Lines

Traditional ferrite phase shifters using magnetic tuning systems are slow, demand high power, and are bulky. Therefore, it is of great importance to develop tunable microwave phase shifters [110–112] for miniature oscillators and phased array antenna systems. In this respect, ME resonators can be useful for the voltage control of the phase by tuning the propagation of magnetostatic waves (MSW) [110], hybrid spin-electromagnetic waves [112], and ferromagnetic resonance [111] in the ferrite.

Figure 8.15a shows an MSW phase shifter that uses a YIG–PZT ME resonator [110]. Two microstrip transducers on an alumina substrate are used to excite and detect the MSW. The resonator is positioned on top of the transducers and subjected to an external magnetic field. A near-linear phase shift of the order of 1 radian was obtained under electric field due to the ME interaction, as shown in Figure 8.15b. A dual-field tunable ME phase shifter based on the propagation of hybrid spin-electromagnetic waves in a bilayer ME composite was also reported [112]. The bilayer composite consists of a barium strontium titanate (BST) slab and a YIG film, as shown in Figure 8.15c. Under an electric field, a continuously variable differential phase shift as high as 650° was achieved (Figure 8.15d), which was attributed to change in the dielectric constant of BST.

Furthermore, the YIG-ferroelectric-based MSW phase shifter as illustrated in Figure 8.15a can also act as a delay line with voltage-controlable delay time [113]. Under an electric field of 8 kV/cm, a decrease in delay time of about 25%

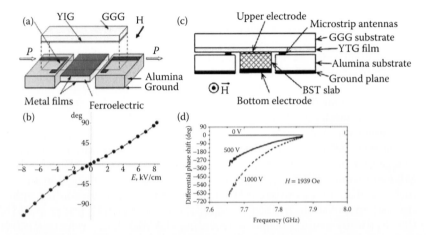

FIGURE 8.15 (a) Schematic of an MSW phase shifter based on YIG–PZT ME resonator. (b) Phase shift at 4 GHz against electric field E across PZT with $H_{dc} = 750$ Oe [110]. (c) Schematic diagram of the YIG-BST-based hybrid spin-electromagnetic wave phase shifter. (d) Electric field–induced differential phase shift of microwave signal. (From Ustinov, A.B. et al., *Appl. Phys. Lett.*, 90, 031913, 2007.)

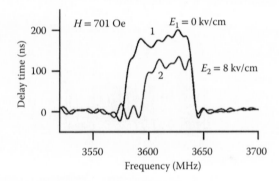

FIGURE 8.16 Delay time as a function of frequency for a YIG/PMN-PT-based delay line under different electric fields. (From Fetisov, Y.K. and Srinivasan, G., *Appl. Phys. Lett.*, 87, 103502, 2005.)

was obtained within the transmission band using YIG/PMN-PT, as shown in Figure 8.16. This tunability is attributed to the change in dielectric constant of PMN-PT and its effect on hybrid spin-electromagnetic waves.

8.3.4 ME Memories

Ferroelectric polarization and magnetization are used to store binary information in ferroelectric random access memory (FeRAM) [114] and magnetic random access memory (MRAM), respectively [115,116]. However, they each have limitations, such as destructive read operation [114], low read/write speed, and high power consumption [117]. A memory device that combines the best of

FeRAM and MRAM is of great importance. In this regard, the ME coupling in multiferroic materials offers unique possibilities for novel device designs.

8.3.4.1 Electrically Assisted Magnetic Recording

In general, commercial magnetic storage devices use a magnetic field to write the data by changing the magnetization direction [115,116]. However, this is usually accompanied by high energy consumption and low speed (due to the need of mechanically relocating the writing head). To enable magnetic recording with a small field, a focused laser beam can be used to locally heat the media to decrease the required magnetic field for magnetization direction change in a heat-assisted magnetic recording (HAMR) [118]. Alternatively, ME nano-composites consisting of magnetic nanopillars embedded in a ferroelectric matrix may also be used to enable low-field magnetic recording with the assistance of an electric field [119,120]. Zavaliche et al. [121] first proposed the electrically assisted magnetic recording (EAMR) device (Figure 8.17a) and experimentally demonstrated its capability using self-assembled $BiFeO_3$–$CoFe_2O_4$ vertical heterostructure epitaxial films. The magnetization of each pillar was selectively switched by combining a weak, uniform magnetic field with the assisting electric field. Compared with HAMR, EAMR device has the advantage of low power consumption, high density, and good stability. On the other hand, instead of using the magnetization, Li et al. [122] demonstrated

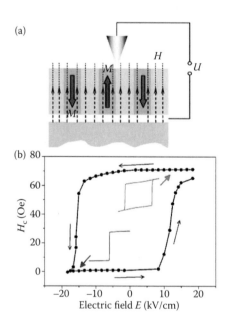

FIGURE 8.17 (a) Schematic of electric field–assisted magnetic recording, where the magnetization up and down directions represent "0" and "1" (From Zavaliche, F. et al,. *Nano Lett.*, 7, 1586–1590, 2007.). (b) Electric-field modulation of the magnetic coercive field H_c; the inset shows two typical Kerr hysteresis loops with two different H_c states. (From Li, Z. et al., *Appl. Phys. Lett.*, 96, 162505, 2010.)

an $FeGa/BiScO_3$-$PbTiO_3$-based ME bilayer memory cell in which the electric field induced two different coercive fields (i.e., low-H_c, high-H_c) that can be used as data bits. Figure 8.17b shows the coercive field variation as a function of applied electric field, which presents a hysteresis loop with two completely different states (low-H_c or high-H_c) at zero field. Thus, an electric-write/magnetic-read memory with two different H_c states was realized.

8.3.4.2 ME Random Access Memories

Multiferroics that possess spontaneous electric polarization (P) and magnetization (M) with a strong coupling between them enable the design of multistate memory devices [117,119,120]. In such a device, the available four states can be defined as (+P,+M), (+P,–M), (–P,+M), and (–P,–M), as shown in Figure 8.18a [114]. However, these states may not be independent of each other and combinations that are independently accessible are either (+P,+M) and (–P,–M)

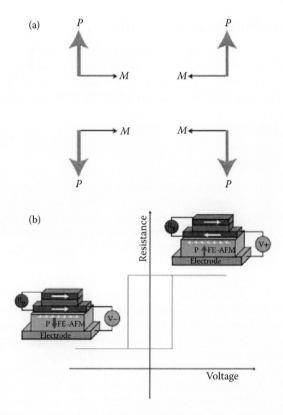

FIGURE 8.18 (a) The four states of a multiferroic tunnel junction. (From Scott, J.F., *Nat. Mater.*, 6, 256–257, 2007.) (b) By combining a magnetic tunnel junction with multiferroic, a novel memory device can be achieved, where the resistance across the magnetic tunnel junction can be controlled by an electric field applied to the multiferroic layer. (From Bibes, M. and Barthélémy, A., *Nat. Mater.*, 7, 425–426, 2008.)

or (+P,–M) and (–P,+M) [114]. Furthermore, there are no known single-phase multiferroic materials so far with larger polarization and magnetization available at room temperature. To address these challenges, various device designs have been proposed. Gajek et al. first demonstrated a four-state resistive memory device by using a multiferroic tunneling junction [123]. In this work, a thin layer of multiferroic $La_{0.1}Bi_{0.9}MnO_3$ (LBMO) was used as the barrier. The LBMO ferromagnetism permits read operations similar to that of MRAM, the electrical switching evokes a ferroelectric write operation. Subsequently, Bibes et al. proposed a different design, in which a spin-valve or magnetic tunnel junction is prepared on a multiferroic layer, as shown in Figure 8.18b [124]. The ME coupling enables an electric field to control the interface exchange coupling, which then controls the magnetization of the ferromagnetic layer, leading to resistance change of the tunnel junction [117,119,120]. A detailed discussion on this topic can be found in Chapter 5 of this book.

8.4 Future Potential Applications

Although prototype devices of ME composite–based magnetic sensor, energy harvesters, transformers, tunable devices, filters, and phase shifters have been demonstrated, the development and practical applications of functional ME devices still have a long way to go. In particular, more attention should be paid to the following three areas: (1) Self-biased ME composites, (2) MEMS scale ME devices, and (3) Flexible ME composites.

8.4.1 Self-Biased ME Devices

Conventional ME composites usually have limited bandwidth near the resonance frequency and need an optimized dc bias (H_{dc}) for maximum coupling coefficient. Small deviation from the resonance frequency or optimized H_{dc} leads to dramatic decrease in ME voltage output and unstable power output. In order to address this issue, great efforts have been devoted to reduce H_{dc}. In certain cases, researchers were able to reduce the optimum H_{dc} from 6.8 kOe to 5 Oe by optimizing the composition and geometry of the magnetostrictive phase [1,17,25]. However, the requirement for H_{dc} still hinders their implementation in miniaturized devices. Therefore, ME composites with large α_{ME} in the absence of H_{dc}, namely "self-biased ME effect," have been developed in various configurations using different strategies. Figure 8.19 illustrates the ME coefficient versus. magnetic field curves of conventional and different types of self-biased ME composites.

The reported self-biased ME composites can be classified into five different groups: (1) functional graded ferromagnetic based [125–129], (2) exchange bias based [130], (3) low-field magnetostriction hysteresis based [76], (4) build-in stress based [131], and (5) nonlinear based [132]. Each system has different working mechanisms and characteristic interactions. However, they all offer considerable potential for future ME-based devices because of the ease of device miniaturization and reduced electromagnetic noise due to the elimination of the H_{dc} source [76,130,133]. Furthermore, each of them also has their unique advantages. For example, the nonlinear-based self-biased

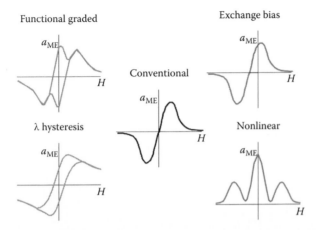

Functional graded

Exchange bias

Conventional

λ hysteresis

Nonlinear

FIGURE 8.19 Schematic diagrams of ME coefficient versus magnetic field curve for different types of self-biased ME composites and conventional ME composite.

ME composite can reduce vibrational and $1/f$ noise by frequency modulation [134,135]; the exchange bias based system can be used as vector magnetometer because of the angular dependence [136]. They can also enhance H_{dc} stability by using tunable self-biased effect [137]. A variety of novel ME devices have been developed including self-biased sensors [135,138], energy harvesters [74,75], and memories [139]. However, a great deal of work still needs to be done to expand the utilization of self-biased ME composites.

8.4.2 MEMS Devices

Most of the ME devices discussed in this chapter are bulky. Only a few have been demonstrated in the form of nanostructured thin films. The investigation of film-based ME devices are critically important in particular for sensors and high-frequency device applications. Among them, ME composites–based microelectromechanical systems (MEMS) is of great importance [136,140]. Recent investigations on ME multilayers and epitaxial heterostructures showed very promising features [119,136,141,142]. Thin-film ME composites benefit from a strong coupling between consecutive phases and reduced size. Furthermore, thin-film cantilevers can be designed to show low resonance frequencies with ultra-high sensitivity [136,142–144].

For example, a 2-2 type thin-film heterostructure FeCoSiB (1.75 μm)/AlN (1.8 μm) was fabricated using magnetron sputtering on Si substrate in the form of a cantilever [145]. After magnetic field annealing, an extremely large ME coefficient of 737 V/cm Oe at the resonance frequency of 753 Hz and 3.1 V/cm Oe at 100 Hz with a H_{dc} of 6 Oe was observed (Figure 8.20a). Such small and low frequency resonant ME sensors can be used in sensor arrays for the detection of low frequency signals such as magnetoencephalography and magnetocardiography. They can be further expanded by using exchange bias–based self-biased ME multilayers on piezoelectric cantilever substrates [136]. Meanwhile,

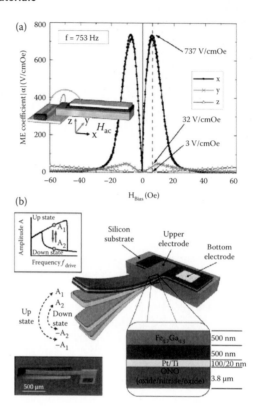

FIGURE 8.20 (a) Schematic and ME coefficient of a cantilever ME laminate composite made of FeCoSiB (1.75 μm)/AlN (1.8 μm). (From Jahns, R. et al., *J. Am. Ceram. Soc.*, 96, 1673–1681, 2013.) (b) Schematic and SEM image of the ME cantilever with mechanical bistable states as a function of driving frequency. (From Onuta, T.-D. et al., *Adv. Mater.*, 27, 202–206, 2014.)

highly efficient miniaturized energy harvesters using thin-film ME structures on Si-MEMS cantilevers have been developed [146]. The devices are built on a silicon oxide/nitride/oxide stack with a free-standing PZT/$Fe_{0.7}Ga_{0.3}$ bilayer heterostructure on top. At the resonant frequency (3.8 kHz) with a load of 12.5 kΩ, this MEMS harvester was able to generate a peak power density of 0.7 mW/cm^3 (RMS) at 1 Oe. Furthermore, using a similar structure, the ME cantilever was demonstrated as a dynamic memory with bistable states that can be reversibly switched under dc magnetic or electric field by operating in the nonlinear regime, as shown in Figure 8.20b [147]. These prototypes demonstrated that the MEMS-based systems offer a variety of multifunctionalities including sensing, actuation, energy conversion devices, and memory.

8.4.3 Flexible ME Devices

Despite the large ME coefficient and high sensitivity achieved in bulk and MEMS-based ME composites, the rigidity and brittleness of the ceramic

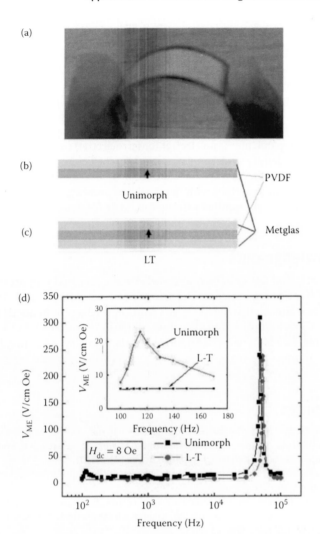

FIGURE 8.21 (a) Picture of a flexible Metglas–PVDF unimorph laminate, schematics of (b) the unimorph configuration and (c) the three layer sandwich configuration. (d) Frequency dependence of the ME voltage coefficient of both laminates. (From Zhai, J. et al., *Appl. Phys. Lett.*, 89, 083507, 2006.)

piezoelectric component is always a concern. It not only hinders the processing of complex ME laminate structures, but also leads to a high cost and is difficult to integrate with other devices. Thus, from a practical point of view, it is important to investigate highly flexible ME composites and their potential applications in wearable devices. Various types of polymer-based ME composites that are highly flexible have been developed [148]. Polyvinylidene fluoride (PVDF) and its copolymer poly(vinylidene fluoride-trifluoroethylene) (P(VDF-TrFE)) are the most widely used piezoelectric polymers due to their relatively large piezoelectric responses and commercial availability [149]. For example, Zhai et al. developed a thin (<100 μm) and flexible Metglas–PVDF ME laminate in both unimorph and bimorph configurations (Figure 8.21a through c) [150]. Besides being flexible, the composite also showed a giant ME voltage coefficient of 7.2 V/cm Oe at low frequencies, and up to 310 V/cm Oe at resonance frequency, leading to high sensitivity, as shown in Figure 8.21d. Alternatively, Le et al. [151] developed a flexible current sensor based on a bilayer structure consisting of Cytop polymer and a magnetic tape filled with magnetically soft particles. The Cytop polymer is an electret (or ferroelectret) material that exhibits a high charge density and large piezoelectric coefficient (25–700 pC/N). It was found to possess a large ME coefficient of 4.58 V/cm Oe at 1 kHz. These preliminary results, together with the flexibility and ease of fabrication make polymer-based ME composites a strong candidate for sensors and actuators.

Acknowledgments

Authors Y.Z. and S.P. gratefully acknowledge the financial support from the Office of Basic Energy Science, Department of Energy (DOE) and Air Force Office of Scientific Research (AFOSR). J.R. acknowledges the financial support from the Global Frontier R&D Program (2013M3A6B1078872) on Center for Hybrid Interface Materials (HIM) funded by the Ministry of Science, ICT and Future Planning of Korea.

References

1. Nan, C. W., M. I. Bichurin, S. Dong, D. Viehland, and G. Srinivasan. 2008. Multiferroic magnetoelectric composites: Historical perspective, status, and future directions. *Journal of Applied Physics* 103:031101.
2. Eerenstein, W., N. D. Mathur, and J. F. Scott. 2006. Multiferroic and magneto-electric materials. *Nature* 442:759–765.
3. Schmid, H. 1994. Multi-ferroic magnetoelectrics. *Ferroelectrics* 162:317.
4. Fiebig, M. 2005. Revival of the magnetoelectric effect. *Journal of Physics D-Applied Physics* 38:R123–R152.
5. Debye, P. 1926. Remark to some new trials on a magneto-electrical direct effect. *Zeitschrift Fur Physik* 36:300–301.
6. Buchel'nikov, V. D., V. S. Romanov, and V. G. Shavrov. 1998. Oscillating polaritons in antiferromagnetics with the magnetoelectric effect. *Journal of Communications Technology and Electronics* 43:80–84.
7. Folen, V. J., G. T. Rado, and E. W. Stalder. 1961. Anisotropy of the magnetoelectric effect in Cr_2O_3. *Physical Review Letters* 6:607–608.
8. Priya, S., R. Islam, S. Dong, and D. Viehland. 2007. Recent advancements in magnetoelectric particulate and laminate composites. *Journal of Electroceramics* 19:149–166.
9. Prellier, W., M. P. Singh, and P. Murugavel. 2005. The single-phase multiferroic oxides: From bulk to thin film. *Journal of Physics Condensed Matter* 17:R803–R832.
10. Spaldin, N. A., and M. Fiebig. 2005. The renaissance of magnetoelectric multiferroics. *Science* 309:391–392.

11. Suchtelen, J. V. 1972. Product properties: A new application of composite materials. *Philips research reports* 27:28–37.
12. Van den Boomgaard, J., D. R. Terrell, R. A. J. Born, and H. Giller. 1974. An *in situ* grown eutectic magnetoelectric composite material. I. Composition and unidirectional solidification. *Journal of Materials Science* 9:1705–1709.
13. Van den Boomgaard, J., A. Vanrun, and J. V. Suchtelen. 1976. Magnetoelectricity in piezoelectric-magnetostrictive composites. *Ferroelectrics* 10:295–298.
14. Van den Boomgaard, J., and R. A. J. Born. 1978. A sintered magnetoelectric composite material $BaTiO_3$-Ni(Co, Mn)Fe_2O_4. *Journal of Materials Science* 13:1538–1548.
15. Ryu, J., A. V. Carazo, K. Uchino, H. E. Kim. 2001. Magnetoelectric properties in piezoelectric and magnetostrictive laminate composites. *Japan Journal of Applied Physics* 40:4948–4951.
16. Ramesh, R., and N. A. Spaldin. 2007. Multiferroics: Progress and prospects in thin films. *Nature Materials* 6:21–29.
17. Srinivasan, G. 2010. Magnetoelectric composites. *Annual Review of Materials Research* 40:153–178.
18. Zhai, J., Z. Xing, S. Dong, J. Li, and D. Viehland. 2008. Magnetoelectric laminate composites: An overview. *Journal of the American Ceramic Society* 91:351–358.
19. Wang, Y. J., D. Gray, D. Berry, J. Q. Gao, M. H. Li, J. F. Li, and D. Viehland. 2011. An extremely low equivalent magnetic noise magnetoelectric sensor. *Advanced Materials* 23:4111–4114.
20. Dong, S., J. Y. Zhai, J. F. Li, and D. Viehland. 2006. Small dc magnetic field response of magnetoelectric laminate composites. *Applied Physics Letters* 88:082907.
21. Cho, K.-H., and S. Priya. 2011. Direct and converse effect in magnetoelectric laminate composites. *Applied Physics Letters* 98:232904.
22. Priya, S., J. Ryu, C. S. Park, J. Oliver, J. J. Choi, and D. S. Park. 2009. Piezoelectric and magnetoelectric thick films for fabricating power sources in wireless sensor nodes. *Sensors (Basel)* 9:6362–6384.
23. Palneedi, H., A. Annapureddy, S. Priya, and J. Ryu, 2016. Status and perspectives of multiferroic magnetoelectric composite materials and applications. *Actuators* 5:9.
24. Dong, S., J. Zhai, F. Bai, J. F. Li, and D. Viehland. 2005. Push-pull mode magnetostrictive/piezoelectric laminate composite with an enhanced magnetoelectric voltage coefficient. *Applied Physics Letters* 87:062502.
25. Dong, S., J. Zhai, J. Li, and D. Viehland. 2006. Near-ideal magnetoelectricity in high-permeability magnetostrictive/piezofiber laminates with a (2-1) connectivity. *Applied Physics Letters* 89:252904.
26. Gao, J. Q., D. Gray, Y. Shen, J. F. Li, and D. Viehland. 2011. Enhanced dc magnetic field sensitivity by improved flux concentration in magnetoelectric laminates. *Applied Physics Letters* 99:153502.
27. Gao, J., L. Shen, Y. Wang, D. Gray, J. Li, and D. Viehland. 2011. Enhanced sensitivity to direct current magnetic field changes in Metglas/Pb($Mg_{1/3}Nb_{2/3}$)O_3–$PbTiO_3$ laminates. *Journal of Applied Physics* 109:074507.
28. Xing, Z., S. Dong, J. Zhai, L. Yan, J. Li, and D. Viehland. 2006. Resonant bending mode of Terfenol-D/steel/Pb(Zr,Ti)O_3 magnetoelectric laminate composites. *Applied Physics Letters* 89:112911.
29. Xing, Z., J. Li, and D. Viehland. 2008. Giant magnetoelectric effect in Pb(Zr,Ti)O_3-bimorph/NdFeB laminate device. *Applied Physics Letters* 93:013505.

30. Liu, G., X. Li, J. Chen, H. Shi, W. Xiao, and S. Dong. 2012. Colossal low-frequency resonant magnetomechanical and magnetoelectric effects in a three-phase ferromagnetic/elastic/piezoelectric composite. *Applied Physics Letters* 101:142904.

31. Busatto, G., R. La Capruccia, F. Iannuzzo, F. Velardi, and R. Roncella. 2003. MAGFET based current sensing for power integrated circuit. *Microelectronics Reliability* 43:577–583.

32. Bai, J. G., G. Q. Lu, and T. Lin. 2003. Magneto-optical current sensing for applications in integrated power electronics modules. *Sensors and Actuators a-Physical* 109:9–16.

33. Dong, S., J. F. Li, and D. Viehland. 2004. Circumferentially magnetized and circumferentially polarized magnetostrictive/piezoelectric laminated rings. *Journal of Applied Physics* 96:3382–3387.

34. Dong, S., J. G. Bai, J. Y. Zhai, J. F. Li, G. Q. Lu, D. Viehland, S. J. Zhang, and T. R. Shrout. 2005. Circumferential-mode, quasi-ring-type, manetoelectric laminate composite—a highly sensitive electric current and/or vortex magnetic field sensor. *Applied Physics Letters* 86:182506.

35. Leung, C. M., S. W. Or, S. Y. Zhang, and S. L. Ho. 2010. Ring-type electric current sensor based on ring-shaped magnetoelectric laminate of epoxy-bonded $Tb_{0.3}Dy_{0.7}Fe_{1.92}$ short-fiber/NdFeB magnet magnetostrictive composite and $Pb(Zr, Ti)O_3$ piezoelectric ceramic. *Journal of Applied Physics* 107:09D918.

36. Wang, Y. J., J. Q. Gao, M. H. Li, Y. Shen, D. Hasanyan, J. F. Li, and D. Viehland. 2014. A review on equivalent magnetic noise of magnetoelectric laminate sensors. *Philosophical Transactions. Series A, Mathematical, Physical, and Engineering Sciences* 372:20120455.

37. Xing, Z. P., J. Y. Zhai, J. F. Li, and D. Viehland. 2009. Investigation of external noise and its rejection in magnetoelectric sensor design. *Journal of Applied Physics* 106:024512.

38. Zhai, J. Y., Z. P. Xing, S. Dong, J. F. Li, and D. Viehland. 2008. Thermal noise cancellation in symmetric magnetoelectric bimorph laminates. *Applied Physics Letters* 93:072906.

39. Israel, C., S. Kar-Narayan, and N. D. Mathur. 2010. Eliminating the temperature dependence of the response of magnetoelectric magnetic-field sensors. *IEEE Sensors Journal* 10:914–917.

40. Xing, Z. P., J. F. Li, and D. Viehland. 2007. Noise and scale effects on the signal-to-noise ratio in magnetoelectric laminate sensor/detection units. *Applied Physics Letters* 91:182902.

41. Gao, J. Q., Y. J. Wang, M. H. Li, Y. Shen, J. F. Li, and D. Viehland. 2012. Quasi-static ($f < 10^{-2}$ Hz) frequency response of magnetoelectric composites based magnetic sensor. *Materials Letters* 85:84–87.

42. Wang, Y., J. Li, and D. Viehland. 2014. Magnetoelectrics for magnetic sensor applications: Status, challenges and perspectives. *Materials Today* 17:269–275.

43. Wang, Y., D. Gray, J. Gao, D. Berry, M. Li, J. Li, D. Viehland, and H. Luo. 2012. Improvement of magnetoelectric properties in Metglas/$Pb(Mg_{1/3}Nb_{2/3})O_3$–$PbTiO_3$ laminates by poling optimization. *Journal of Alloys and Compounds* 519:1–3.

44. Xiang, Y., R. Zhang, and W. Cao. 2011. Poling field versus piezoelectric property for [001] c oriented 91%$Pb(Zn_{1/3}Nb_{2/3})O_3$–9%$PbTiO_3$ single crystals. *Journal of Materials Science* 46:1839–1843.

45. Wang, Y., D. Gray, D. Berry, M. Li, J. Gao, J. Li, and D. Viehland. 2012. Influence of interfacial bonding condition on magnetoelectric properties in piezofiber/Metglas heterostructures. *Journal of Alloys and Compounds* 513:242–244.

46. Li, M. H., D. Berry, J. Das, D. Gray, J. F. Li, and D. Viehland. 2011. Enhanced sensitivity and reduced noise floor in magnetoelectric laminate sensors by an improved lamination process. *Journal of the American Ceramic Society* 94:3738–3741.

47. Apo, D. J., M. Sanghadasa, and S. Priya. 2014. Vibration modeling of arc-based cantilevers for energy harvesting applications. *Energy Harvesting and Systems* 1:1–12.

48. Beeby, S. P., M. J. Tudor, and N. M. White. 2006. Energy harvesting vibration sources for microsystems applications. *Measurement Science and Technology* 17:R175–R195.

49. Paradiso, J. A., and T. Starner. 2005. Energy scavenging for mobile and wireless electronics. *IEEE Pervasive Computing* 4:18–27.

50. Anton, S. R., and H. A. Sodano. 2007. A review of power harvesting using piezoelectric materials (2003–2006). *Smart Materials and Structures* 16:R1–R21.

51. Sodano, H. A., D. J. Inman, and G. Park. 2004. A review of power harvesting from vibration using piezoelectric material. *The Shock and Vibration Digest* 36:197–205.

52. Roundy, S., and P. K. Wright. 2004. A piezoelectric vibration based generator for wireless electronics. *Smart Materials and Structures* 13:1131–1142.

53. Tadesse, Y., S. J. Zhang, and S. Priya. 2009. Multimodal energy harvesting system: Piezoelectric and electromagnetic. *Journal of Intelligent Material Systems and Structures* 20:625–632.

54. Bai, X., Y. Wen, J. Yang, P. Li, J. Qiu, and Y. Zhu. 2012. A magnetoelectric energy harvester with the magnetic coupling to enhance the output performance. *Journal of Applied Physics* 111:07A938.

55. Zhu, Y., and J. W. Zu. 2012. A magnetoelectric generator for energy harvesting from the vibration of magnetic levitation. *IEEE Transactions on Magnetics* 48:3344–3347.

56. Xing, X., J. Lou, G. M. Yang, O. Obi, C. Driscoll, and N. X. Sun. 2009. Wideband vibration energy harvester with high permeability magnetic material. *Applied Physics Letters* 95:134103.

57. Wang, L., and F. G. Yuan. 2008. Vibration energy harvesting by magnetostrictive material. *Smart Materials and Structures* 17:045009.

58. Khaligh, A., P. Zeng, and C. Zheng. 2010. Kinetic energy harvesting using piezoelectric and electromagnetic technologies-state of the art. *IEEE Transactions on Industrial Electronics* 57:850–860.

59. Li, P., Y. Wen, J. Chaobo, and L. Xinshen. 2011. A magnetoelectric composite energy harvester and power management circuit. *IEEE Transactions on Industrial Electronics* 58:2944–2951.

60. Li, P., Y. M. Wen, and L. X. Bian. 2007. Enhanced magnetoelectric effects in composite of piezoelectric ceramics, rare-earth iron alloys, and ultrasonic horn. *Applied Physics Letters* 90:022503.

61. Gao, J. Q., D. Hasanyan, Y. Shen, Y. J. Wang, J. F. Li, and D. Viehland. 2012. Giant resonant magnetoelectric effect in bi-layered Metglas/Pb(Zr,Ti)O$_3$ composites. *Journal of Applied Physics* 112:104101.

62. Dong, S., J. Y. Zhai, J. F. Li, D. Viehland, and S. Priya. 2008. Multimodal system for harvesting magnetic and mechanical energy. *Applied Physics Letters* 93:103511.

63. Ryu, J., J.-E. Kang, Y. Zhou, S.-Y. Choi, W.-H. Yoon, D.-S. Park, J.-J. Choi, B.-D. Hahn, C.-W. Ahn, J.-W. Kim, Y.-D. Kim, S. Priya, S. Y. Lee, S. Jeong, and D.-Y. Jeong. 2015. Ubiquitous magneto-mechano-electric generator. *Energy and Environmental Science* 8:2402–2408.

64. Priya, S. 2007. Advances in energy harvesting using low profile piezoelectric transducers. *Journal of Electroceramics* 19:167–184.

65. Dai, X., Y. Wen, P. Li, J. Yang, and G. Zhang. 2009. Modeling, characterization and fabrication of vibration energy harvester using Terfenol-D/PZT/Terfenol-D composite transducer. *Sensors and Actuators a-Physical* 156:350–358.

66. Dai, X., Y. Wen, P. Li, J. Yang, and M. Li. 2011. Energy harvesting from mechanical vibrations using multiple magnetostrictive/piezoelectric composite transducers. *Sensors and Actuators a-Physical* 166:94–101.

67. Li, M., Y. Wen, P. Li, J. Yang, and X. Dai. 2011. A rotation energy harvester employing cantilever beam and magnetostrictive/piezoelectric laminate transducer. *Sensors and Actuators a-Physical* 166:102–110.

68. Yang, J., Y. Wen, and P. Li. 2011. Magnetoelectric energy harvesting from vibrations of multiple frequencies. *Journal of Intelligent Material Systems and Structures* 22:1631–1639.

69. Yang, J., Y. Wen, P. Li, and X. Bai. 2011. A magnetoelectric-based broadband vibration energy harvester for powering wireless sensors. *Science China Technological Sciences* 54:1419–1427.

70. Moss, S. D., J. E. McLeod, I. G. Powlesland, and S. C. Galea. 2012. A bi-axial magnetoelectric vibration energy harvester. *Sensors and Actuators A-Physical* 175:165–168.

71. Ju, S., S. H. Chae, Y. Choi, S. Lee, H. W. Lee, and C.-H. Ji. 2013. A low frequency vibration energy harvester using magnetoelectric laminate composite. *Smart Materials and Structures* 22:115037.

72. Lafont, T., L. Gimeno, J. Delamare, G. A. Lebedev, D. I. Zakharov, B. Viala, O. Cugat, N. Galopin, L. Garbuio, and O. Geoffroy. 2012. Magnetostrictive-piezoelectric composite structures for energy harvesting. *Journal of Micromechanics and Microengineering* 22:094009.

73. Kambale, R. C., W.-H. Yoon, D.-S. Park, J.-J. Choi, C.-W. Ahn, J.-W. Kim, B.-D. Hahn, D.-Y. Jeong, B. Chul Lee, G.-S. Chung, and J. Ryu. 2013. Magnetoelectric properties and magnetomechanical energy harvesting from stray vibration and electromagnetic wave by $Pb(Mg_{1/3}Nb_{2/3})O_3$-$Pb(Zr,Ti)O_3$ single crystal/Ni cantilever. *Journal of Applied Physics* 113:204108.

74. Zhou, Y., D. J. Apo, and S. Priya. 2013. Dual-phase self-biased magnetoelectric energy harvester. *Applied Physics Letters* 103:192909.

75. Patil, D. R., Y. Zhou, J. E. Kang, N. Sharpes, D. Y. Jeong, Y. D. Kim, K. H. Kim, S. Priya, and J. Ryu. 2014. Anisotropic self-biased dual-phase low frequency magneto-mechano-electric energy harvesters with giant power densities. *APL Materials* 2:046102.

76. Zhou, Y., S. C. Yang, D. J. Apo, D. Maurya, and S. Priya. 2012. Tunable self-biased magnetoelectric response in homogenous laminates. *Applied Physics Letters* 101:232905.

77. Dong, S., J. F. Li, and D. Viehland. 2004. Voltage gain effect in a ring-type magnetoelectric laminate. *Applied Physics Letters* 84:4188–4190.

78. Dong, S., J. F. Li, D. Viehland, J. Cheng, and L. E. Cross. 2004. A strong magnetoelectric voltage gain effect in magnetostrictive-piezoelectric composite. *Applied Physics Letters* 85:3534–3536.

79. Islam, R. A., H. Kim, S. Priya, and H. Stephanou. 2006. Piezoelectric transformer based ultrahigh sensitivity magnetic field sensor. *Applied Physics Letters* 89:152908.

80. Kim, H., R. A. Islam, and S. Priya. 2007. Working principle of voltage controlled differential magnetic field sensor. *Applied Physics Letters* 90:012909.

81. Jia, Y. M., H. S. Luo, X. Y. Zhao, and F. F. Wang. 2008. Giant magnetoelectric response from a piezoelectric/magnetostrictive laminated composite combined with a piezoelectric transformer. *Advanced Materials* 20:4776–4779.

82. Dong, S., J. Y. Zhai, S. Priya, J. F. Li, and D. Viehland. 2009. Tunable features of magnetoelectric transformers. *IEEE Transactions on Ultrasonics Ferroelectrics and Frequency Control* 56:1124–1127.

83. Leung, C. M., S. W. Or, F. F. Wang, and S. L. Ho. 2011. Dual-resonance converse magnetoelectric and voltage step-up effects in laminated composite of long-type $0.71Pb(Mg_{1/3}Nb_{2/3})O_3$-$0.29PbTiO_3$ piezoelectric single-crystal transformer and $Tb_{0.3}Dy_{0.7}Fe_{1.92}$ magnetostrictive alloy bars. *Journal of Applied Physics* 109:104103.

84. Dong, S., J. Zhai, J. F. Li, D. Viehland, and M. I. Bichurin. 2006. Magnetoelectric gyration effect in $Tb_{1-x}Dy_xFe_{2-y}/Pb(Zr,Ti)O_3$ laminated composites at the electromechanical resonance. *Applied Physics Letters* 89:243512.

85. Zhai, J., J. Li, S. Dong, D. Viehland, and M. I. Bichurin. 2006. A quasi (unidirectional) Tellegen gyrator. *Journal of Applied Physics* 100:124509.

86. Zhou, Y., Y. Yan, and S. Priya. 2014. Co-fired magnetoelectric transformer. *Applied Physics Letters* 104:232906.

87. Priya, S., S. Ural, H. W. Kim, K. Uchino, and T. Ezaki. 2004. Multilayered unipoled piezoelectric transformers. *Japanese Journal of Applied Physics* 43:3503–3510.

88. Priya, S. 2006. High power universal piezoelectric transformer. *IEEE Transactions on Ultrasonics Ferroelectrics and Frequency Control* 53:23–29.

89. Kamentsev, K. E., Y. K. Fetisov, and G. Srinivasana. 2006. Low-frequency nonlinear magnetoelectric effects in a ferrite-piezoelectric multilayer. *Applied Physics Letters* 89:142510.

90. Ma, J., Z. Li, Y. H. Lin, and C. W. Nan. 2011. A novel frequency multiplier based on magneto electric laminate. *Journal of Magnetism and Magnetic Materials* 323:101–103.

91. Zhang, W. H., G. Yin, J. W. Cao, J. M. Bai, and F. L. Wei. 2012. Frequency multiplying behavior in a magnetoelectric unimorph. *Applied Physics Letters* 100:032903.

92. Fetisov, L. Y., Y. K. Fetisov, G. Sreenivasulu, and G. Srinivasan. 2013. Nonlinear resonant magnetoelectric interactions and efficient frequency doubling in a ferromagnetic-ferroelectric layered structure. *Journal of Applied Physics* 113:116101.

93. Sundaresan, V. B., J. Atulasimha, and J. Clarke. 2013. Magnetoelectric surgical tools for minimally invasive surgery. edited by U. S. Patent. USA: Virginia Commonwealth University (Richmond, VA).

94. Vishnu Baba Sundaresan, J. A. 2009. Characterization of ME cantilever for use as an ablation tool in minimally invasive surgery. In *ASME 2009 Conference on Smart Materials, Adaptive Structures and Intelligent Systems* 1: Active Materials, Mechanics and Behavior; Modeling, Simulation and Control. Oxnard, California, USA.

95. Sun, N. X., and G. Srinivasan. 2012. Voltage control of magnetism in multiferroic heterostructures and devices. *Spin* 02:1240004.

96. Lou, J., M. Liu, D. Reed, Y. Ren, and N. X. Sun. 2009. Giant electric field tuning of magnetism in novel multiferroic FeGaB/lead zinc niobate–lead titanate (PZN-PT) heterostructures. *Advanced Materials* 21:4711–4715.

97. Liu, M., O. Obi, J. Lou, Y. Chen, Z. Cai, S. Stoute, M. Espanol, M. Lew, X. Situ, K. S. Ziemer, V. G. Harris, and N. X. Sun. 2009. Giant electric field tuning of magnetic properties in multiferroic ferrite/ferroelectric heterostructures. *Advanced Functional Materials* 19:1826–1831.

98. Fetisov, Y. K., and G. Srinivasan. 2008. Nonlinear electric field tuning characteristics of yttrium iron garnet–lead zirconate titanate microwave resonators. *Applied Physics Letters* 93:033508.

99. Ustinov, A. B., B. A. Kalinikos, V. S. Tiberkevich, A. N. Slavin, and G. Srinivasan. 2008. Q factor of dual-tunable microwave resonators based on yttrium iron garnet and barium strontium titanate layered structures. *Journal of Applied Physics* 103:063908.

100. Ustinov, A. B., V. S. Tiberkevich, G. Srinivasan, A. N. Slavin, A. A. Semenov, S. F. Karmanenko, B. A. Kalinikos, J. V. Mantese, and R. Ramer. 2006. Electric field tunable ferrite-ferroelectric hybrid wave microwave resonators: Experiment and theory. *Journal of Applied Physics* 100(9):093905.

101. Ustinov, A. B., G. Srinivasan, and B. A. Kalinikos. 2008. High-Q active ring microwave resonators based on ferrite-ferroelectric layered structures. *Applied Physics Letters* 92:093905.

102. Semenov, A. A., S. F. Karmanenko, V. E. Demidov, B. A. Kalinikos, G. Srinivasan, A. N. Slavin, and J. V. Mantese. 2006. Ferrite-ferroelectric layered structures for electrically and magnetically tunable microwave resonators. *Applied Physics Letters* 88:033503.

103. Tatarenko, A. S., and M. I. Bichurin. 2012. Microwave magnetoelectric devices. *Advances in Condensed Matter Physics* 2012:1–10.

104. Vaz, C. A. F. 2012. Electric field control of magnetism in multiferroic heterostructures. *Journal of Physics-Condensed Matter* 24:333201.

105. Fetisov, Y. K., and G. Srinivasan. 2006. Electric field tuning characteristics of a ferrite-piezoelectric microwave resonator. *Applied Physics Letters* 88:143503.

106. Srinivasan, G., A. S. Tatarenko, and M. I. Bichurin. 2005. Electrically tunable microwave filters based on ferromagnetic resonance in ferrite-ferroelectric bilayers. *Electronics Letters* 41:596–598.

107. Tatarenko, A. S., V. Gheevarughese, and G. Srinivasan. 2006. Magnetoelectric microwave bandpass filter. *Electronics Letters* 42:540–541.

108. Adam, J. D., L. E. Davis, G. F. Dionne, E. F. Schloemann, and S. N. Stitzer. 2002. Ferrite devices and materials. *IEEE Transactions on Microwave Theory and Techniques* 50:721–737.

109. Pettiford, C., S. Dasgupta, L. Jin, S. D. Yoon, and N. X. Sun. 2007. Bias field effects on microwave frequency behavior of PZT/YIG magnetoelectric bilayer. *IEEE Transactions on Magnetics* 43:3343–3345.

110. Fetisov, Y. K., and G. Srinivasan. 2005. Ferrite/piezoelectric microwave phase shifter: Studies on electric field tunability. *Electronics Letters* 41:1066–1067.

111. Tatarenko, A. S., G. Srinivasan, and M. I. Bichurin. 2006. Magnetoelectric microwave phase shifter. *Applied Physics Letters* 88:183507.

112. Ustinov, A. B., G. Srinivasan, and B. A. Kalinikos. 2007. Ferrite-ferroelectric hybrid wave phase shifters. *Applied Physics Letters* 90:031913.

113. Fetisov, Y. K., and G. Srinivasan. 2005. Electrically tunable ferrite-ferroelectric microwave delay lines. *Applied Physics Letters* 87:103502.

114. Scott, J. F. 2007. Data storage: Multiferroic memories. *Nature Materials* 6:256–257.

115. Prinz, G. A. 1998. Magnetoelectronics. *Science* 282:1660–1663.

116. Chappert, C., A. Fert, and F. N. Van Dau. 2007. The emergence of spin electronics in data storage. *Nature Materials* 6:813–23.

117. Roy, A., R. Gupta, and A. Garg. 2012. Multiferroic memories. *Advances in Condensed Matter Physics* 2012:1–12.

118. Terry, W. M. 2005. Ultimate limits to thermally assisted magnetic recording. *Journal of Physics: Condensed Matter* 17:R315.

119. Ma, J., J. Hu, Z. Li, and C. W. Nan. 2011. Recent progress in multiferroic magneto-electric composites: From bulk to thin films. *Advanced Materials* 23:1062–1087.
120. Wang, Y., J. M. Hu, Y. H. Lin, and C. W. Nan. 2010. Multiferroic magnetoelectric composite nanostructures. *NPG Asia Materials* 2:61–68.
121. Zavaliche, F., T. Zhao, H. Zheng, F. Straub, M. P. Cruz, P. L. Yang, D. Hao, and R. Ramesh. 2007. Electrically assisted magnetic recording in multiferroic nano-structures. *Nano Letters* 7:1586–1590.
122. Li, Z., J. Wang, Y. Lin, and C. W. Nan. 2010. A magnetoelectric memory cell with coercivity state as writing data bit. *Applied Physics Letters* 96:162505.
123. Gajek, M., M. Bibes, S. Fusil, K. Bouzehouane, J. Fontcuberta, A. Barthelemy, and A. Fert. 2007. Tunnel junctions with multiferroic barriers. *Nature Materials* 6:296–302.
124. Bibes, M., and A. Barthélémy. 2008. Multiferroics: Towards a magnetoelectric memory. *Nature Materials* 7:425–426.
125. Mandal, S. K., G. Sreenivasulu, V. M. Petrov, and G. Srinivasan. 2010. Flexural deformation in a compositionally stepped ferrite and magnetoelectric effects in a composite with piezoelectrics. *Applied Physics Letters* 96:192502.
126. Yang, S. C., C. S. Park, K. H. Cho, and S. Priya. 2010. Self-biased magnetoelectric response in three-phase laminates. *Journal of Applied Physics* 108:093706.
127. Mandal, S. K., G. Sreenivasulu, V. M. Petrov, and G. Srinivasan. 2011. Magnetization-graded multiferroic composite and magnetoelectric effects at zero bias. *Physical Review B* 84:014432.
128. Sreenivasulu, G., S. K. Mandal, S. Bandekar, V. M. Petrov, and G. Srinivasan. 2011. Low-frequency and resonance magnetoelectric effects in piezoelectric and func-tionally stepped ferromagnetic layered composites. *Physical Review B* 84:144426.
129. Laletin, U., G. Sreenivasulu, V. M. Petrov, T. Garg, A. R. Kulkarni, N. Venkataramani, and G. Srinivasan. 2012. Hysteresis and remanence in mag-netoelectric effects in functionally graded magnetostrictive-piezoelectric lay-ered composites. *Physical Review B* 85:104404.
130. Lage, E., C. Kirchhof, V. Hrkac, L. Kienle, R. Jahns, R. Knochel, E. Quandt, and D. Meyners. 2012. Exchange biasing of magnetoelectric composites. *Nature Materials* 11:523–529.
131. Yan, Y. K., Y. Zhou, and S. Priya. 2013. Giant self-biased magnetoelectric cou-pling in co-fired textured layered composites. *Applied Physics Letters* 102:052907.
132. Shen, Y., J. Gao, Y. Wang, P. Finkel, J. Li, and D. Viehland. 2013. Piezomagnetic strain-dependent non-linear magnetoelectric response enhancement by flux concentration effect. *Applied Physics Letters* 102:172904.
133. Lage, E., F. Woltering, E. Quandt, and D. Meyners. 2013. Exchange biased magnetoelectric composites for vector field magnetometers. *Journal of Applied Physics* 113:17C725.
134. Petrie, J., D. Viehland, D. Gray, S. Mandal, G. Sreenivasulu, G. Srinivasan, and A. S. Edelstein. 2011. Enhancing the sensitivity of magnetoelectric sensors by increasing the operating frequency. *Journal of Applied Physics* 110:124506.
135. Zhuang, X., M. L. C. Sing, C. Cordier, S. Saez, C. Dolabdjian, L. G. Shen, J. F. Li, M. H. Li, and D. Viehland. 2011. Evaluation of applied axial field modu-lation technique on ME sensor input equivalent magnetic noise rejection. *IEEE Sensors Journal* 11:2266–2272.
136. Jahns, R., A. Piorra, E. Lage, C. Kirchhof, D. Meyners, J. L. Gugat, M. Krantz, M. Gerken, R. Knöchel, E. Quandt, and D. J. Green. 2013. Giant magnetoelec-tric effect in thin-film composites. *Journal of the American Ceramic Society* 96:1673–1681.

137. Zhou, Y., and S. Priya. 2014. Near-flat self-biased magnetoelectric response in geometry gradient composite. *Journal of Applied Physics* 115:104107.
138. Lu, C. J., P. Li, Y. M. Wen, A. C. Yang, C. Yang, D. C. Wang, W. He, and J. T. Zhang. 2014. Zero-biased magnetoelectric composite $Fe_{73.5}Cu_1Nb_3Si_{13.5}B_9$/Ni/Pb(Zr$_{1-x}$,Ti$_x$)$O_3$ for current sensing. *Journal of Alloys and Compounds* 589:498–501.
139. Wu, J., Z. Shi, J. Xu, N. Li, Z. Zheng, H. Geng, Z. Xie, and L. Zheng. 2012. Synthesis and room temperature four-state memory prototype of $Sr_3Co_2Fe_{24}O_{41}$ multiferroics. *Applied Physics Letters* 101:122903.
140. Stephan, M., J. Robert, G. Henry, Q. Eckhard, K. Reinhard, and W. Bernhard. 2012. MEMS magnetic field sensor based on magnetoelectric composites. *Journal of Micromechanics and Microengineering* 22:065024.
141. Martin, L. W., and R. Ramesh. 2012. Overview No. 151 Multiferroic and magnetoelectric heterostructures. *Acta Materialia* 60:2449–2470.
142. Kambale, R. C., D. Y. Jeong, and J. Ryu. 2012. Current status of magnetoelectric composite thin/thick films. *Advances in Condensed Matter Physics* 2012:824643.
143. Nan, T., Y. Hui, M. Rinaldi, and N. X. Sun. 2013. Self-biased 215 MHz magnetoelectric NEMS resonator for ultra-sensitive DC magnetic field detection. *Scientific Reports* 3:1985.
144. Piorra, A., R. Jahns, I. Teliban, J. L. Gugat, M. Gerken, R. Knöchel, and E. Quandt. 2013. Magnetoelectric thin film composites with interdigital electrodes. *Applied Physics Letters* 103:032902.
145. Greve, H., E. Woltermann, H.-J. Quenzer, B. Wagner, and E. Quandt. 2010. Giant magnetoelectric coefficients in $(Fe_{90}Co_{10})_{78}Si_{12}B_{10}$-AlN thin film composites. *Applied Physics Letters* 96:182501.
146. Onuta, T.-D., Y. Wang, C. J. Long, and I. Takeuchi. 2011. Energy harvesting properties of all-thin-film multiferroic cantilevers. *Applied Physics Letters* 99:203506.
147. Onuta, T.-D., Y. Wang, S. E. Lofland, and I. Takeuchi. 2014. Multiferroic operation of dynamic memory based on heterostructured cantilevers. *Advanced Materials* 27:202–206.
148. Martins, P., and S. Lanceros-Méndez. 2013. Polymer-based magnetoelectric materials. *Advanced Functional Materials* 23:3371–3385.
149. Baur, C., Y. Zhou, J. Sipes, S. Priya, and W. Voit. 2014. Organic, flexible, polymer composites for high-temperature piezoelectric applications. *Energy Harvesting and Systems* 1:167–177.
150. Zhai, J., S. Dong, Z. Xing, J. Li, and D. Viehland. 2006. Giant magnetoelectric effect in Metglas/polyvinylidene-fluoride laminates. *Applied Physics Letters* 89:083507.
151. Le, M. Q., F. Belhora, A. Cornogolub, P. J. Cottinet, L. Lebrun, and A. Hajjaji. 2014. Enhanced magnetoelectric effect for flexible current sensor applications. *Journal of Applied Physics* 115:194103.

Section III

Theoretical Approaches
in Multiferroic Study

Landau Theory of Multiferroics

Chuanwei Huang and Lang Chen

Contents

9.1 Introduction to Landau Theory

All materials, whether crystalline or not, possess a certain kind of symmetry. Phase transitions are ubiquitous in nature and usually accompanied by changes of crystal symmetry and various structural variables. Generally, the concept of symmetry and symmetry breaking plays an important role in all fields of physics. Based solely on symmetry considerations, the Landau theory provides a powerful tool to describe a great variety of phase transitions and measurable physical quantities near transitions [1]. This is achieved by introducing an important physical quantity, that is, order parameter, which is zero in the high-symmetry phase, and changes to a finite value for lower-symmetry phases. The total free energy in the vicinity of the transition can be expressed as a power series in the order parameter, and the stable state can be determined by minimizing the free energy with respect to the order parameter. Specific thermodynamic function/quantities can be obtained by differentiating the free energy. The Landau approach establishes straightforwardly the relationship between crystal symmetry and physical properties [2]. In a word, the Landau theory is a symmetry-based analysis of equilibrium physical behaviors in the vicinity of phase transitions.

In this chapter, Section 9.2 introduces the basic concepts of the Landau theory, initially proposed by L. D. Landau, and adapted by A. F. Devonshire and I. E. Dzialoshinskii to describe ferroelectric and ferromagnetic systems, respectively. Next, the applications of the Landau theory to both type I and type II multiferroics are reviewed in Section 9.3. We end the chapter with summary and outlooks in Section 9.4.

The Landau theory offers a reliable description of a system's equilibrium behaviors near the phase transition using input parameters obtained from either experiments or first-principles calculations. Symmetry considerations in the Landau theory facilitate the determination of crystal symmetry and provide selection rules for allowed phase evolutions. The physical properties such as order parameters, manifesting themselves in different analytic forms of the correlation functions at different temperatures, can be modeled at the macroscopic level by writing the thermodynamic free energy as a polynomial expansion with relevant coefficients. The phase transitions and associated properties are phenomena at the macroscopic scale, which can be described by averaging over various distributed variables in the Landau approach. The derivatives (with respect to temperature or other field) of the free energy lead to thermodynamic information about the phase transition such as specific heat, susceptibility, and so on.

There are four essential steps to analyze a phase transition using the Landau theory and to deal with the order parameter near the phase transition of a system.

1. Define an order parameter φ: This is the quantity that changes from zero to a finite value across the phase transition. As listed in Table 9.1, different materials or phase transitions have different order parameters.

2. Construct a free-energy functional: The stable phase can be determined by minimizing the free-energy functional: $G = G_0 + G_L(T, \varphi)$, where G_0 is the thermodynamic free energy of the parent phase at a high temperature, and G_L includes all information about the order parameter φ. This free energy is a symmetry-related polynomial expansion in φ that characterizes the transformation.

3. Construct G_L: The Landau functional is assumed to be a polynomial expansion of φ that obeys all possible symmetries associated with φ.

TABLE 9.1 Broken Symmetries and Order Parameters Involved in Different Phase Transitions

Phase Transition	Symmetry Broken	Order Parameter
Order–disorder	Reflection (Z2)	$\rho_1 - \rho_2$
Uniaxial ferromagnets	Reflection (Z2)	M
Liquid–gas	Reflection (Z2)	$\rho_1 - \rho_g$
Heisenberg ferromagnets	Rotation [SO(3)]	M
Superfluid	U1 gauge	Superfluid wavefunction $\Psi\, e^{i\phi}$
Superconductivity	U1 gauge	Gap $\Delta\, e^{i\psi}$
Nematic liquid crystal	SO(3)/Z2 rotation	Orientation (3nn−1)
Ferroelectrics	Lattice symmetry	P

4. Temperature dependence: The fundamental idea of the Landau theory is the temperature-driven symmetry breaking accompanying the phase transition. The spontaneous order parameter and related physical properties arise at temperatures lower than the transition point.

$$G_L(T,\varphi) = \int dV\left[\frac{1}{2}a_0(T-T_0)\varphi^2 + \cdots\right] \tag{9.1}$$

After the construction of the Landau functional and minimizing it over φ with temperatures, the nature of the transition and related physical quantities could be describable in terms of temperature.

9.2 Landau Theory of Ferroic Materials

Nowadays, there is a growing demand for materials exhibiting multiple functionalities with enhanced cross couplings, desirable for a variety of applications such as sensing, actuation, and data storage. Among them, ferroic materials such as ferroelectrics, ferroelastics, and ferromagnets have been attracting much interest [3–5]. A fundamental basis for understanding the intriguing properties of ferroic materials is the symmetry evolution as functions of external stimuli, such as temperature, stress/pressure, and electrical/magnetic field, which have been well described by the phenomenological Landau theory.

9.2.1 Classification of Ferroic Materials

In condensed matter physics, inversion and time reversal symmetries are particularly important as electrons carry both spin and charge. Lattice deformation and charge displacements break the inversion symmetry, whereas long-range order of magnetic moments (induced by electron spin or orbital movement) and current break the time reversal symmetry. Such symmetry breaking leads to ferroic orders. Based on which symmetry is broken, ferroic materials can be classified into ferromagnets, ferroelectrics, ferroelastics, and ferrotoridics with different spontaneous order parameters. The characteristic of ferromagnets is the presence of spontaneous magnetization M_s (an axial vector that breaks time reversal symmetry). Similarly, ferroelectricity is related to a spontaneous polarization P_s that breaks spatial inversion symmetry. Ferroelastics has a spontaneous strain μ_s that breaks rotational symmetry and ferrotoroidics has a tordization T with both the spatial inversion and time-reversal symmetries broken.

Taking iron as an example, it is in the nonmagnetic phase with no net magnetic moment, namely, paramagnetic phase, at high temperatures. As the temperature decreases, it experiences a phase transition to magnetic phase with the appearance of a spontaneous magnetization (order parameter). The spontaneous magnetization has two or more preferred directions, and can be switched among them by an external magnetic field, leading to the hysteresis loop. Domains, within which the magnetic moments align in the same direction, usually form in a ferromagnet. The boundary between adjacent domains is called a domain wall. In analogy to ferromagnets, materials exhibiting spontaneous polarizations, which can be reversed by an external electrical field, are called ferroelectrics. They undergo a transition from paraelectric to a low-symmetry ferroelectric phase (the temperature-driven breaking of spatial inversion symmetry). Domain structure also arises with different orientations of the polarization. The analysis can be extended to ferroelastic materials, which in the lower symmetry state have spontaneous strains μ_s, accompanied by domain structure and stress-strain hysteresis loops.

On the basis of these characteristics, Aizu coined a general name "ferroics" for these materails [3]. They exhibit several common features, including (1) spontaneous breaking of symmetry across the transition temperature; (2) subsequent appearance of order parameter and domain structure; and (3) hysteretic behavior. Another feature of ferroic material is the strongly temperature-dependent macroscopic properties near the ferroic phase transition. These features are the foundation for various practical applications of ferroic materials.

9.2.2 Landau Theory of Ferroelectrics

One of the advantages of the Landau theory is its ability to constitute a direct and exact relationship between the crystal symmetry and the associated physical properties of the system. In 1937, Landau proposed a theory to describe the second-order phase transitions of magnetism and superconductivity. Based on the Landau theory, any change of the system symmetry can be described by the order parameter φ. The magnitude of φ represents the degree of deviation from the high-symmetry state. If φ is small around a phase transition, then the free-energy functional G can be expressed as a power series in φ:

$$G = G_0(T) + a\varphi + \frac{A}{2}\varphi^2 + \frac{B}{3}\varphi^3 + \frac{C}{4}\varphi^4 + \qquad (9.2)$$

G_0 is the thermodynamic free energy of the high-temperature parent phase. The equilibrium value of φ at a given temperature or pressure can be determined by minimizing $G(T, P, \varphi)$, that is, the first derivative of G with respect to φ must be zero, and the second derivative is positive.

$$\left(\frac{\partial G}{\partial \varphi}\right)_{T,P} = a + A\varphi + 3B\varphi^2 + C\varphi^3 + \cdots = 0 \qquad (9.3)$$

and

$$\left(\frac{\partial^2 G}{\partial^2 \varphi}\right) = A + 6B\varphi + 3C\varphi^2 + \cdots > 0 \qquad (9.4)$$

The order parameter, for example, the spontaneous polarization P in ferroelectrics, magnetization M in ferromagnets or strain μ in ferroelastics, is a physical observable quantity. According to Ehrenfest's scheme, the order of a transition is the lowest order differential of G that shows a discontinuity. One feature of a second-order phase transition is the continuity of order parameter φ at the transition point. Thus, the parameter a in Equation 9.3 must be equal to zero ($a = 0$). Meanwhile, B is restricted to 0 to stabilize the system with the conditions (at the critical point with $\varphi = 0$, $A = 0$). Since the order parameter φ is small near the phase transition, the expansion to fourth order is enough for a good approximation of the total free energy. Furthermore, only even power terms of Equation 9.2 are allowed because the free energy remains unchanged if the spontaneous polarization P in ferroelectrics changes to $-P$.

$$G = G_0(T) + \frac{A}{2}P^2 + \frac{C}{4}P^4 \qquad (9.5)$$

When $A > 0$, there is only one minimum at $P = 0$. When $A < 0$, C must be positive so that the free energy has two minima with $P \neq 0$ as plotted in Figure 9.1.

However, the ferroic phase transitions are first order in most cases. There is a jump of order parameter φ at the transition point. Since the basic assumption of the Landau theory is that the transition from the high- to low-temperature phases takes place continuously at the transition temperature T_c, in principle, the Landau expansion in Equation 9.2 ceases to be valid. Thus, it is necessary to modify the theory to incorporate the possibility of having a metastable phase, that is, a local minimum in the free energy, by adding higher terms in the expansion (i.e., the sixth or higher term) [6]:

$$G = G_0(T, P) + \frac{A}{2}\varphi^2 + \frac{C}{4}\varphi^4 + \frac{D}{6}\varphi^6 \qquad (9.6)$$

Then the system will undergo a first-order phase transition if $A > 0$, $C < 0$, and $D > 0$.

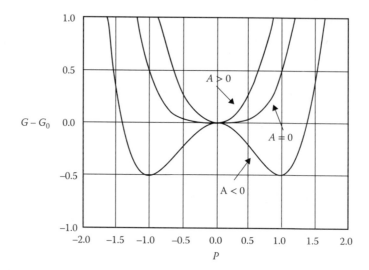

FIGURE 9.1 The free energy difference $G-G_0$ across a ferroelectric–paraelectric phase transition as a function of polarization P for different coefficient A.

So far, the Landau theory has been widely employed to investigate phase transitions and related physical properties in proper ferroelectrics, where the order parameter is the spontaneous polarization. Taking $BaTiO_3$ as an example, the order parameter is the spontaneous polarization. Though there also exists a spontaneous deformation during the transition from the cubic to the tetragonal phase, it is a consequence of the spontaneous polarization, and thus it is a secondary effect. However, there is a different class of system called improper ferroelectrics, where the order parameter is not the spontaneous polarization [7]. The polarization in improper ferroelectrics is only an inducted secondary or higher-order effect accompanying the structure change. In this case, the expression of free energy in order parameter φ and polarization P will contain not only even powers of φ and P, but also cross terms, such as φP, if φ and P have the same symmetry:

$$G = \frac{A}{2}\varphi^2 + \frac{C}{4}\varphi^4 + \frac{\omega}{2}P^2 + \frac{\kappa}{4}P^4 - \xi\varphi P \ldots \tag{9.7}$$

So far, we have discussed the first- and second-order phase transitions using the homogeneous Landau theory. However, domains usually form in ferroic materials, which lead to inhomogeneous magnetization, polarization and/ or strain. Thus, a gradient term should be added to the Landau expression to incorporate the nonuniform order parameter. Meanwhile, depolarization field energy in ferroelectrics is inevitable, if the polarization is normal to the surface. Though it can be significantly reduced with the deposition of electrodes, a spatial charge distribution still exists near the interface between the electrode and the ferroelectric film, resulting in a residual depolarization field. Thus, the depolarization field term should be incorporated into the Landau theory [8,9]. Furthermore, due to the strong coupling between polarization and strain, pressure and biaxial strain affect the physical properties of ferroelectric film significantly [10,11]. Therefore, it is essential to add the strain-dependent elastic energy to the Landau equation in ferroelectric and multiferroic films. In these cases, the total free energy of ferroelectric films can be written as [12–14]

$$G = G_{bulk} + G_{grad} + G_{depo} + G_{elas} \tag{9.8}$$

where G_{bulk}, G_{grad}, G_{depo}, and G_{elas} are the bulk-free energy, gradient energy, depolarization energy and elastic energy, respectively. The explicit expressions of these terms have been addressed in References [14,15].

9.2.3 Landau Theory of Magnets

In contrast to the static electrical charge density $\rho(r)$ that characterizes ferroelectrics, the description of magnetics requires a different parameter, the current density $j(r)$. Under the time-reversal operator [defined as $R(t) = -t$], static charges remain unchanged [$R\rho(r) = \rho(r)$], while $Rj(r) = -j(r)$. This suggests that the nonmagnetic phases, for example, paramagnetic and diamagnetic phases, are characterized by $j(r) = 0$. On the other hand, $j(r) \neq 0$ in ferromagnetic and antiferromagnetic phases, and $M = \int j dv$ is the average magnetization over the whole volume. Accordingly, the free energy of ferromagnets can be described in the form of Equation 9.2 with the order parameter M.

Despite many similarities between ferroelectrics and ferromagnets [16], there are three major differences in the Landau treatments for them. (1) The interaction between magnetic dipoles is nearly negligible, in sharp contrast to the long-range interaction between electric dipoles. Thus, the Landau theory is more reliable within a broader temperature range for ferroelectrics than for ferromagnets. (2) The presence of surface charges in ferroelectrics can dramatically suppress the stability of ferroelectricity. However, it does not have a magnetic analog due to the absence of free magnetic monopoles. (3) The strong coupling between the polarization and the lattice in ferroelectrics leads to the much thinner domain walls in ferroelectrics than that in ferromagnets.

9.3 Multiferroics

From the symmetry point of view, there are 31 ferroelectric (P), 31 ferromagnetic (M) and 31 ferrrotoroidic (υ) point groups, and four intersections (MP, Mυ, Pυ, and PMυ) of their ensembles (Figure 9.2) [17], which stand for magnetoelectric (MP), magnetotoroidic (Mυ), electrotoroidic (Pυ) and magnetoelectrotoroidic (PMυ). Specifically, 13 groups allow two kinds of primary ferroic orders, and only 9 allow all three kinds of primary ferroic orders. If a system possesses two or more ferroic orders simultaneously, it is termed multiferroic.

9.3.1 Magnetoelectric Coupling and Multiferroics

The coupling between electric and magnetic orders is known as magnetoelectric (ME) effect, which is an induction of polarization by a magnetic field (direct ME effect) or a magnetization by an electric field (converse ME effect).

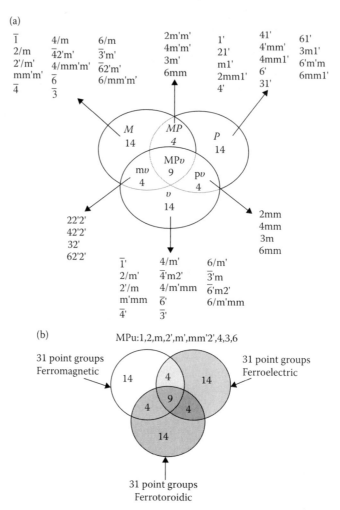

FIGURE 9.2 (a) Seven ensembles of point groups, produced by the intersection of 31 ferroelectrics, 31 ferromagnets and 31 ferrotoroidics point groups. *P, M,* and *υ* stand for spontaneous polarization, magnetization and velocity, respectively. (b) Simplified representation of (a). (From Hans, S., *J. Phys.: Condens. Matter*, 20, 434201, 2008).

The ME effect was first predicted by Curie in 1894 [18], which is then studied both theoretically and experimentally [19,20]. So far, more than 100 compounds have been discovered to exhibit ME effect. Thermodynamically, the ME effect can also be understood within the framework of the Landau theory by the expansion of free energy:

$$G(E,H) = G_0 - P_i^s E_i - M_i^s H_i - \frac{1}{2}\varepsilon_0\varepsilon_{ij}E_iE_j - \frac{1}{2}\mu_0\mu_{ij}H_iH_j$$
$$- \alpha_{ij}E_iH_j - \frac{1}{2}\beta_{ijk}E_iH_jH_k - \frac{1}{2}\gamma_{ijk}H_iE_jE_k \dots$$

(9.9)

Differentiations of free energy lead to polarization P and magnetization M:

$$P_i(E,H) = -\frac{\partial G}{\partial E_i} = P_i^s + \varepsilon_0\varepsilon_{ij}E_j + \alpha_{ij}H_j + \frac{1}{2}\beta_{ijk}H_jH_k + \gamma_{ijk}H_iE_j\ldots \quad (9.10)$$

$$M_i(E,H) = -\frac{\partial G}{\partial H_i} = M_i^s + \mu_0\mu_{ij}H_j + \alpha_{ij}E_j + \beta_{ijk}E_iH_j + \frac{1}{2}\gamma_{ijk}E_jE_k \quad (9.11)$$

where E and H are the electric and magnetic fields, respectively, P^s and M^s stand for the spontaneous polarization and magnetization, ε_{ij} and μ_{ij} stand for the second-order tensors of electric and magnetic susceptibilities, α_{ij} is the linear ME effect, which corresponds to an induced polarization by a magnetic field or a magnetization by an electrical field, and β_{ijk} and γ_{ijk} are the higher third-order tensors of ME effect. Generally speaking, most of current researches focus on the linear ME effect and overlook the higher coupling terms (Table 9.2). It is shown that the maximum linear ME response α_{ij} is limited by the relation [21]:

$$a_{ij}^2 < \chi_{ii}^e\chi_{jj}^m \quad (9.12)$$

where χ^e and χ^m are the electric and magnetic susceptibilities, respectively. Thus, large ME coupling is expected in single-phase materials with ferroelectric and ferromagnetic orders simultaneously, which are usually referred to as multiferroics. However, due to the different symmetry restrictions and

TABLE 9.2 The 58 Point Groups That Permit the Linear Magnetoelectric Effect [17,24]

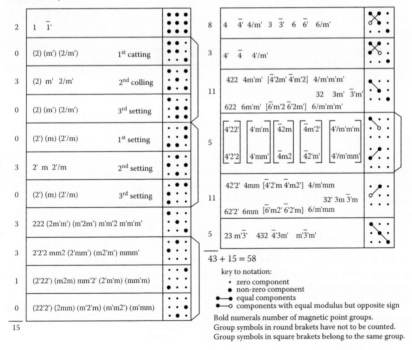

2	1 $\bar{1}'$			8	4 $\bar{4}'$ 4/m' 3 $\bar{3}'$ 6 $\bar{6}'$ 6/m'	
0	(2) (m') (2/m')	1st catting		3	4' $\bar{4}$ 4'/m'	
3	(2) m' 2/m'	2nd colling		11	422 4m'm' [$\bar{4}$'2m' $\bar{4}$'m'2] 4/m'm'm'	
					32 3m' $\bar{3}$'m'	
0	(2) (m') (2/m')	3rd setting			622 6m'm' [$\bar{6}$'m'2 $\bar{6}$'2m'] 6/m'm'm'	
0	(2') (m) (2'/m)	1st setting		5	[4'22' 4'm'm $\bar{4}$2m $\bar{4}$m'2' 4'/m'm'm]	
3	2' m 2'/m	2nd setting			[4'2'2 4'mm' $\bar{4}$m2 $\bar{4}$2'm' 4'/m'mm']	
0	(2') (m) (2'/m)	3rd setting		11	42'2' 4mm [$\bar{4}$'2'm $\bar{4}$'m2'] 4/m'mm	
					32' 3m $\bar{3}$'m	
3	222 (2m'm') (m'2m') m'm'2 m'm'm'				62'2' 6mm [$\bar{6}$'m2' $\bar{6}$'2'm] 6/m'mm	
3	2'2'2 mm2 (2'mm') (m2'm') mmm'			5	23 m$\bar{3}'$ 432 $\bar{4}$'3m' m$\bar{3}'$m'	
1	(2'22') (m2m) mm'2' (2'm'm) (mm'm)					
0	(22'2') (2mm) (m'2'm) (m'm2') (m'mm)					

$43 + 15 = 58$

15

key to notation:
- · zero component
- • non-zero component
- •—• equal components
- •—○ components with equal modulus but opposite sign

Bold numerals number of magnetic point groups.
Group symbols in round brakets have not to be counted.
Group symbols in square brakets belong to the same group.

TABLE 9.3	Properties of Two Types of Multiferroics				
Multiferroic	**FE**	**FM**	**ME Effect**	**_H_-Induced FE**	**_E_-Induced FM**
Type I	Strong	Weak	Weak (or no)	Weak (or no)	Weak (or no)
Type II	Weak	Weak	Strong	Strong	Weak (or no)

seemingly mutual exclusion between the ferroelectricity and ferromagnetism, only a few single-phase multiferroics have been discovered [22–25].

To achieve larger ME couplings, multiferroic composites made of ferroelectric (e.g., $BaTiO_3$) and ferromagnetic (e.g., $CoFe_2O_4$) materials have been introduced [23,26]. ME coupling in the composites is mediated by strain transfer across the interface, and thus requires that the two phases have good contact between them. Detailed discussion about ME composites can be found in other chapters of this book.

In general, crystals are called multiferroic when two or three of primary ferroic properties are presented in the same phase. Here we mainly focus on multiferroics that simultaneously possess ferroelectric and (anti)-ferromagnetic orders. They can be divided into two groups with distinctive properties as summarized in Table 9.3 [27]. Type I multiferroics are the materials that have different sources for ferroelectricity and magnetism with the two effects being quite independent of each other. In type I multiferroics, the magnetic order parameter that breaks time reversal symmetry and the ferroelectric order parameter that breaks spatial inversion symmetry coexist, and thus exhibit a certain degree of coupling between them. Typical type-I multiferroics include $BiFeO_3$ [28–31], and $BiMnO_3$ [22,32–36], where the ferroelectric and magnetic orders originate from the A-site and B-site ions, respectively. It also includes other systems, for example, hexagonal manganites $RMnO_3$ (R = Ho–Lu, Y) [37–40], which have their polarizations as the by-product of a complex lattice distortion.

In 2003, Kimura et al. reported a multiferroic material with spontaneous polarization originates from magnetic order [41]. This discovery was followed by the observations of similar effects in $REMn_2O_5$ [42,43]. Other materials, such as $FeVO_4$ and $Ni_3V_2O_8$ [44–47], also exhibit similar phenomena. These are referred to as type-II multiferroics, where ferroelectricity is induced by a magnetic order [41,48–50] and results in a very strong intrinsic coupling between ferroelectric and magnetic order parameters. However, the polarization of type II multiferroics is generally small (less than 1 $\mu C/cm^2$).

9.3.2 Landau Theory of Type-I Multiferroics

The Landau theory provides a general framework for understanding the ferroic behaviors and the ME coupling effect in multiferroics. Due to the additional ME coupling, the total free energy of a multiferroic materials can be written as $G = G_m + G_e + G_{inter}$, where G_m, and G_e are the free energy terms due to magnetic and electric orders, respectively. G_{inter} is the interaction term. For type I multiferroics, the free energy can be expressed as

$$G = G_0 + \frac{\alpha}{2}P^2 + \frac{\beta}{4}P^4 - PE + \frac{a}{2}M^2 + \frac{b}{4}M^4 - MH + \frac{\gamma}{2}P^2M^2 \quad (9.13)$$

where the term P^2M^2 represents the ME interaction.

$BiFeO_3$, as the only known room temperature single-phase multiferroic material, has been extensively studied. Scott et al. first described the ME coupling between magnetization M and polarization P in the domain walls and calculated the domain wall properties in multiferroic $BiFeO_3$ [51,52]. To simplify the illustration, the order parameter is restricted to one dimension and the model is established based only on 180° domains.

$$G = G_0 + \frac{\alpha}{2}P^2 + \frac{\beta}{4}P^4 + \frac{a}{2}M^2 + \frac{b}{4}M^4 + \frac{\gamma}{2}P^2M^2 + \frac{\kappa}{2}(\nabla P)^2 + \frac{\lambda}{2}(\nabla M)^2 \quad (9.14)$$

The domain wall energy of $BiFeO_3$ can be obtained by minimizing the difference between the system with and without the domain wall. If the domain wall is restricted to the $z-y$ plane, then Equation 9.15 can be rewritten as

$$
\begin{aligned}
G = {}& \frac{\alpha}{4P_0^2}\left(P(x)^2 - P_0^2\right)^2 + \frac{a}{2}M(x)^2 + \frac{b}{4}M(x)^4 \\
&+ \frac{\gamma}{2}P(x)^2 M(x)^2 + \frac{\kappa}{2}\left(\frac{dP(x)}{dx}\right)^2 + \frac{\lambda}{2}\left(\frac{dM(x)}{dx}\right)^2
\end{aligned}
\quad (9.15)
$$

The shapes of the free energy, polarization, and magnetization cross the domain wall of $BiFeO_3$ are shown in Figure 9.3.

Furthermore, in thin films, the misfit strain can alter the phase diagram significantly, which has been intensively investigated based on the Landau theory [23,53,54]. The free energy of $BiFeO_3$ thin film can be expressed as

$$
\begin{aligned}
G = {}& a_1^*\left(P_1^2 + P_2^2\right) + a_3^*P_3^2 + a_{11}^*\left(P_1^4 + P_2^4\right) + a_{33}^*P_3^4 \\
&+ a_{13}^*\left(P_1^2 P_3^2 + P_2^2 P_3^2\right) + a_{12}^*P_1^2 P_2^2 + \frac{u_m^2}{S_{11} + S_{22}}
\end{aligned}
\quad (9.16)
$$

With the following renormalized coefficients:

$$a_1^* = a_1 - u_m \frac{Q_{11} + Q_{12}}{S_{11} + S_{12}}, a_3^* = a_1 - u_m \frac{2Q_{12}}{S_{11} + S_{12}}$$

$$a_{11}^* = a_{11} + \frac{1}{2}\frac{1}{S_{11}^2 - S_{12}^2}\left[\left(Q_{11}^2 + Q_{12}^2\right)S_{11} - 2Q_{11}Q_{12}S_{12}\right]$$

$$a_{33}^* = a_{11} + \frac{Q_{12}^2}{S_{11} + S_{12}}, a_{13}^* = a_{12} + \frac{Q_{12}(Q_{11} + Q_{12})}{S_{11} + S_{12}}$$

$$a_{12}^* = a_{12} - \frac{1}{2}\frac{1}{S_{11}^2 - S_{12}^2}\left[\left(Q_{11}^2 + Q_{12}^2\right)S_{11} - 2Q_{11}Q_{12}S_{12}\right] + \frac{Q_{44}^2}{2S_{44}}$$

where a_1 is the dielectric stiffness, a_{ij} and a_{ijk} are the higher-order stiffness coefficients under constant stress, Q_{ij} are the electrostrictive coefficients, and S_{ij} are the elastic compliances of the film.

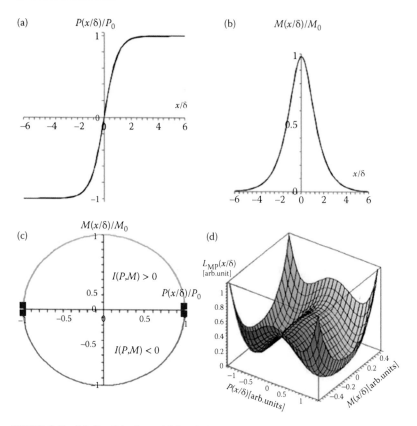

FIGURE 9.3 (a) $P = 0$ in the middle of the domain wall but $P \neq 0$ in the domains. (b) $M \neq 0$ in the middle of the domain wall but $M = 0$ in the domains. (c) $M(x)$ and $P(x)$ are constrained on an ellipse. (d) Free-energy landscape as functions of **P** and **M**. (From Daraktchiev, M. et al., *Phys. Rev. B*, 81, 224118, 2010).

The strain–temperature phase diagram of BiFeO3 films can be obtained by minimizing the total free energy. Due to the lack of sufficient input parameters for higher-order terms, the expansion of the Landau equation is up to the fourth order [55]. There are four stable phases for strained BiFeO3 films: the paraelectric cubic phase ($P_1 = P_2 = P_3 = 0$); the tetragonal phase ($P_1 = P_2 = 0$, $P_3 \neq 0$); the monoclinic phase ($P_1 = P_2 \neq 0$, $P_3 \neq 0$); and the orthorhombic phase ($P_1 = P_2 \neq 0$, $P_3 = 0$). It is shown in Figure 9.4 that there are several phase transitions at different temperatures and misfit strains. With the increase of in-plane strain, the stable phase of BiFeO$_3$ film first changes from tetragonal to monoclinic, and then changes to orthorhombic. Recently, a more detailed analysis revealed that there are other lower symmetry phases such as M_A, M_B, and M_C in epitaxial BiFeO$_3$ films [54,56]. Furthermore, it is proven that the tetragonal phase can be stabilized with large in-plane compressive strain of -4.0% at ambient temperatures [54,56]. Accompanying the phase transition from tetragonal to orthorhombic, the spontaneous polarization of BiFeO$_3$ film

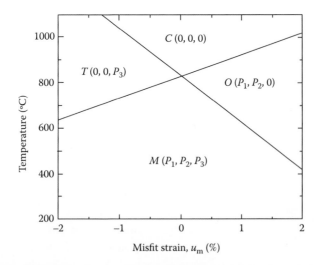

FIGURE 9.4 Misfit strain–temperature phase diagram of BiFeO₃ thin films. (From Ma, H. et al. *Appl. Phys. Lett.*, 92, 182902, 2008).

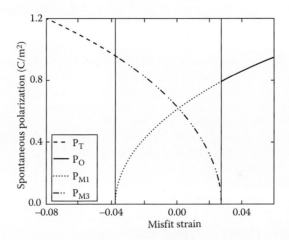

FIGURE 9.5 Misfit strain-dependent spontaneous polarization *P* of BiFeO₃ thin films [54]. (From Huang, C.W. et al., *Appl. Phys. Lett.*, 97, 152901, 2010.)

changes from out-of-plane to in-plane. Compared to the bulk counterpart, the polarization of BiFeO3 films is also enhanced by the misfit strain, with large P (i.e., 1.5 C/m² in Figure 9.5) in the "tetragonal" phase [54,57,58].

Effects of normal epitaxial strain on the phase diagram of BiFeO₃ film have been well addressed. What is less discussed in the literature is the effect of shear strain. It has been shown that shear strain from the substrates could shift the phase-transition points and induce novel properties in ferroelectric PbTiO₃ films [59,60]. Meanwhile, it is worthy to note that the deposition temperature of BiFeO₃ is usually lower than its Curie temperature (T_c ~ 1100 K), which

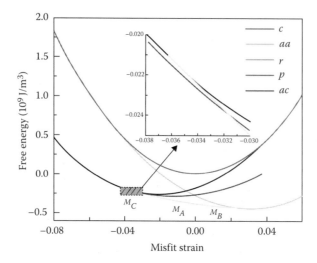

FIGURE 9.6 The free energies of BiFeO$_3$ films plotted as a function of misfit strain for different phases with shear strain μ_{xy} = 0.01. The inset is the enlarged section of the shaded area. (From Chen, Z. et al., *Adv. Funct. Mater.*, 21, 133–138, 2011.)

means that the films are rhombohedral phases during the deposition [61–64]. For BiFeO$_3$, the rhombohedral phase has pseudocubic lattice parameters of a_{pc} = 0.396 nm and a_{pc} = 89.4°, which results in a shear strain μ_{xy} between the film and the substrate. If the substrate has a cubic structure, the magnitude of μ_{xy} can be calculated as μ_{xy} = π (a_{pc}/90–1)/2 = –1.0%. Under such a condition, the phase diagram of BiFeO$_3$ is clearly changed as shown in Figure 9.6. A new stable ac ($P_1 \neq 0$, $P_2 = 0$, $P_3 \neq 0$) phase arises between the T phase and M_A phase with normal compressive strain of about –4%, which agrees well with the experimental observation of a monoclinic M_C phase for epitaxial BiFeO$_3$ thin films deposited on LaAlO$_3$ [65].

9.3.3 Landau Theory of Type-II Multiferroics: RMnO$_3$ and Spiral Magnets

In contrast to type-I multiferroics, type-II multiferroics generally have low magnetic and ferroelectric transition temperatures. However, there is a great deal of interest to better understand the mechanism on magnetically induced ferroelectricity. The Landau theory, which allows us to describe various macroscopic quantities and to interpret the coupling mechansim based on symmetry analysis, is once again widely employed to investigate the phase transition, ferroic order parameters and ME coupling in this type of multiferroic materials, where the ferroelectricity is induced a complex magnetic state [46,66–70]. Recently, a unified phenomenological Landau model was developed [66], which considers simultaneously the temperature and composition effects on the phase diagrams of type-II multiferroic RMnO$_3$ (R = Eu, Gd, Tb, Dy, and Ho) and related solid solutions. The model can produce the observed phase diagrams and the polar properties for given orthorhombic rare-earth compounds or solid solutions. As plotted in Figure 9.7, the specific phase sequences (PM → L-INC → Cycl-XY) and temperature-dependent polarizations and modulation wave vectors are effectively illustrated for TbMnO$_3$ and DyMnO$_3$ based on the Landau theory.

To investigate the ME coupling and the induced electric polarization in spin spiral multiferroics, besides many microscopic mechanisms that have been proposed [71–73], Mostovoy et al. [74] have introduced a simple but effective ME coupling term in Landau free energy:

$$G_{ME}(\mathbf{P},\mathbf{M}) = \gamma \mathbf{P}g\left[\mathbf{M}(\nabla \cdot \mathbf{M}) - (\mathbf{M}\cdot\nabla)\mathbf{M} + \cdots\right] \qquad (9.17)$$

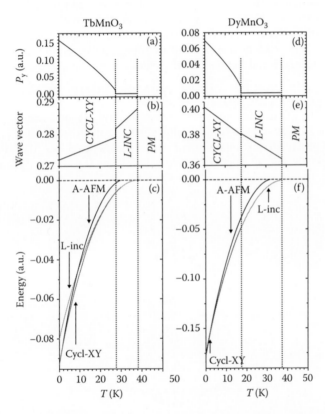

FIGURE 9.7 Polarization, modulation wave vector, and free energy of different phases for TbMnO$_3$ (a–c) and DyMnO$_3$ (d–f) as functions of temperature [66]. L-inc stands for longitudinal incommensurate phase, and Cycl-XY stands for cycloidal modulated phase with the primary longitudinal modulation σ_x and a secondary orthogonal modulation σ_{yA} of the y component of Mn spin. (From Ribeiro, J. L. et al., *Phys. Rev. B* 82, 064410, 2010.)

where P, and M are the polarization and magnetization vectors, respectively; γ is the coupling coefficient. Minimizing the free energy with respect to P yields

$$P = \gamma\chi_e\left[(M\cdot\nabla)M - M(\nabla\cdot M)\right] \quad (9.18)$$

where χ_e is the dielectric susceptibility. So the electric polarization is induced by magnetic ordering and the occurrence of ferroelectricity correlates with the transition to a spiral spin density wave (SDW) state. For a SDW state with the wave vector Q,

$$M = M_1 e_1 \cos Q\cdot x + M_2 e_2 \sin Q\cdot x + M_3 e_3 \quad (9.19)$$

If only M_1 or M_2 is nonzero, Equation 9.19 describes a sinusoidal wave. If both M_1 and M_2 are nonzero, Equation 9.19 describes a helix with the spin rotation axis e_3. Additionally, the helix is conical when M_1, M_2, and M_3 are nonzero simultaneously. Thus, the spiral spin order can spontaneously break both the time-reversal symmetry and spatial inversion symmetry, which could result in the presence of polarization (as described in Figure 9.8). According to Equations 9.18 and 9.19, one can find an average polarization transverse both to e_3 and Q, and is independent of M_3, which can be expressed as

$$\bar{P} = \frac{1}{V}\int d^3x P = \gamma\chi_e M_1 M_2\left[e_3 \times Q\right] \quad (9.20)$$

Now, the phase transition of ferroelectric SDW magnets can be obtained using the Landau approach. Generally, it is estimated that the energy gain of induced polarization is smaller than that of magnetic energy. Thus, the magnetic field–dependent polarization merely reflects the changes in magnetic

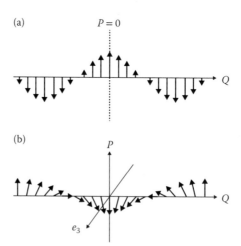

FIGURE 9.8 (a) The sinusoidal spin-density wave does not induce an electric polarization. (b) For a helicoidal spin-density wave, P is orthogonal to both the spin rotation axis e_3 and the wave vector Q. (From Mostovoy, M., *Phys. Rev. Lett.*, 96, 067601, 2006.)

ordering. In this scenario, the total Landau free energy of this system can be described as

$$G_M = \sum_{i=x,y,z} \frac{a_i}{2}(M_i)^2 + \frac{b}{4}\mathbf{M}^4 + \frac{c}{2}\mathbf{M}\left(\frac{d^2}{dx^2} + Q^2\right)^2 \mathbf{M}$$

$$+ b_{xy}(M_x)^2(M_y)^2 + b_{yz}(M_y)^2(M_z)^2 + b_{xy}'(M_x)^2\left(\frac{dM_y}{dx}\right)^2$$

(9.21)

The third term of Equation 9.21 represents the periodic SDW ordering with the wave vector $\mathbf{Q}$ along the x axis. The last three terms of Equation 9.21 relates to the flops of the anisotropic rare earth spins, coupled to Mn spins. Based on this phenomenological thermodynamic potential, typical phase diagrams of spiral magnets are calculated as shown in Figure 9.9, which are similar to the experimental results of TbMnO₃ [75].

9.4 Conclusions and Outlook

The main characteristics and developments of the phenomenological Landau theory for (multi-)ferroic materials are discussed. After briefly addressing the similarities and differences in the Landau treatments for ferroelectric and ferromagnetic, we focus more on multiferroics such as $BiFeO_3$ and $TbMnO_3$. To advance this active field of research, several challenges remain to be addressed.

1. Strain effects on magnetism in multiferroic thin films.

   Due to the strong coupling between the polarization and the crystal lattices, mechanical boundary conditions can drastically change the physical properties of ferroelectrics [11,76,77]. Although the approach has been extended to investigate the ferroelectric-related properties in multiferroic $BiFeO_3$ films [53–56], the effect of strains on magnetism in multiferroic systems remain largely unexplored, particularly from a phenomenological standpoint.

2. Breakdown of the Landau theory

   Generally, the validity of the Landau theory has been addressed following the Levanyuk–Ginzburg criterion [78]. More recently, the reliability of Landau–Ginzburg theory in ferroelectrics has been elaborated [79]. However, the corresponding issue remains largely unexplored for multiferroics, particularly in ultrathin multiferroic films.

3. Accuracy of the coefficients of Landau expansion.

   One of the advantages of Landau approach is that it is directly linked to measurable thermodynamic quantities in the vicinity of a phase transition. However, the accuracy of the calculated physical

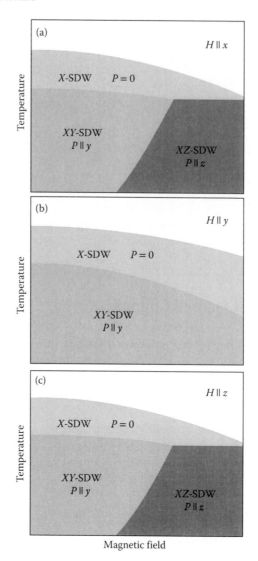

FIGURE 9.9 Typical phase diagrams of sinusoidal–helicoidal transition included for $H \parallel x$ (Panel a), y (Panel b), and z (Panel c). X-SDW (green) denotes the sinusoidal SDW state with spins along the x axis, while XY-SDW (pink) and XZ-SDW (blue) denote the spiral states with spins rotating in xy and xz planes, respectively. (From Mostovoy, M., *Phys. Rev. Lett.*, 96, 067601, 2006.)

quantity is highly sensitive to the coefficients of the series-expansion terms of the free energy. Only when the input parameters can be precisely determined from experiments or first-principle calculations is the Landau approach able to successfully predict the phase diagram and related physical quantities near the transition.

References

1. Landau, L. D., and E. M. Lifshitz. 1959. *Statistical Physics*. Oxford, UK: Pergamon.

2. Toledano, J.-C., and P. Toledano. 1987. *The Landau Theory of Phase Transitions*. Singapore: World Scientific.

3. Aizu, K. 1969. Possible species of "Ferroelastic" crystals and of simultaneously ferroelectric and ferroelastic crystals. *Journal of the Physical Society of Japan* 27:387–396.

4. Taganstev, A. K., L. E. Cross, and J. Fousek, eds. 2010. *Domains in Ferroic Materials and Thin Films*. New York: Springer.

5. Wadhawan, V. K. 2000. *Introduction to Ferroic Materials*. UK: Gordon & Breach.

6. Lines, M. E., and A. M. Glass. 1977. *Principles and Applications of Ferroelectrics and Related Materials*. Oxford, UK: Oxford University Press.

7. Levanyuk, A. P., and G. S. Daniil. 1974. Improper ferroelectrics. *Soviet Physics Uspekhi* 17:199–214.

8. Wurfel, P., and I. P. Batra. 1973. Depolarization-field-induced instability in thin ferroelectric films-experiment and theory. *Physical Review B* 8:5126–5133.

9. Batra, I. P., P. Wurfel, and B. D. Silverman. 1973. Phase transition, stability, and depolarization field in ferroelectric thin films. *Physical Review B* 8:3257–3265.

10. Schlom, D. G., L.-Q. Chen, C.-B. Eom, K. M. Rabe, S. K. Streiffer, and J.-M. Triscone. 2007. Strain tuning of ferroelectric thin films. *Annual Review of Materials Research* 37:589–626.

11. Pertsev, N. A., A. G. Zembilgotov, and A. K. Tagantsev. 1998. Effect of mechanical boundary conditions on phase diagrams of epitaxial ferroelectric thin films. *Physical Review Letters* 80:1988–1991.

12. Zhang, J. X., G. Sheng, and L. Q. Chen. 2010. Large electric field induced strains in ferroelectric islands. *Applied Physics Letters* 96:132901.

13. Qiu, Q. Y., S. P. Alpay, and V. Nagarajan. 2010. Phase diagrams, dielectric response, and piezoelectric properties of epitaxial ultrathin (001) lead zirconate titanate films under anisotropic misfit strains. *Journal of Applied Physics* 107:114105.

14. Li, Y. L., S. Y. Hu, and L. Q. Chen. 2005. Ferroelectric domain morphologies of (001) $PbZr_{1-x}Ti_xO_3$ epitaxial thin films. *Journal of Applied Physics* 97:034112.

15. Qiu, Q. Y., V. Nagarajan, and S. P. Alpay. 2008. Film thickness versus misfit strain phase diagrams for epitaxial $PbTiO_3$ ultrathin ferroelectric films. *Physical Review B* 78:064117.

16. Spaldin, N. 2007. Analogies and differences between ferroelectrics and ferromagnets. In *Physics of Ferroelectrics*. Berlin: Springer.

17. Hans, S. 2008. Some symmetry aspects of ferroics and single phase multiferroics. *Journal of Physics: Condensed Matter* 20:434201.

18. Curie, P. 1894. Sur la *symétrie* dans les phénomènes physiques, symétrie d'un champ électrique et d'un champ magnétique. *Journal of Theoretical and Applied Physics* 3:393–415.

19. Dzyaloshinskii, I. E. 1959. On the magneto-electrical effect in antiferromagnetics. *Soviet Physics—JETP* 10:628–629.

20. Astrov, D. N. 1960. The magnetoelectric effect in antiferromagnetics. *Soviet Physics—JETP* 11:708–709.

21. O'Dell, T. H. 1963. The field invariants in a magneto-electric medium. *Philosophical Magazine* 8:411–418.

22. Kimura, T., S. Kawamoto, I. Yamada, M. Azuma, M. Takano, and Y. Tokura. 2003. Magnetocapacitance effect in multiferroic $BiMnO_3$. *Physical Review B* 67:180401.

23. Ma, J., J. Hu, Z. Li, and C.-W. Nan. 2011. Recent progress in multiferroic magnetoelectric composites: from bulk to thin films. *Advanced Materials* 23:1062–1087.

24. Rivera, J. P. 2009. A short review of the magnetoelectric effect and related experimental techniques on single phase (multi-)ferroics. *The European Physical Journal B* 71:299–313.

25. Hill, N. A. 2000. Why are there so few magnetic ferroelectrics? *The Journal of Physical Chemistry B* 104:6694–6709.

26. Zheng, H., J. Wang, S. E. Lofland, Z. Ma, L. Mohaddes-Ardabili, T. Zhao, L. Salamanca-Riba, S. R. Shinde, S. B. Ogale, F. Bai, D. Viehland, Y. Jia, D. G. Schlom, M. Wuttig, A. Roytburd, and R. Ramesh. 2004. Multiferroic $BaTiO_3$-$CoFe_2O_4$ Nanostructures. *Science* 303:661–663.

27. Jeroen van den, B., and I. K. Daniel. 2008. Multiferroicity due to charge ordering. *Journal of Physics: Condensed Matter* 20:434217.

28. Lebeugle, D., D. Colson, A. Forget, and M. Viret. 2007. Very large spontaneous electric polarization in $BiFeO_3$ single crystals at room temperature and its evolution under cycling fields. *Applied Physics Letters* 91:022907.

29. Lebeugle, D., A. Mougin, M. Viret, D. Colson, and L. Ranno. 2009. Electric field switching of the magnetic anisotropy of a ferromagnetic layer exchange coupled to the multiferroic compound $BiFeO_3$. *Physical Review Letters* 103:257601.

30. Wang, J., J. B. Neaton, H. Zheng, V. Nagarajan, S. B. Ogale, B. Liu, D. Viehland, V. Vaithyanathan, D. G. Schlom, U. V. Waghmare, N. A. Spaldin, K. M. Rabe, M. Wuttig, and R. Ramesh. 2003. Epitaxial $BiFeO_3$ multiferroic thin film heterostructures. *Science* 299:1719–1722.

31. Catalan, G., and J. F. Scott. 2009. Physics and applications of bismuth ferrite. *Advanced Materials* 21:2463–2485.

32. Moreira dos Santos, A., A. K. Cheetham, T. Atou, Y. Syono, Y. Yamaguchi, K. Ohoyama, H. Chiba, and C. N. R. Rao. 2002. Orbital ordering as the determinant for ferromagnetism in biferroic $BiMnO_3$. *Physical Review B* 66:064425.

33. Solovyev, I. V., and Z. V. Pchelkina. 2010. Magnetic-field control of the electric polarization in $BiMnO_3$. *Physical Review B* 82:094425.

34. De Luca, G. M., D. Preziosi, F. Chiarella, R. Di Capua, S. Gariglio, S. Lettieri, and M. Salluzzo. 2013. Ferromagnetism and ferroelectricity in epitaxial $BiMnO_3$ ultra-thin films. *Applied Physics Letters* 103:062902.

35. Hill, N. A., and K. M. Rabe. 1999. First-principles investigation of ferromagnetism and ferroelectricity in bismuth manganite. *Physical Review B* 59:8759–8769.

36. Schmidt, R., W. Eerenstein, and P. A. Midgley. 2009. Large dielectric response to the paramagnetic-ferromagnetic transition $BiMnO_3$ epitaxial thin films. *Physical Review B* 79:214107.

37. Ismailza, I. G., and S. A. Kizhaev. 1965. Determination of the Curie point of the ferroelectrics $YMnO_3$ and $YbMnO_3$. *Soviet Physics-Solid State* 7:236–238.

38. Katsufuji, T., S. Meri, M. Masaki, Y. Moritomo, N. Yamamoto, and H. Takagi. 2001. Dielectric and magnetic anomalies and spin frustration in hexagonal $RMnO_3$ (R = Y, Yb, and Lu). *Physics Reivew B* 64:104419.

39. Lorenz, B., A. P. Litvinchuk, M. M. Gospodinov, and C. W. Chu. 2004. Field-induced Reentrant Novel phase and a ferroelectric-magnetic order coupling in $HoMnO_3$. *Physics Review Letter* 92:087204.

40. Lottermoser, T., T. Lonkai, U. Amann, D. Hohlwein, J. Ihringer, and M. Fiebig. 2004. Magnetic phase control by an electric field. *Nature* 430:541–544.

41. Kimura, T., T. Goto, H. Shintani, K. Ishizaka, T. Arima, and Y. Tokura. 2003. Magnetic control of ferroelectric polarization. *Nature* 426:55–58.

42. Chapon, L. C., G. R. Blake, M. J. Gutmann, S. Park, N. Hur, P. G. Radaelli, and S. W. Cheong. 2004. Structural anomalies and multiferroic behavior in magnetically frustrated $TbMn_2O_5$. *Physical Review Letters* 93:177402.

43. Shukla, D. K., R. Kumar, S. Mollah, R. J. Choudhary, P. Thakur, S. K. Sharma, N. B. Brookes, and M. Knobel. 2010. Swift heavy ion irradiation induced magnetism in magnetically frustrated $BiMn_2O_5$. *Physical Review B* 82:174432.

44. Lawes, G., A. B. Harris, T. Kimura, N. Rogado, R. J. Cava, A. Aharony, O. Entin-Wohlman, T. Yildirim, M. Kenzelmann, C. Broholm, and A. P. Ramirez. 2005. Magnetically driven ferroelectric order in $Ni_3V_2O_8$. *Physical Review Letters* 95:087205.

45. Fabrizi, F., H. C. Walker, L. Paolasini, F. de Bergevin, T. Fennell, N. Rogado, R. J. Cava, T. Wolf, M. Kenzelmann, and D. F. McMorrow. 2010. Electric field control of multiferroic domains in $Ni_3V_2O_8$ imaged by x-ray polarization-enhanced topography. *Physical Review B* 82:024434.

46. Dixit, A., G. Lawes, and A. B. Harris. 2010. Magnetic structure and magneto-electric coupling in bulk and thin film $FeVO_4$. *Physical Review B* 82:024430.

47. Kundys, B., C. Martin, and C. Simon. 2009. Magnetoelectric coupling in poly-crystalline $FeVO_4$. *Physical Review B* 80:172103.

48. Kimura, T. 2007. Spiral magnets as magnetoelectrics. *Annual Review of Materials Research* 37:387–413.

49. Mochizuki, M., and N. Furukawa. 2010. Theory of magnetic switching of ferroelectricity in spiral magnets. *Physical Review Letters* 105:187601.

50. Rubi, D., C. de Graaf, C. J. M. Daumont, D. Mannix, R. Broer, and B. Noheda. 2009. Ferromagnetism and increased ionicity in epitaxially grown $TbMnO_3$ films. *Physical Review B* 79:014416.

51. Daraktchiev, M., G. Catalan, and J. F. Scott. 2008. Landau theory of ferroelectric domain walls in magnetoelectrics. *Ferroelectrics* 375:122–131.

52. Daraktchiev, M., G. Catalan, and J. F. Scott. 2010. Landau theory of domain wall magnetoelectricity. *Physical Review B* 81:224118.

53. Zhang, J. X., Y. L. Li, Y. Wang, Z. K. Liu, L. Q. Chen, Y. H. Chu, F. Zavaliche, and R. Ramesh. 2007. Effect of substrate-induced strains on the spontaneous polarization of epitaxial $BiFeO_3$ thin films. *Journal of Applied Physics* 101:114105.

54. Huang, C. W., Y. H. Chu, Z. H. Chen, J. Wang, T. Sritharan, Q. He, R. Ramesh, and L. Chen. 2010. Strain-driven phase transitions and associated dielectric/piezoelectric anomalies in $BiFeO_3$ thin films. *Applied Physics Letters* 97:152901.

55. Ma, H., L. Chen, J. Wang, J. Ma, and F. Boey. 2008. Strain effects and thickness dependence of ferroelectric properties in epitaxial $BiFeO_3$ thin films. *Applied Physics Letters* 92:182902.

56. Zeches, R. J., M. D. Rossell, J. X. Zhang, A. J. Hatt, Q. He, C.-H. Yang, A. Kumar, C. H. Wang, A. Melville, C. Adamo, G. Sheng, Y.-H. Chu, J. F. Ihlefeld, R. Erni, C. Ederer, V. Gopalan, L. Q. Chen, D. G. Schlom, N. A. Spaldin, L. W. Martin, and R. Ramesh. 2009. A strain-driven morphotropic phase boundary in $BiFeO_3$. *Science* 326:977–980.

57. Zhang, J. X., Q. He, M. Trassin, W. Luo, D. Yi, M. D. Rossell, P. Yu, L. You, C. H. Wang, C. Y. Kuo, J. T. Heron, Z. Hu, R. J. Zeches, H. J. Lin, A. Tanaka,

C. T. Chen, L. H. Tjeng, Y. H. Chu, and R. Ramesh. 2011. Microscopic origin of the giant ferroelectric polarization in tetragonal-like $BiFeO_3$. *Physical Review Letters* 107:147602.

58. Ederer, C., and N. A. Spaldin. 2005. Effect of epitaxial strain on the spontaneous polarization of thin film ferroelectrics. *Physical Review Letters* 95:257601.

59. Zembilgotov, A. G., U. Bottger, and R. Waser. 2008. Effect of in-plane shear strain on phase states and dielectric properties of epitaxial ferroelectric thin films. *Journal of Applied Physics* 104:054118.

60. Zembilgotov, A. G., N. A. Pertsev, U. Bottger, and R. Waser. 2005. Effect of aniso-tropic in-plane strains on phase states and dielectric properties of epitaxial fer-roelectric thin films. *Applied Physics Letters* 86:052903.

61. Farag, N., M. Bobeth, W. Pompe, A. E. Romanov, and J. S. Speck. 2005. Modeling of twinning in epitaxial (001)-oriented $La_{0.67}Sr_{0.33}MnO_3$ thin films. *Journal of Applied Physics* 97:113516.

62. Arnold, D. C., K. S. Knight, F. D. Morrison, and P. Lightfoot. 2009. Ferroelectric-paraelectric transition in $BiFeO_3$: Crystal structure of the orthorhombic beta phase. *Physical Review Letters* 102:027602.

63. Palai, R., R. S. Katiyar, H. Schmid, P. Tissot, S. J. Clark, J. Robertson, S. A. T. Redfern, G. Catalan, and J. F. Scott. 2008. Beta phase and gamma—beta metal–insulator transition in multiferroic $BiFeO_3$. *Physical Review B* 77:014110.

64. Sandiumenge, F., J. Santiso, L. Balcells, Z. Konstantinovic, J. Roqueta, A. Pomar, J. P. Espinós, and B. Martínez. 2013. Competing misfit relaxation mechanisms in epitaxial correlated oxides. *Physical Review Letters* 110:107206.

65. Chen, Z., Z. Luo, C. Huang, Y. Qi, P. Yang, L. You, C. Hu, T. Wu, J. Wang, C. Gao, T. Sritharan, and L. Chen. 2011. Low-symmetry monoclinic phases and polariza-tion rotation path mediated by epitaxial strain in multiferroic $BiFeO_3$ thin films. *Advanced Functional Materials* 21:133–138.

66. Ribeiro, J. L., and L. G. Vieira. 2010. Landau model for the phase diagrams of the orthorhombic rare-earth manganites $RMnO_3$ (R = Eu, Gd, Tb, Dy, Ho). *Physical Review B* 82:064410.

67. Lawes, G., A. B. Harris, T. Kimura, N. Rogado, R. J. Cava, A. Aharony, O. Entin-Wohlman, T. Yildirim, M. Kenzelmann, C. Broholm, and A. P. Ramirez. 2005. Magnetically driven ferroelectric order in $Ni_3V_2O_8$. *Physical Review Letters* 95:087205.

68. Harris, A. B., M. Kenzelmann, A. Aharony, and O. Entin-Wohlman. 2008. Effect of inversion symmetry on the incommensurate order in multiferroic RMn_2O_5 R = rare earth). *Physical Review B* 78:014407.

69. Kenzelmann, M., G. Lawes, A. B. Harris, G. Gasparovic, C. Broholm, A. P. Ramirez, G. A. Jorge, M. Jaime, S. Park, Q. Huang, A. Y. Shapiro, and L. A. Demianets. 2007. Direct transition from a disordered to a multiferroic phase on a triangular lattice. *Physical Review Letters* 98:267205.

70. Artyukhin, S., K. T. Delaney, N. A. Spaldin, and M. Mostovoy. 2013. Landau theory of topological defects in multiferroic hexagonal manganites. *Nature Material* 13:42–49.

71. Hu, J. 2008. Microscopic origin of magnetoelectric coupling in noncollinear multiferroics. *Physical Review Letters* 100:077202.

72. Katsura, H., N. Nagaosa, and A. V. Balatsky. 2005. Spin current and magneto-electric effect in noncollinear magnets. *Physical Review Letters* 95:057205.

73. Sergienko, I. A., and E. Dagotto. 2006. Role of the Dzyaloshinskii–Moriya inter-action in multiferroic perovskites. *Physical Review B* 73:094434.

74. Mostovoy, M. 2006. Ferroelectricity in spiral magnets. *Physical Review Letters* 96:067601.
75. Kimura, T., G. Lawes, T. Goto, Y. Tokura, and A. P. Ramirez. 2005. Magneto-electric phase diagrams of orthorhombic $RMnO_3$ (R = Gd, Tb, and Dy). *Physical Review B* 71:224425.
76. Pertsev, N. A., A. K. Tagantsev, and N. Setter. 2000. Phase transitions and strain-induced ferroelectricity in $SrTiO_3$ epitaxial thin films. *Physical Review B* 61:R825–R829.
77. Ban, Z. G., and S. P. Alpay. 2002. Phase diagrams and dielectric response of epi-taxial barium strontium titanate films: A theoretical analysis. *Journal of Applied Physics* 91:9288–9296.
78. Levanyuk, A. P. 1959. Contribution to the theory of light scattering near the second-order phase-transition points. *Sovient Physics-JETP* 9:571–576.
79. Chandra, P., and P. Littlewood. 2007. A Landau primer for ferroelectrics. In *Physics of Ferroelectrics*. Berlin: Springer.

First-Principles Calculations for Multiferroic BiFeO₃

Jian-Xin Zhu

Contents

10.1 Introduction to First-Principles Calculations

Alongside the phenomenological approach toward the understanding of multiferroic materials, first-principles calculations have made a significant stride in providing much more microscopic details, enabling a direct comparison of theoretical predictions with experimental results. In a broader scope, the development of reliable methods for the quantitative calculations of properties in real materials is one of the most important challenges in modern condensed matter and materials physics. To set the stage, we start with a Hamiltonian for both electrons and ions in materials as follows:

$$H = -\sum_i \frac{1}{2m_e} \nabla_i^2 - \sum_{i,I} \frac{Z_I e^2}{|r_i - R_I|} + \frac{1}{2} \sum_{i,j} \frac{e^2}{|r_i - r_j|} - \sum_I \frac{1}{2M_I} \nabla_I^2 + \frac{1}{2} \sum_{I,J} \frac{Z_I Z_J e^2}{|R_I - R_J|} \quad (10.1)$$

In Equation 10.1, the first three terms represent electron kinetic energy, the Coulomb potential of electrons in a nuclear environment, and electron–electron Coulomb repulsion, while the last two terms represent the nucleus kinetic energy and the nucleon–nucleon Coulomb repulsion. Here, m_e and M_I are the masses of individual electrons and nuclei; the symbol -e stands for the electron charge and $Z_I e$ denotes the nucleus charge; r_i and R_I are the coordinates of electronic and nuclear degrees of freedom. Since the Hamiltonian involves enormous degrees of freedom, it is forbiddingly difficult to solve the entire problem quantum mechanically and approximations must be introduced. In the Oppenheimer approximation, because the mass of a nucleus is much larger than that of an electron, the nuclei move much slower than the electrons such that one can focus on the electronic degrees of freedom described by the Schrödinger equation:

$$H_{el} = -\sum_i \frac{1}{2m_e}\nabla_i^2 - \sum_{i,I}\frac{Z_I e^2}{|r_i - R_I|} + \frac{1}{2}\sum_{i,j}\frac{e^2}{|r_i - r_j|}$$ (10.2)

as if the nuclei are fixed in motion. The motion of the latter will be treated classically in an effective medium with electronic contribution through the Feynman–Hellman theorem. Even so, the electronic degrees of freedom are still highly entangled and it is a challenging problem to solve the electronic-related Schrödinger Equation 10.2. Within the density functional theory (DFT), the original many-electron problem can be mapped onto a different auxiliary system. In the Kohn–Sham ansatz, the ground state density of the original interacting electron system is equal to a much simplified Kohn–Sham particle system [1]. The motion of these Kohn–Sham particles is described by the single-particle Schrödinger equation:

$$\left(-\frac{\hbar^2}{2m_e}\nabla^2 + V_{eff}(r)\right)\psi_v^\sigma(r) = \varepsilon_v\psi_v^\sigma(r)$$ (10.3)

where

$$V_{eff}(r) = V_{ext}(r) + V_{Hartree}(r) + V_{xc}^\sigma(r)$$ (10.4)

consists of the external potential arising from the nuclei and any other external fields $V_{ext}(r)$, the Hartree potential $V_{Hartree}(r) = \int \frac{n(r')}{|r-r'|}dr'$, and the exchange-correlation potential $V_{xc}(r)$. Here the total electron density is in turn given by the solution to the Eigen-Equation 10.3: $n(r) = \sum_\sigma n_\sigma(r) = \sum_\sigma \sum_{v=1}^{N_\sigma}|\psi_v^\sigma(r)|^2$ for the total $N = N_\uparrow + N_\downarrow$ electrons occupying N lowest orbitals. As such, the above Kohn–Sham equation should be solved self-consistently. We note that the external applied field can also be spin dependent when the spin Zeeman interaction in the presence of an applied magnetic field becomes important in some situations. In the Kohn–Sham ansatz, the crucial quantity is the exchange-correlation energy, the functional derivative of which determines the exchange-correlation potential $v_{xc}^{\sigma}(r)$. Currently, the two most widely used potentials are based on the local density approximation (LDA) [2] and the generalized gradient approximation (GGA) [3]. For the past several decades, the LDA or GGA in the DFT framework has been very successful in describing structural, electronic, and optical properties of many good metals and several semiconductors, where the electronic correlations are rather weak. However, when the LDA-based DFT method is applied to complex materials like transition metal oxides (TMOs) and heavy-fermion materials with open d or f shell electrons, the description is not adequate. For example, TMOs like MnO, NiO, La$_2$CuO$_4$ are known Mott insulators, while the LDA predicts that they are metals. There are several attempts to improve the LDA approach including the self-interaction correction (SIC) method [4], the GW approximation [5,6], the hybrid DFT [7], the LDA + U method [8], and the more recently developed combination of LDA with dynamical mean-field theory (DMFT), the so-called LDA + DMFT method [9,10]. We note that the SIC method usually gives incorrect location of occupied d-bands in these TMOs and is also numerically expensive, while the GW method in its standard implementation works reasonably well for weakly interacting semiconductors. The LDA + DMFT method is a truly quantum many-body approach, which captures dynamical correlation effects and the localization-delocalization transition in strongly correlated electron systems. When one is in particular interested in the electronic and magnetic behaviors in the strongly correlated insulators, the LDA + U method turns out to be appealing with its computational efficiency and relatively transparent physical interpretation. The LDA + U total energy functional has proven to give large but reasonable corrections for a number of magnetic insulators. Therefore, we describe this method here with more technical details.

In the LDA + U method, an appropriate set of localized orbitals, for example, 3d in TMOs or 4f in rare-earth heavy fermion systems are first identified. A strong intra-atomic interaction term is then introduced and treated in a Hartree-Fock manner. As such, the method includes the orbital-dependent, self-energy operators, missing in the Kohn–Sham-based potential. The total energy functional in the LDA + U can be written as [11]

$$E_{\text{LDA+U}}\left[n_{\sigma}(r),\{n_{h,\sigma}\}\right]=E^{\text{LDA}}\left[n_{\sigma}(r)\right]+E^{U}\left[\{n_{h,\sigma}\}\right]-E_{dc}\left[\{n_{h,\sigma}\}\right] \quad (10.5)$$

where the Hubbard-U energy functional for the localized orbitals (denoted with symbol "h") in the spherically averaged density–density coupling approximation is given by

$$E^{U}\left[\{n_{h,\sigma}\}\right]=\frac{U}{2}\sum_{m,m',\sigma}n_{h,m\sigma}n_{h,m'\sigma}+\frac{U-J}{2}\sum_{m\neq m',\sigma}n_{h,m\sigma}n_{h,m'\sigma} \quad (10.6)$$

with the double-counting term

$$E_{dc}\left[\{n_{h,\sigma}\}\right]=\frac{1}{2}UN_{h}(N_{h}-1)-\frac{1}{2}J\sum_{\sigma}N_{h,\sigma}(N_{h,\sigma}-1) \quad (10.7)$$

in the fully localized limit approximation. Here, $N_h = N_{h,\uparrow} + N_{h,\downarrow}$. In a solid, the screened, effective Coulomb interaction parameter U and exchange parameter J can be obtained either from the constrained density functional calculations [12] or by inferring from the spectroscopy measurement together with the atomic Hartree-Fock calculations [13]. Nowadays, the LDA-based *ab initio* method, that is, solving the Kohn–Sham single-particle Schrödinger equations, has been implemented in the fully plane-wave basis, full-potential linearized augmented plane wave, and full-potential linear muffin-tin orbital basis codes, which are available as open sources or commercially.

To describe the properties of multiferroic materials, the two most important physical quantities are the magnetic and ferroelectric moments. The magnetic moment can be readily evaluated with the spin-polarized Kohn–Sham wave functions as

$$M = \mu_B \int_\Omega d^3r \sum_{\sigma,v}^{occ} \sigma \left| \psi_v^\sigma(r) \right|^2 \tag{10.8}$$

where μ_B and Ω are the Bohr magneton and the volume of the primitive magnetic unit cell. However, the calculation of electric polarization in crystalline solids is tricky. In the early years, the question as to whether the quantity in crystalline solids is well defined has been a hotly debated topic. It gradually became clear that the change in polarization between a polar and nonpolar reference system rather than the absolute value is well defined and can be compared with experimental measurements [14]. The controversy was thoroughly resolved about two decades ago, when the modern theory of polarization was introduced [15–19]. The theory shows that the electric polarization in a crystalline solid is a lattice rather than a vector. The total polarization for a given crystalline structure is the sum of ionic and electronic contributions:

$$P = P_{ion} + P_{el} \tag{10.9}$$

Here, the ionic contribution P_{ion} is given as

$$P_{ion} = \frac{e}{\Omega} \sum_i Z_i r_i \tag{10.10}$$

where r_i is the ionic potion in the unit cell and eZ_i is the nominal charge in the understanding that Z_i is either the valence atomic number in the pseudopotential/fixed-core context or the full atomic number when the core electrons are taken into account in the sum over occupied bands for the electronic contribution. The electronic contribution is given as a Berry phase multiplied by a prefactor, that is,

$$P_{el} = -\frac{ie}{(2\pi)^3} \sum_{\sigma,n}^{occ} \int_{BZ} d^3k \left\langle u_{nk}^\sigma \left| \frac{\partial u_{nk}^\sigma}{\partial k} \right. \right\rangle \tag{10.11}$$

where u_{nk}^σ is the periodic part of the Bloch function:

$$u_{nk}^\sigma(r) = e^{-ik\cdot r} \psi_{nk}^\sigma(r) \tag{10.12}$$

and BZ stands for the Brillouin zone corresponding to the primitive unit cell.

10.2 Applications of the *Ab Initio* Method to Multiferroic Materials

Hereafter, we will give an overview of the DFT-based first-principles electronic structure calculations of real multiferroic materials. In particular, we will focus on the TMO-based systems.

10.2.1 Ferroelectric, Magnetic, and Electronic Properties of BiFeO$_3$ in Bulk R3c and Strained P4mm Structures

As mentioned in previous chapters, bulk BiFeO$_3$ is ferroelectric with a Curie temperature of $T_C \sim 1100$ K and antiferromagnetic with a Neel temperature of about $T_N \sim 640$ K. This may be the only single-phase TMO, exhibiting both ferroelectricity and weak net magnetism (due to the incomplete compensation of a spin canting structure) and the strong coupling between these two order parameters at room temperature [20]. It is in contrast with other TMOs like BiMnO$_3$ (with $T_C \sim 450$ K and ferromagnetic $T_N \sim 100$ K) and YMnO$_3$ (with $T_C \sim 570$–900 K and antiferromagnetic $T_N \sim 70$–130 K).

At room temperature, bulk BiFeO$_3$ has a highly distorted perovskite structure with rhombohedral symmetry and R3c space group ($a = b = c = 5.63$ Å, $\alpha = \beta = \gamma = 59.4°$) [21]. The R3c crystal structure is shown in Figure 10.1a in a hexagonal conventional unit cell containing six formula units of BiFeO$_3$. The primitive R3c unit cell contains two formula units of BiFeO$_3$. Early experimental studies showed that although bulk single crystal has large atomic displacements and very high ferroelectric Curie temperature, the spontaneous

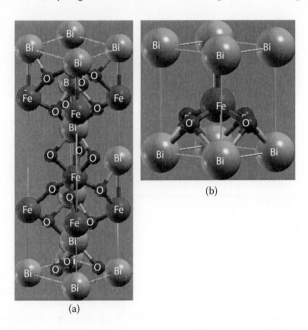

(a)

(b)

FIGURE 10.1 (a) Rhombohedral and (b) tetragonal structures of BiFeO$_3$.

polarization measured was surprisingly small with a value of about 6.1 μC/cm^2 along the easy-axis [111] direction (in a pseudocubic lattice) at liquid nitrogen temperature [22]. The application of first-principles-based modern theory of ferroelectric polarization to the bulk R3c BFO was first reported by Wang et al. [23]. The total electric polarization lattice for the bulk R3c BFO was found to be: $p_+ = (6.6 + n184.2)\mu$C/cm^2. Similarly, the polarization for its enantiomorphic counterpart is $p_- = (-6.6 + m184.2)\mu$C/cm^2. If one evaluated the spontaneous polarization as the difference of absolute polarization between the positively oriented +(R3c) and its enantiomorphic counterpart −(R3c) by naively taking $n = m$, the obtained value would be $p_s = (p_+ - p_-)/2 = 6.6\mu$C/cm^2. Although this value almost agrees perfectly with the experimental measurement, the theoretical value is obtained incorrectly. Later on, Neaton et al. [24] re-examined the ferroelectricity in the bulk R3c BiFeO$_3$. They showed that a correct way of calculating P_s should be followed by first identifying the so-called switching path and taking the difference of polarization between two endpoints only on the same lattice branch. Figure 10.2 shows an example for BiFeO$_3$ the absolute polarization for the endpoints and several intermediate structures along an idealized "switching path" connecting +(R3c) and −(R3c) through the centrosymmetric cubic perovskite structure. Therefore, it is clear from Figure 10.2 that the polarization values with $n = 0$ for +(R3c) and with $m = -1$ for −(R3c) are on the same lattice branch, which leads to the spontaneous polarization $P_s = 98.7$ μC/cm^2. The same value can be obtained by noting the absolute polarization for the centrosymmetric cubic structure that has the values of $(n - 1/2)92.1$ μC/cm^2. Although this value is much larger than the experimental value originally measured for the bulk R3c BiFeO$_3$, it is more

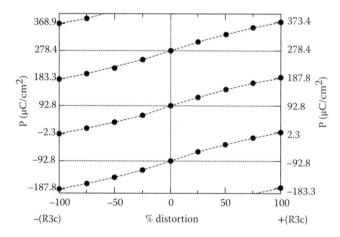

FIGURE 10.2 Electric polarization along a deformation path from the original +(R3c) through the centrosymmetric cubic structure to the inverted −(R3c). The LDA + U method with $U_{eff} = 2$ eV is used to ensure that the system stays insulating in all intermediate structures. Here, an ideal rhombohedral angle $\alpha = 60°$ is assumed and the polarization quantum is 185.6 μC/cm^2. (From Neaton, J. et al. *Phys. Rev. B*, 71, 014113, 2005.)

close to the expectation with such a higher Curie temperature and also agrees well with measurements on (111) oriented thin films [25].

The first-principles simulations have also been carried out to study the effect of onsite Hubbard interaction U_{eff} on the polarization of the bulk BiFeO$_3$. The results are shown in Table 10.1. As can be seen, the Hubbard-U only slightly decreases the calculated value of spontaneous polarization. We note that in all previously mentioned calculations of electric polarization, a rock-salt G-type antiferromagnetic homogeneous collinear order has been assumed. The local spin-density approximation (LSDA)–obtained local magnetic moment at the Fe site is about 3.3 μ_B, comparable with the experimental value of 3.75 μ_B [26]. The inclusion of Hubbard-U in the LSDA + U method improves the agreement with the experiment, increasing the Fe-site magnetic moment to 3.8 μ_B for $U_{eff} = 2$ eV and 4.0 μ_B for $U_{eff} = 4$ eV. A typical band structure based on LSDA and LSDA + U is shown in Figure 10.3. The band gap is indirect and has the value of about 0.40, 1.3, and 1.9 eV for $U_{eff} = 0, 2,$ and 4 eV, respectively, in comparison with the experimental value of 2.5 eV [27] and 2.75 eV [28] from independent measurements.

Experimentally, a long-wavelength spiral spin structure [29] and a small out-of-plane canting induced weak ferromagnetism [30] have been reported. The former, which is detrimental to the observation of macroscopic magnetization, can be easily suppressed through doping or strain engineering. For the latter, first-principles calculations have indeed shown [31] the occurrence of weak ferromagnetism of the Dzyaloshinskii–Moriya [32,33] type, when the Fe magnetic moments are oriented within the pseudocubic (111) planes and the spin-orbit coupling is included in the calculations. The magnitude of the measurable magnetization from the canting is about 0.1 μ_B per Fe site and is slightly reduced with the Hubbard-U interaction strength.

TABLE 10.1 Hubbard-U Dependence of the Calculated Structural Parameters and Spontaneous Polarization for the R3c Bulk BiFeO$_3$

U_{eff} (eV)		0	2	4
Bi (2a)	x	0	0	0
Fe (2a)	x	0.231	0.228	0.227
O (6b)	x	0.542	0.542	0.542
	y	0.943	0.942	0.943
	z	0.398	0.396	0.397
a_{rh} (Å)		5.46	5.50	5.52
α (deg)		60.36	59.99	59.84
Ω (Å^3)		115.98	117.86	118.34
P_s (μC/cm^2)		98.7	94.0	89.7

Source: Neaton, J. et al. Phys. Rev. B, 71, 014113, 2005.
Note: The Wyckoff positions 2a and 6b are referred to the rhombohedral system and are Bi(x,x,x), Fe(x,x,x), and O(x,y,z)

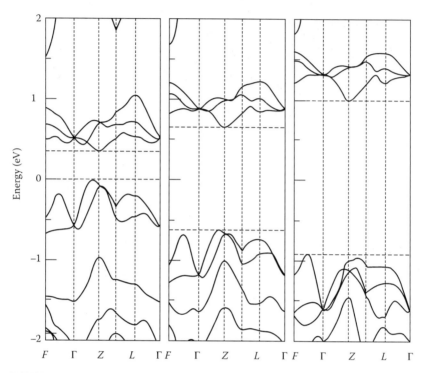

FIGURE 10.3 Band structure of R3c BiFeO$_3$ for U_{eff} = 0 (LSDA, left panel), 2 (middle panel), and 4 eV (right panel). The zero energy denotes the Fermi level. (From Neaton, J. et al. *Phys. Rev. B*, 71, 014113, 2005.)

The situation is more complicated for thin films. As is well known in the study of TMOs, strain, which is ubiquitous in thin films, is effective in tuning the properties through the modification of hybridization and bonding. It can also stabilize the otherwise metastable structures by varying the lattice mismatch between the film and the substrate. Experimentally, BiFeO$_3$ thin films on (001)-oriented LaAlO$_3$ and YAlO$_3$ substrate are revealed to have a tetragonal-like phase with the symmetry close to P4mm, which is depicted in Figure 10.1b. For BiFeO$_3$, DFT-based first-principles calculations have obtained a systematic evolution of the *c/a* ratio with the in-plane lattice constant [34–36], as shown in Figure 10.4, and predicted an unusually large *c/a* ratio for the stabilized P4mm phase. This evolution provides guidance on the right substrate material that can be used to tune the *c/a* ratio.

In reality, the films with even a small in-plane compressive strain are not strictly rhombohedral, and various values of *c/a* ratio have been reported. Noticeably, spontaneous polarization P_s on the order of 100 μC/cm^2 in the high-quality thin films has been experimentally observed [23,25,37]. This magnitude is in reasonable agreement with the theoretical value obtained for the bulk R3c structure, suggesting its robustness against the structure variation in BiFeO$_3$. On the other hand, a magnetic moment as large as 1 μB for the 70 nm thin film has been reported in the pioneering work of Reference [23], which

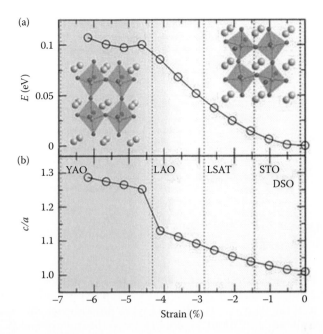

FIGURE 10.4 Evolution of the energy and *c/a* ratio as a function of in-plane lattice constant for BiFeO$_3$. (From Zeches, R. et al., *Science* 326, 977–980, 2009.)

caused a debate as to whether it was intrinsic or not [38–40]. Albrecht et al. [41] have performed first-principles calculations to investigate the ferromagnetism in BiFeO$_3$ thin films. They have obtained findings similar to those of Reference [29]. Regardless of being thick or ultrathin, the systems all possess a weak ferromagnetism of 0.027 μ_B arising from a spin canting. Such findings are different from the suggestion that the coupling between magnetic dipoles and mismatch strain could be responsible for the previously reported large values. Furthermore, earlier GGA or GGA + U–based electronic structure calculations [35] on the theoretically predicted super-tetragonal phase with a very large *c/a* ratio, have obtained an almost full magnetic moment ~5.0 μ_B, but a metallic state for U_{eff} = 0 eV and an insulating state with a direct gap of only 0.55 eV even for U_{eff} = 4 eV, in contrast with an indirect gap of 1.8 eV for the R3c phase with U_{eff} = 4 eV. These results are also at odds with the blue shifted absorption spectrum of BiFeO$_3$ thin film on (110) plane YAlO$_3$ substrate.

One notes that, although it was not explicitly mentioned, the band structure calculations in Reference [35] were carried out by assuming a ferromagnetic order in the original P4mm crystal structure. To give a further understanding of this difference, we have applied the DFT as implemented in the full-potential linearized augmented plane wave (FP-LAPW) WIEN2k code [42] to a P4mm phase of BiFeO$_3$ (Figure 10.5). The lattice structure parameters as previously determined in Reference [23] have been used. Electronic charge self-consistency is iterated with a 10 × 10 × 10 k-mesh in the entire Brillouin zone. Spin-polarized calculations are performed with

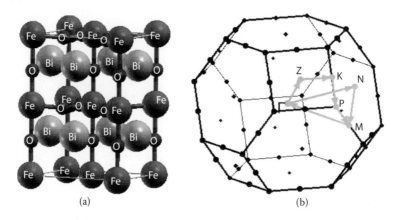

FIGURE 10.5 (a) Body-centered tetragonal crystal structure and (b) the corresponding Brillouin zone.

the GGA exchange-correlation functional. In the calculations, the muffin tin radius of $2.31a_0$ for Bi, $1.95a_0$ for Fe, and $1.73a_0$ for O is used with $RK_{max} = 8.0$. Here a_0 is the Bohr radius. Also a G-type antiferromagnetic ordering is assumed in all the calculations. Figure 10.6 shows the energy band dispersion for $U_{eff} = 0, 2, 4$ eV. When $U_{eff} = 0$ eV (that is, the usual spin-polarized GGA), the tetragonal $BiFeO_3$ in the G-type antiferromagnetic state is metallic with the Fermi energy cutting both the top of the valence band and the bottom of the conduction band. The observation of metallicity in the spin-polarized GGA and its replacement by insulating characteristic with the incorporation of Coulomb interaction reflects the generic inadequacy of conventional LSDA in treating electronic correlations for the narrow band behaviors in strongly correlated electronic materials. By applying the heavy-band visualization, one can identify that the lowest conduction band at Z point is derived mainly from the Bi-6p character and it is so low lying as to cross the Fermi level, giving rise to a gapless behavior of quasiparticles. This behavior is different from the insulating behavior predicted for the R3c bulk phase of $BiFeO_3$, where the conduction bands immediately above the Fermi energy are primarily of Fe-3d character. However, if one considers the band gap at the zone center, the gap is as large as about 1.4 eV already in the spin-polarized GGA calculations. The inclusion of onsite Hubbard repulsion on the Fe-3d orbitals pushes the occupied Fe-3d bands down, while those unoccupied Fe-3d bands are moved further up away from the Fermi energy. This leads to the increase of the gap at the zone center up to about 2.5 eV and 3.0 eV for $U_{eff} = 2$ and 4 eV, respectively. On the other hand, the conduction band bottom at Z point is not up shifted very rapidly, which gives rise to the opening of an indirect energy gap of about 0.35 eV for $U_{eff} = 2$ eV and about 0.7 eV for $U_{eff} = 4$ eV.

Figure 10.7 shows the single-particle density of states for the tetragonal $BiFeO_3$ in the G-type antiferromagnetic state. The evolution of the density of states characteristic near the Fermi energy with the onsite Coulomb interaction U_{eff} reinforces the conclusion made from the band dispersion.

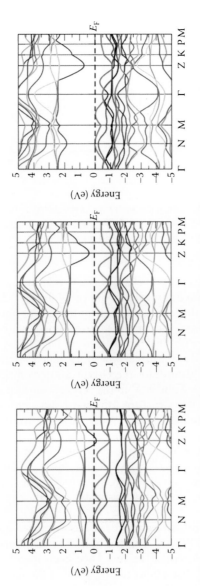

FIGURE 10.6 Band structure of the tetragonal $BiFeO_3$ in the G-type antiferromagnetic state for $U_{eff} = 0$ eV (left panel), 2 eV (middle panel), and 4 eV (right panel). The Fermi energy is denoted by a horizontal dashed line. The path along the high-symmetry k-points is marked in Figure 10.5b. Note that the band structure for the spin up and down channels is identical.

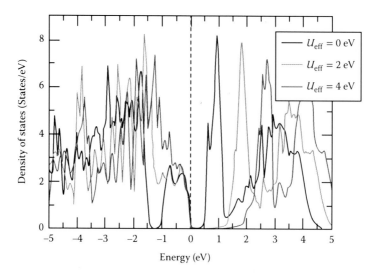

FIGURE 10.7 Electronic density of states for various values of Hubbard-U strength for the P4mm tetragonal $BiFeO_3$ in the G-type antiferromagnetic state. We note that the total DOS for spin up and down in the body-centered unit cell (see Figure 10.5a) is identical.

For $U_{eff} = 0$ eV, the dominant Fe-3d character peak is located about 1 eV above the Fermi energy. It is shifted upward to about 1.9 eV for $U_{eff} = 2$ eV and 2.80 eV for $U_{eff} = 4$ eV. Therefore, the results obtained for the G-type antiferromagnetic state are quite different from those for the ferromagnetic state though for the same tetragonal structure of $BiFeO_3$. For the ferromagnetic state, the opened gap even with $U_{eff} = 4eV$ is of the direct type and is much smaller than that for the bulk R3c structure. Instead, for the G-type antiferromagnetic state, the opened gap for both $U_{eff} = 2$ and 4 eV in the vertical transition sense is larger than those values for the R3c structure. This may point a promising direction toward a proper description of optical properties in the strained $BiFeO_3$ thin films. In addition, from our calculations for the G-type antiferromagnetic state, we have obtained the local magnetic moment on one of the Fe-sites in the BCT unit cell of 3.68 μ_B for $U_{eff} = 0$ eV, 3.91 μ_B for $U_{eff} = 2.0$ eV, and 4.05 μ_B for $U_{eff} = 4.0$ eV, which are smaller than those previously calculated for the ferromagnetic state but are compared reasonably with the experimental value of the bulk $BiFeO_3$ [26].

The DFT-based first-principles approach has also been applied to study the lattice dynamics in $BiFeO_3$. For the bulk R3c structure, two phonon modes with Eigen displacement vectors have been identified to strongly overlap with the atomic distortions taking place at the ferroelectric structural phase transition [43]. The observation suggests a transition of a displacive character. The calculated Raman and infrared reflectivity spectra related to these vibration modes can provide benchmark theoretical inputs for a correct assignment of experimental spectra [44]. In addition, the phonon contribution to the heat capacity has also been calculated using the first-principles approach [45]. Significant deviation from the Debye-like T^3 behavior has been observed down to a very low temperature, casting a constraint on the contribution of the gapped magnon modes to the heat capacity. For the P4mm BiFeO$_3$, the blue shift of some phonon mode has also been predicted [33].

10.2.2 Heterostructures of BiFeO$_3$ with Other Transition Metal Perovskites

Although BiFeO$_3$ has been unique in multiferroic materials because it exhibits both ferroelectricity and antiferromagnetism at room temperature as well as a robust coupling between the polar and magnetic orders, the antiferromagnetic nature, being insensitive to external magnetic field, has cast a severe limitation on technical applications of single-phase BiFeO$_3$. Recently, complex oxide heterostructures have provided an unprecedented platform to control the charge, spin, and orbital degrees of freedom at such interfaces. This control has given rise to emergent phenomena including interface-mediated metallicity [46], superconductivity [47], as well as magnetic effects [48].

Early works have been focused on the electronic reconstruction at the interface between LaAlO$_3$ and SrTiO$_3$, or the charge transfer–driven orbital ordering and ferromagnetism in heterostructures made of high-temperature cuprate YBa$_2$Cu$_3$O$_7$ and doped manganite La$_{0.67}$Ca$_{0.33}$MnO$_3$ [49,50]. The formation of a novel ferromagnetic state in the antiferromagnetic BiFeO$_3$ at the interface with ferromagnetic La$_{0.7}$Sr$_{0.3}$MnO$_3$ is another important observation [51,52]. To understand the interface effect, we have investigated the magnetic structure of (La$_{0.7}$Sr$_{0.3}$MnO$_3$)$_6$/(BiFeO$_3$)$_5$ by performing DFT-based first-principles calculations using the plane-wave basis set and the projected-augmented-wave method as implemented in the Vienna simulation package (VASP) code [53]. Our calculations are performed within the LSDA plus onsite Hubbard repulsion (LSDA + U) on d-orbitals of Mn and Fe. We have fixed the Hubbard-U on Mn-3d orbitals U_{eff}(Mn) = 4.0 eV while varying the value of U_{eff} on Fe-3d orbitals. With the starting spins of ferromagnetic La$_{0.7}$Sr$_{0.3}$MnO$_3$ and G-type antiferromagnetic BiFeO$_3$ aligned along x and z directions, respectively, we have been able to obtain a converged self-consistent solution. As shown in Figure 10.8, noticeable magnetization along the x-direction is formed in BiFeO$_3$ near the interface. The magnitude of induced ferromagnetism on Fe sites is sensitive to the size of band gap in bulk BiFeO$_3$. As we discussed in the previous subsection, the calculated band gap of bulk BiFeO$_3$, as exemplified by the density of states shown in the inset of Figure 10.8, depends on the value of Hubbard-U. The induced ferromagnetism at the interface is inversely proportional to the band gap of BiFeO$_3$, suggesting a magnetic proximity effect from the La$_{0.7}$Sr$_{0.3}$MnO$_3$ layer. Quantitatively, the obtained interfacial ferromagnetism of >~0.3 μ_B/Fe is much larger than the canting-induced

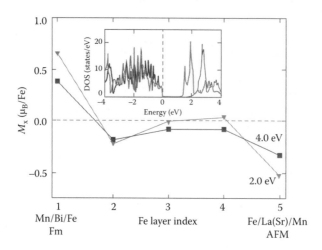

FIGURE 10.8 Induced local magnetic moments on Fe sites for a $(La_{0.7}Sr_{0.3}MnO_3)_6/$
$(BiFeO_3)_5$ superlattice obtained from *ab initio* calculations. The first Fe layer is
interfaced with an Mn layer through the Bi atomic layer, while the last Fe layer is
interfaced with Mn through the La/Sr atomic layer. (From Singh, S. et al., *Phys.
Rev. Lett.*, 113, 047204, 2014.)

moment of 0.03 μ_B/Fe in bulk $BiFeO_3$. This result explains the observation
of a significant ferromagnetic moment in $BiFeO_3$ layers of $La_{0.7}Sr_{0.3}MnO_3/$
$BiFeO_3$ superlattice [54]. In addition, we have also predicted from our first-
principles calculations that the exchange coupling between the Fe and the
Mn moments across the interface is ferromagnetic when they are separated
by Bi atomic layers, but it is antiferromagnetic when they are separated
by Sr/La atomic layers. The latter type of exchange coupling is consistent
with the experimental measurements on $BiFeO_3/La_{0.7}Sr_{0.3}MnO_3$ bilayer heterostructures and recent refined neutron scattering measurements on the
$La_{0.7}Sr_{0.3}MnO_3/BiFeO_3$ superlattice [55].

Recently, we have also performed first-principles calculations on the
$BiFeO_3/YBa_2Cu_3O_7$ heterostructure [56]. We consider various Fe-spin configurations near the interface, as schematically shown in Figure 10.9. The total
energy for these configurations is then computed. We then fit the energy to
a spin-exchange model, $E = \dfrac{1}{2}\sum_{i \neq j} J_{ij} S_i \cdot S_j$, where J_{ij} is the exchange interaction
strength between spins S_i and S_j. In this way, the interfacial exchange interac-
tion in the real system can be quantified. The calculation based on the resultant
exchange model enables us to find the true ground state with a ferromagnetic
spin ordering in $BiFeO_3$ near the interface. The mechanism for the emergence
of interfacial ferromagnetism in the present heterostructure is not of the
exchange-bias type. Instead, the charge transfer could play an important role.
Our theoretical observation of ferromagnetism is consistent with the magnetic
and ultrafast optical spectroscopy measurements in the same system [56].

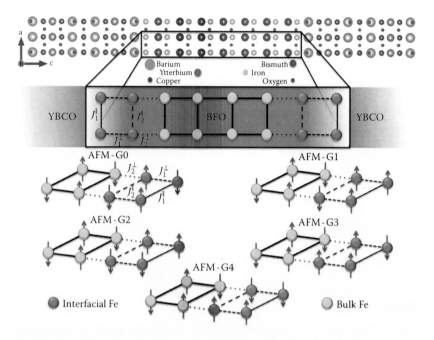

FIGURE 10.9 Schematic drawing of the BiFeO$_3$(BFO)/YBa$_2$Cu$_3$O$_7$(YBCO) hetero-structure (top panel) and five representative spin configurations on Fe atoms in BiFeO$_3$. The spin configuration AFM-G0 is the same as that for the G-type anti-ferromagnetic bulk state. For the spin configurations AFM-G1 through AFM-G4, the spin alignment deviates from that of the AFM-G0 in the first two layers of BiFeO$_3$ near the interface. (From Zhu, J.-X. et al., *Sci. Rep.*, 4, 5368, 2014.)

10.3 Conclusions and Outlook

In summary, the DFT-based first-principles approach has been reviewed and its applications to bulk BiFeO$_3$ as well as its heterostructures have been discussed. Its power of providing microscopic insight and connecting the physical properties with experimental measurements have been exemplified. Several remarks are in order. First, the possible origin of robust ferromagnetism in BiFeO$_3$ thin films should still be explored, especially along the scenario of possible mixed Fe^{2+}/Fe^{3+} valence due to oxygen vacancy [57]. The idea is now borne out in the double perovskite Bi$_2$FeMnO$_6$, where the unequal 3d electrons on Fe and Mn atoms serve naturally as a mixed-valence model for the single-phase BiFeO$_3$. Second, there have also been extensive first-principles investigations on the spin spiral–driven RMnO$_3$ [58,59] and exchange striction–based RMn$_2$O$_5$ [60] with R = Y, Tb, Dy, and so on. It has become clear that the spin-spiral state originates from the spin-orbit coupling, while the ionic displacement from their centrosymmetric positions plays a crucial role in determining the ferroelectric polarization direction and magnitude. Third, recently, the LDA + DMFT methodology has been applied to metallic ferromagnets and dynamical electronic correlation effects are found to be important for a large magnetocrystalline anisotropy energy [61]. Although its application to the insulating TMO systems may be an overkill, the LDA + DMFT method could be important in describing the localization/delocalization behavior at interfaces of heterostructures, including those with multiferroic TMOs.

Acknowledgments

The author would like to acknowledge his stimulating discussions with Dr. Quanxi Jia, Prof. Jason Haraldsen, Prof. Elbert Chia, Dr. Towfiq Ahmed, Dr. Mike Fitsimmons, and Dr. X.-D. Wen. The author would also like to thank Prof. Junling Wang for being an extremely patient and understanding editor, and the publishers and copyright holders of all figures from previous sources used in this chapter, which have been referenced in the relevant figure captions. This work was supported by U.S. DOE at LANL under Contract No. DEAC52-06NA25396 and the LANL LDRD-DR Program.

References

1. Martin, R. M. 2004. *Electronic Structure: Basic Theory and Practical Methods.* Cambridge: Cambridge University Press.
2. Khon, W., and L. Sham. 1965. Self-consistent equations including exchange and correlation effects. *Physical Review* 140:A1133–A1138.
3. Perdew, J. P., K. Burke, and M. Ernzerhof. 1996. Generalized gradient approximation made simple. *Physical Review Letters* 77:3865–3868.
4. Svane, A., and O. Gunnarsson. 1990. Transition-metal oxides in the self-interaction–corrected density-functional formalism. *Physical Review Letters* 65:1148–1151.
5. Hedin, L. 1965. New method for calculating the one-particle Green's function with application to the electron-gas problem. *Physical Review* 139:A796–A823.
6. Hedin, L., and S. Lundqvist. 1970. Effects of electron–electron and electron–phonon interactions on the one-electron states of solids. *Solid State Physics* 23:1–181.
7. Perdew, J. P., M. Ernzerhof, and K. Burke. 1996. Rationale for mixing exact exchange with density functional approximations. *The Journal of Chemical Physics* 105:9982–9985.
8. Anisimov, V. I., F. Aryasetiawan, and A. Lichtenstein. 1997. First-principles calculations of the electronic structure and spectra of strongly correlated systems: The LDA + U method. *Journal of Physics: Condensed Matter* 9:767–808.
9. Georges, A., G. Kotliar, W. Krauth, and M. J. Rozenberg. 1996. Dynamical mean-field theory of strongly correlated fermion systems and the limit of infinite dimensions. *Reviews of Modern Physics* 68:13–125.
10. Kotliar, G., S. Y. Savrasov, K. Haule, V. S. Oudovenko, O. Parcollet, and C. Marianetti. 2006. Electronic structure calculations with dynamical mean-field theory. *Reviews of Modern Physics* 78:865–951.
11. Liechtenstein, A., V. Anisimov, and J. Zaanen. 1995. Density-functional theory and strong interactions: Orbital ordering in Mott-Hubbard insulators. *Physical Review B* 52:R5467–R5470.

12. Gunnarsson, O., O. Andersen, O. Jepsen, and J. Zaanen. 1989. Density-functional calculation of the parameters in the Anderson model: Application to Mn in CdTe. *Physical Review B* 39:1708–1722.

13. Czyżyk, M., and G. Sawatzky. 1994. Local-density functional and on-site correlations: The electronic structure of La$_2$CuO$_4$ and LaCuO$_3$. *Physical Review B* 49:14211–14228.

14. Baroni, S., P. Giannozzi, and A. Testa. 1987. Green's-function approach to linear response in solids. *Physical Review Letters* 58:1861–1864.

15. King-Smith, R., and D. Vanderbilt. 1993. Theory of polarization of crystalline solids. *Physical Review B* 47:1651–1654.

16. Resta, R. 1993. Macroscopic electric polarization as a geometric quantum phase. *EPL (Europhysics Letters)* 22:133–138.

17. Vanderbilt, D., and R. King-Smith. 1993. Electric polarization as a bulk quantity and its relation to surface charge. *Physical Review B* 48:4442–4455.

18. Resta, R. 1994. Macroscopic polarization in crystalline dielectrics: The geometric phase approach. *Reviews of Modern Physics* 66:899–915.

19. Spaldin, N. A. 2012. A beginner's guide to the modern theory of polarization. *Journal of Solid State Chemistry* 195:2–10.

20. Smolenskiĭ, G., and I. Chupis. 1982. Ferroelectromagnets. *Soviet Physics Uspekhi* 25:475–493.

21. Kubel, F., and H. Schmid. 1990. Structure of a ferroelectric and ferroelastic monodomain crystal of the perovskite BiFeO$_3$. *Acta Crystallographica Section B: Structural Science* 46:698–702.

22. Teague, J. R., R. Gerson, and W. J. James. 1970. Dielectric hysteresis in single crystal BiFeO$_3$. *Solid State Communications* 8:1073–1074.

23. Wang, J., J. Neaton, H. Zheng, V. Nagarajan, S. Ogale, B. Liu, D. Viehland, V. Vaithyanathan, D. Schlom, and U. Waghmare. 2003. Epitaxial BiFeO$_3$ multiferroic thin film heterostructures. *Science* 299:1719–1722.

24. Neaton, J., C. Ederer, U. Waghmare, N. Spaldin, and K. Rabe. 2005. First-principles study of spontaneous polarization in multiferroic BiFeO$_3$. *Physical Review B* 71:014113.

25. Bai, F., J. Wang, M. Wuttig, J. Li, N. Wang, A. P. Pyatakov, A. K. Zvezdin, L. E. Cross, and D. Viehland. 2005. Destruction of spin cycloid in (111)c-oriented BiFeO$_3$ thin films by epitiaxial constraint: Enhanced polarization and release of latent magnetization. *Applied Physics Letters* 86:032511.

26. Sosnowska, I., W. Schäfer, W. Kockelmann, K. Andersen, and I. Troyanchuk. 2002. Crystal structure and spiral magnetic ordering of BiFeO$_3$ doped with manganese. *Applied Physics A* 74:S1040–S1042.

27. Gao, F., Y. Yuan, K. Wang, X. Chen, F. Chen, J. Liu, and Z. Ren. 2006. Preparation and photoabsorption characterization of BiFeO$_3$ nanowires. *Applied Physics Letters* 89:102506.

28. Ihlefeld, J., N. Podraza, Z. Liu, R. Rai, X. Xu, T. Heeg, Y. Chen, J. Li, R. Collins, and J. Musfeldt. 2008. Optical band gap of BiFeO$_3$ grown by molecular-beam epitaxy. *Applied Physics Letters* 92:142908.

29. Sosnowska, I., T. P. Neumaier, and E. Steichele. 1982. Spiral magnetic ordering in bismuth ferrite. *Journal of Physics C: Solid State Physics* 15:4835–4846.

30. Vorob'ev, G., A. Zvezdin, A. Kadomtseva, Y. F. Popov, V. Murashov, and D. Rakov. 1995. Possible coexistence of weak ferromagnetism and spatially modulated spin structure in ferroelectrics. *Physics of the Solid State* 37:1793–1795.

31. Ederer, C., and N. A. Spaldin. 2005. Weak ferromagnetism and magnetoelectric coupling in bismuth ferrite. *Physical Review B* 71:060401.

32. Dzialoshinskii, I. 1957. Thermodynamic theory of weak ferromagnetism in anti-ferromagnetic substances. *Sovient Physics JETP-USSR* 5:1259–1272.

33. Moriya, T. 1960. Anisotropic superexchange interaction and weak ferromagnetism. *Physical Review* 120:91–98.

34. Ricinschi, D., K.-Y. Yun, and M. Okuyama. 2006. A mechanism for the 150 μC cm^{-2} polarization of BiFeO$_3$ films based on first-principles calculations and new structural data. *Journal of Physics: Condensed Matter* 18:L97–L105.

35. Tütüncü, H., and G. Srivastava. 2008. Electronic structure and lattice dynamical properties of different tetragonal phases of BiFeO$_3$. *Physical Review B* 78:235209.

36. Zeches, R., M. Rossell, J. Zhang, A. Hatt, Q. He, C.-H. Yang, A. Kumar, C. Wang, A. Melville, and C. Adamo. 2009. A strain-driven morphotropic phase boundary in BiFeO$_3$. *Science* 326:977–980.

37. Yun, K. Y., D. Ricinschi, T. Kanashima, M. Noda, and M. Okuyama. 2004. Giant ferroelectric polarization beyond 150 μC/cm^2 in BiFeO$_3$ thin film. *Japanese Journal of Applied Physics* 43:L647–L648.

38. Béa, H., M. Bibes, A. Barthélémy, K. Bouzehouane, E. Jacquet, A. Khodan, J.-P. Contour, S. Fusil, F. Wyczisk, and A. Forget. 2005. Influence of parasitic phases on the properties of BiFeO$_3$ epitaxial thin films. *Applied Physics Letters* 87:072508.

39. Béa, H., M. Bibes, S. Petit, J. Kreisel, and A. Barthélémy. 2007. Structural distortion and magnetism of BiFeO3 epitaxial thin films: A Raman spectroscopy and neutron diffraction study. *Philosophical Magazine Letters* 87:165–174.

40. Catalan, G., and J. F. Scott. 2009. Physics and applications of bismuth ferrite. *Advanced Materials* 21:2463–2485.

41. Albrecht, D., S. Lisenkov, W. Ren, D. Rahmedov, I. A. Kornev, and L. Bellaiche. 2010. Ferromagnetism in multiferroic BiFeO3 films: A first-principles-based study. *Physical Review B* 81:140401.

42. Blaha, P., K. Schwarz, G. Madsen, D. Kvasnicka, and J. Luitz. Wien2k, An augmented plane wave local orbitals program for calculating crystal properties (Technical University of Wien, Austria, 2001). ISBN 3-9501031-1-2.

43. Hermet, P., M. Goffinet, J. Kreisel, and P. Ghosez. 2007. Raman and infrared spectra of multiferroic bismuth ferrite from first principles. *Physical Review B* 75:220102.

44. Tütüncü, H., and G. Srivastava. 2008. Electronic structure and zone-center phonon modes in multiferroic bulk BiFeO$_3$. *Journal of Applied Physics* 103:083712.

45. Wang, Y., J. E. Saal, P. Wu, J. Wang, S. Shang, Z.-K. Liu, and L.-Q. Chen. 2011. First-principles lattice dynamics and heat capacity of BiFeO$_3$. *Acta Materialia* 59:4229–4234.

46. Ohtomo, A., and H. Hwang. 2004. A high-mobility electron gas at the LaAlO3/SrTiO3 heterointerface. *Nature* 427:423–426.

47. Reyren, N., S. Thiel, A. Caviglia, L. F. Kourkoutis, G. Hammerl, C. Richter, C. Schneider, T. Kopp, A.-S. Rüetschi, and D. Jaccard. 2007. Superconducting interfaces between insulating oxides. *Science* 317:1196–1199.

48. Brinkman, A., M. Huijben, M. Van Zalk, J. Huijben, U. Zeitler, J. Maan, W. Van der Wiel, G. Rijnders, D. Blank, and H. Hilgenkamp. 2007. Magnetic effects at the interface between non-magnetic oxides. *Nature Materials* 6:493–496.

49. Chakhalian, J., J. Freeland, H.-U. Habermeier, G. Cristiani, G. Khaliullin, M. Van Veenendaal, and B. Keimer. 2007. Orbital reconstruction and covalent bonding at an oxide interface. *Science* 318:1114–1117.

50. Chakhalian, J., J. Freeland, G. Srajer, J. Strempfer, G. Khaliullin, J. Cezar, T. Charlton, R. Dalgliesh, C. Bernhard, and G. Cristiani. 2006. Magnetism at the interface between ferromagnetic and superconducting oxides. *Nature Physics* 2:244–248.

51. Yu, P., J.-S. Lee, S. Okamoto, M. Rossell, M. Huijben, C.-H. Yang, Q. He, J. Zhang, S. Yang, and M. Lee. 2010. Interface ferromagnetism and orbital reconstruction in BiFeO$_3$–La$_{0.7}$Sr$_{0.3}$MnO$_3$ Heterostructures. *Physical Review Letters* 105:027201.

52. Wu, S., S. A. Cybart, P. Yu, M. Rossell, J. Zhang, R. Ramesh, and R. Dynes. 2010. Reversible electric control of exchange bias in a multiferroic field-effect device. *Nature Materials* 9:756–761.

53. Kresse, G., and J. Furthmuller. 1996. Efficient iterative schemes for *ab initio* total-energy calculations using a plane-wave basis set. *Physical Review B* 54:11169–11186.

54. Singh, S., J. Haraldsen, J. Xiong, E. Choi, P. Lu, D. Yi, X.-D. Wen, J. Liu, H. Wang, and Z. Bi. 2014. Induced magnetization in La$_{0.7}$Sr$_{0.3}$MnO$_3$/BiFeO$_3$ superlattices. *Physical Review Letters* 113:047204.

55. Prashant, J., Q. Wang, M. Roldan, A. Glavic, V. Lauter, C. Urban, Z. Bi, T. Ahmed, J.-X. Zhu, M. Varela, Q. Jia, and M. Fitzsimmons. 2015. Synthetic magnetoelectric coupling in a nanocomposite multiferroic. *Scientific Reports* 5:9089.

56. Zhu, J.-X., X.-D. Wen, J. Haraldsen, M. He, C. Panagopoulos, and E. E. Chia. 2014. Induced ferromagnetism at BiFeO$_3$/YBa$_2$Cu$_3$O$_7$ interfaces. *Scientific Reports* 4:5368.

57. Wang, J., A. Scholl, H. Zheng, S. B. Ogale, D. Viehland, D. G. Schlom, N. A. Spaldin, K. M. Rabe, M. Wuttig, L. Mohaddes, J. Neaton, U. Waghmare, T. Zhao, and R. Ramesh. 2005. Response to comment on "Epitaxial BiFeO$_3$ multiferroic thin film heterostructures". *Science* 307:1203.

58. Malashevich, A., and D. Vanderbilt. 2008. First principles study of improper ferroelectricity in TbMnO$_3$. *Physical Review Letters* 101:037210.

59. Xiang, H. J., S.-H. Wei, M. H. Whangbo, and J. L. F. Da Silva. 2008. Spin-orbit coupling and ion displacements in multiferroic TbMnO$_3$. *Physical Review Letters* 101:037209.

60. Wang, C., G.-C. Guo, and L. He. 2008. First-principles study of the lattice and electronic structure of TbMn$_2$O$_5$. *Physical Review B* 77:134113.

61. Zhu, J.-X., M. Janoschek, R. Rosenberg, F. Ronning, J. D. Thompson, M. A. Torrez, E. D. Bauer, and C. D. Batista. 2014. LDA + DMFT Approach to magnetocrystalline anisotropy of strong magnets. *Physical Review X* 4:021027.

11

Interface Coupling in Multiferroic Heterojunctions

Jason T. Haraldsen

Contents

11.1 Introduction

Multiferroic materials that possess ferromagnetic and ferroelectric properties simultaneously have been investigated intensively over the last decade due to the anticipation of strong magnetoelectric coupling effects [1–10], which could be utilized in future spintronics [11,12]. However, obtaining multiferroicity and strong magnetoelectric coupling in a single-phase material is not very simple. This is due to the very nature of ferro-states. Ferroelectricity in perovskite oxides, for example, prefers to have empty d-orbitals that will help in shifting the atomic positions to produce an overall dipole moment [13]. However, ferromagnetism is produced through the coupling of magnetic moments, which typically requires partially filled d-orbitals [1]. This makes coexistence and coupling of electric polarization and magnetic order difficult, because the ferro-states have to couple through orbital and lattice degrees of freedom. This disassociation weakens the magnetoelectric coupling in many single-phase multiferroic materials [14–16],

Single-phase multiferroic

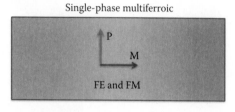

Multiferroic heterojunction

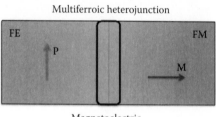

Magnetoelectric
coupling

FIGURE 11.1 Single-phase multiferroics (top panel) and multiferroic composites or heterstructures (bottom panel). The interfaces are crucial in multiferroic composites or heterostructures.

even though a large electric polarization or magnetic order may exist. This is usually the case for type-I multiferroic materials, including lone-pair and charge-ordered systems [17]. On the other hand, in the so-called type-II multiferroics (e.g., $Ni_3V_2O_8$, $TbMnO_3$, and $CuFeO_2$), the ferroelectric polarization is produced through a strong coupling to incommensurate magnetic ordering [18–20]. Here, magnetic order (typically a spin spiral) breaks the inversion symmetry of the system and produces an electric polarization through the converse Dzyloshinskii–Moriya interaction [18,19]. In these type-II multiferroics, there is a strong coupling between magnetism and ferroelectricity. However, the net polarization is typically orders of magnitude smaller than that of type-I multiferroics and the Curie temperatures are usually much lower than the room temperature.

To obtain multiferroics with strong magnetoelectric coupling at room temperature that can be used for spintronic devices, multiferroic composites or heterostructures have been investigated (Figure 11.1) [11,21–24]. In this case, individual components with large electric polarization and magnetization at room temperature can be selected to have the best of both. Because of the multiple components involved, interfaces are inevitable. The quality of the interface determines the magnetoelectric coupling efficiency in such composites or heterostructures.

11.2 Interfaces in Multiferroic Heterojunctions

Advancement in film-growth technologies through the last decade has significantly improved our capability to create various complex material interfaces. The ability to couple multiple degrees of freedom has brought about new and exciting opportunities for both fundamental physics study and applications

[21,25,26]. In particular, multiferroic composites or heterostructures provide a unique ground for the mixing of ferroic properties [25]. However, there are various interfacial phenomena that both complicate and facilitate magnetoelectric coupling in such systems. Typically, these come in the form of orbital reconstruction, spin interaction, interlayer mixing, and strain due to lattice mismatch.

Orbital reconstruction is the change of electron orbital direction and energy in response to variations in composition and crystal lattice across the interface [1]. In case of a multiferroic heterostructure, it produces a direct coupling between the magnetic order and electric polarization. This type of configuration allows for the easy manipulation of the magnetoelectric coupling through applied strain or external fields, and may produce effects not accessible in the bulk compounds.

Here, we start by looking into the direct coupling between magnetization and electric polarization at the interface using a standard Ginzburg–Landau formulation. We end by discussing how strain/stress can be used to tune the coupling through the heterostructure.

11.2.1 Ferromagnetism versus Electric Polarization

Through recent years, there have been a number of theoretical studies on the microscopic role of strain and lattice distortion in the magnetoelectric coupling at the interface in a multiferroic heterostructure [27,28]. To answer this question, it is important to first investigate the magnetoelectric coupling arising strictly from interfacial symmetry breaking.

Magnetism and ferroelectricity have different symmetry-breaking mechanisms [29]. Due to inversion symmetry breaking at the interface ($z \neq -z$), the free energy will gain an interaction coupling that is linear with respect to the electric polarization P and quadratic with respect to the magnetization M, λPM^2, where λ is the coupling coefficient. This is illustrated in Figure 11.2 showing a multiferroic heterojunction consisting of distinct ferroelectric and ferromagnetic components. It is assumed that there exists an interface region in each component that extends into the bulk by the correlation lengths ξ_P and ξ_M, respectively. This is typically achieved microscopically through either orbital reconstruction or interlayer mixing at the interface [21,30]. It should be noted that this is not the only type of coupling terms that can exist at the interface. There are many higher order terms to describe this kind of coupling. Here, we discuss only the first order approximation.

Due to the directionality of electric polarization, the symmetry breaking at the interface of a heterojunction produces a coupling that is linear with respect to the z-direction, where $P = (0, 0, P_z)$. The total free energy of the heterostructure is integrated over the bulk (B) and interface (I) regions of a total length $2L$, with the interface regions defined by the correlation lengths of $-\xi_P$ and ξ_M [21]. The bulk and interface can coexist up to the Thomas–Fermi screening length, which is assumed to be much smaller than the interfacial correlation length. We can then define the depth-dependent total free energy as

$$E(z)_{\text{tot}} = E(z)_{\text{B}} + E(z)_{\text{I}} \tag{11.1}$$

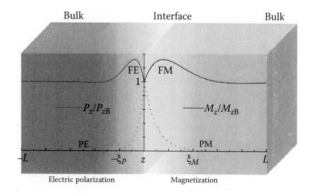

FIGURE 11.2 Illustration of the interfacial magnetoelectric interaction of λPM^2 between ferroelectric (FE) and ferromagnetic (FM) materials. The interface (lighter color) is a relatively small region from $-\xi_P$ to ξ_M. The bulk (darker color) region is the volume that is essentially unaffected by the interfacial interactions. The magnetoelectric interaction with an ordered state produces a fluctuation of the adjacent order parameter. When interacting with a nonordered state, the ordered state induces interfacial ordering. Therefore, an FM state will produce a modulation of the FE state or an interfacial polarization in a paraelectric (PE) state or vice versa. An ordered state is required for this interaction. Therefore, paraelectric–paramagnetic (PE–PM) interfaces will not produce any P or M. (Data from Haraldsen, J., and Balatsky, A., *Mater. Res. Lett.*, 1, 39–44, 2013.)

where

$$E(z)_B = E(z)_{B,P} + E(z)_{B,M} = \alpha_B \left| M_{zB} \right|^2 + \frac{\beta_B}{2} \left| M_{zB} \right|^4 + \gamma_B \left| P_{zB} \right|^2 + \frac{\eta_B}{2} \left| P_{zB} \right|^4 \qquad (11.2)$$

and

$$E(z)_I = E(z)_{I,P} + E(z)_{I,M} + E(z)_{I,\text{int}}$$

$$= \alpha_I \left| M_{zI} \right|^2 + \frac{\beta_I}{2} \left| M_{zI} \right|^4 + g_{M_I} \left(\nabla \left| M_{zI} \right| \right)^2 + \gamma_I \left| P_{zI} \right|^2 + \frac{\eta_I}{2} \left| P_{zI} \right|^4 + g_{P_I} \left(\nabla \left| P_{zI} \right| \right)^2 + \lambda P_{z,I} \left| M_{z,I} \right|^2 \qquad (11.3)$$

Here, $\alpha_I = a(T - T_c)$ describes the ferromagnetic state when $\alpha_I < 0$, since the free energy will produce a nonzero minimum value [21]. Similarly, $\gamma_I = b(T - T_{fe})$ defines the ferroelectric state when $\gamma_I < 0$ [31,32]. As is normal, T_c and T_{fe} are the transition temperatures for the magnetic and ferroelectric ordered states. Here, we mainly consider states that are either ordered (ferroelectric–ferromagnetic) or nonordered (paraelectric–paramagnetic). The overall free energy can be increased or decreased depending on positive or negative P_z at the interface, respectively. Here, g is a positive constant related to the contribution from the gradient of order parameters, where the gradient defines the interface limits at $z = 0$.

As shown in Figure 11.3, the general effect of magnetoelectric coupling can be easily understood by looking at the free energy plots for each order parameter. Since the coupling term is linear with respect to P, it induces a tilt in the energy plot for both the ferroelectric and paraelectric states. This leads to a

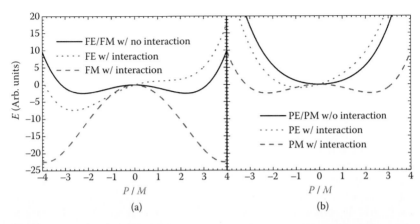

FIGURE 11.3 Effects of the interaction term λPM^2 on the (a) ferroelectric (FE) and ferromagnetic (FM) states as well as (b) paraelectric (PE) and paramagnetic (PM) states. Since the term is linear with respect to polarization, the FE and PE states are shifted to one direction. For magnetization, the interaction term is quadratic, which simply deepens the potential well for the FM state at the interface. For the PM state, the addition of the interaction term makes it possible for the polarization to drive the system into a magnetic state. (Data from Haraldsen, J. and Balatsky, A., *Mater. Res. Lett.*, 1, 39–44, 2013.)

preferred polarization direction in the ferroelectric state (Figure 11.3a) and induces a net polarization in the paraelectric state (Figure 11.3b). The situation is totally different for the magnetization term. Since the coupling term has to be quadratic with regard to the magnetization, due to the requirement of being invariant under time-reversal operation ($M \rightarrow -M$), the interaction enhances the ferromagnetic state near the interface (Figure 11.3a). If the magnetic component is in the paramagnetic state, the magnetoelectric coupling may force the system into a ferromagnetic ordered-state (Figure 11.3b). This clearly shows that the interface coupling in a heterostructure induces changes to the order parameters, which can be treated as perturbations to the parent states.

To further examine the effects of magnetoelectric coupling, we can obtain an analytical solution for the order parameters. Here, the higher-order terms can be ignored, since we are discussing small perturbations produced by the magnetoelectric coupling. These terms will of course be more relevant if polarization and/or magnetization fluctuations become larger. Ignoring the higher-order terms helps the minimization of free energy, and we can determine the equations of motion for $P_{z'I}$ and $M_{z'I}$ through a Fourier transformation, which produces the following conditions:

$$\left(\gamma_I + g_{P_I} k_{P_I}^2 \right) P_{k_I} + \lambda |M_0|^2 = 0 \tag{11.4}$$

and

$$\left(\alpha_I + g_{M_I} k_{M_I}^2 \right) M_{k_I} + 2\lambda P_0 M_0 = 0 \tag{11.5}$$

where P_0 and M_0 are the effective order parameters at the interface ($z = 0$), which is considered to have a finite thickness of the Thomas–Fermi screening length [21]. Since the polarization and magnetization will be small at the interface, we assume that β_I and η_I can be ignored.

Here, k is obtained through the standard Fourier transformation of the spatial gradients. Under these conditions, there are only a few combinations of order parameters at the interface to be considered: ferroelectric–ferromagnetic, ferroelectric–paramagnetic, or paraelectric–ferromagnetic. The paraelectric–paramagnetic combination will not produce any effect since both order parameters at the interface are zero.

Ferroelectric–Ferromagnetic Interface: If both materials are below their Curie temperatures, then the system contains ferroelectric–ferromagnetic interfaces, where γ_I and α_I are both negative. Through the equations of motion for P_{kI} and M_{kI} and Fourier transformation into real space, the order parameters within the interface region can be determined for the direct coupling. For completeness, the electric polarization needs to be written as a sum of the bulk and interface components $P_z = P_{zI} + P_{zB}$ for $z < 0$ (as shown in Figure 11.2). The polarization is written as

$$P_z = \frac{\lambda M_0^2}{|\gamma_I| \xi_P} \sin\left(\frac{|z|}{\xi_P}\right) e^{|z|/\xi_P} + P_{ZB} \tag{11.6}$$

where P_{zB} is the bulk electric polarization. For the magnetization, we define $M_z = M_{zI} + M_{zB}$. Using the same Fourier transformation for $z > 0$, the solution is given by

$$M_z = \frac{2\lambda P_0 M_0}{|\alpha_I| \xi_M} \sin\left(\frac{|z|}{\xi_M}\right) e^{|z|/\xi_M} + M_{ZB} \tag{11.7}$$

Here, M_{zB} is the bulk magnetization [21]. The coupling between the ferroelectric and ferromagnetic states produces a modulation of the order parameters, which decay over a length on the order of $\xi_P = (g_{PI}/|\gamma_I|)^{1/2}$ for the electric polarization, and $\xi_M = (g_{MI}/|\alpha_I|)^{1/2}$ for the magnetization, respectively. The interface constraints are $P_0 = P_{zB}$ and $M_0 = M_{zB}$, which is illustrated in Figure 11.2.

Paraelectric–Ferromagnetic or Ferroelectric–Paramagnetic Interfaces: At temperatures between the Curie temperatures of the two components, the combinations of paraelectric–ferromagnetic or ferroelectric–paramagnetic appear at the interface, where one of γ_I and α_I (but not both) becomes negative. For the paraelectric–ferromagnetic case, the inclusion of the linear coupling term induces an interfacial polarization in the paraelectric phase at $z = 0$, which has a decay length of ξ_P. This is a similar effect to that predicted in case of paraelectric and superconducting coupling as discussed in Reference [13].

Using the same Fourier transformation, the electric polarization for $z < 0$ is given by

$$P_z = \frac{\lambda M_0^2}{|\gamma_I| \xi_P} e^{-2|z|/\xi_P} \tag{11.8}$$

Since the bulk has no net polarization, $P_{zB} = 0$, the polarization around the interface is produced solely from the magnetoelectric coupling, which decays to zero in the bulk. Therefore, the interface constraints are different from that in the ferroelectric–ferromagnetic case. Here, $M_0 = M_{zB}$ and $P_0 = \lambda M_{zB}{}^2/|\gamma_1|\xi_P$. This leads to the induced polarization being dependent on the magnetization and coupling strength. As shown in Figure 11.2, the polarization has a finite value at $z = 0$ and decays to zero in the bulk. This in turn produces a modulation of the magnetization in the ferromagnetic phase through a feedback mechanism.

For the ferroelectric–paramagnetic case, the interface polarization is equal to that of the bulk ($P_0 = P_{zB}$). Therefore, the magnetization transition temperature is given by

$$\hat{\alpha}_1 = \alpha_1 + g_{M_1} k_{M_1}^2 + \lambda P_0 = a_M \left(T - T_0 \right) \tag{11.9}$$

which means that depending on the polarization and magnetoelectric coupling, there could be a phase transition from paramagnetic state to ferromagnetic at the interface. In the Ginzburg–Landau approximation (as shown in Figure 11.3d), this shift can be represented by

$$T_c = T_c^0 + \left(\frac{g_{M_1} k_{M_1}^2 + \lambda P_0}{a_M} \right) \tag{11.10}$$

which will shift the magnetic transition temperature depending on the combined effect of coupling and electric polarization locally at the interface.

To determine the initial magnetization analytically, we reintroduce the quartic term $\beta_1|M|^4$ to the free energy, and assume the previously obtained solution for the electric polarization. This leads to $M_0 = (\hat{\alpha}_1/\beta_1)^{1/2}$, which is the standard Ginzburg–Landau solution. Similar to the paraelectric–ferromagnetic case, the magnetization as a function of z is given by

$$M_z = \frac{2\lambda P_0 M_0}{|\alpha_1|\xi_M} e^{-2|z|/\xi_M} \tag{11.11}$$

This provides a constraint to the polarization at $z = 0$ of $P_{zB} = \alpha_1 \xi_M/2\lambda$ in order to induce a magnetization [21]. A similar feedback produces a modulation of the electric polarization as described in the ferroelectric–ferromagnetic case.

Therefore, the coupling between an ordered state and a nonordered state will induce either a polarization or magnetization, depending on which state is ordered. The induced polarization or magnetization at the interface, in turn, will generate spatial modulation of the order parameter in the parent phase. Since we are using a mean-field approach, it is not sensitive enough to distinguish between metallic and insulator states. The technique assumes general interactions through the interface. However, it should be noted that ferroelectric materials are typically insulators and mobile carriers from a magnetic metallic state could increase the microscopic interactions.

11.2.2 Strain-Induced Magnetoelectric Coupling

Up to this point, we have only considered the magnetoelectric coupling that is produced through fundamental interactions at the interface, which are likely to be through orbital ordering or reconstruction. However, the interface in a heterojunction produces strain due to lattice mismatch between the layers.

Multiferroic Materials

This strain can mediate a magnetoelectric coupling of its own. In a phenomenological model, this adds another term to the free energy.

Strain-induced magnetoelectric coupling is typically analyzed based on the magnetostrictive and piezoelectric properties of the magnetic and ferroelectric phases of the heterostructure. Because of elastic continuity at the interface, structural changes on one side of the interface can easily be transferred to the other. If the magnetism and polarization are coupled to these structural changes, then we achieve the ability to control magnetization (polarization) with an electric field (magnetic field) [33,34]. In perovskite oxide-based heterostructures, this type of coupling is usually highly directional due to its connection with Jahn–Teller distortions, which is very sensitive to epitaxial strain [35,36]. Overall, this means that electronic and magnetic structure in a multiferroic heterostructure can be manipulated through either the application of mechanical strain or the magnetic and/or electric fields (Figure 11.4). We can thus describe the magnetoelectric effect as follows:

$$ME = \frac{P}{S} \otimes \frac{S}{H} \tag{11.12}$$

where H is the magnetic field, P is the electric polarization, and S is the mechanical strain [35]. This can also occur in the converse manner, where the application of an electric field induces change in the magnetization:

$$ME = \frac{M}{S} \otimes \frac{S}{E} \tag{11.13}$$

Here, M is the magnetization and E the electric field.

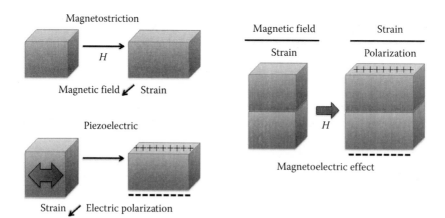

FIGURE 11.4 The strain-mediated magnetoelectric effect. A magnetic field induces strain in the magnetic material, which is transferred to the ferroelectric across the interface and produces electric polarization. The converse effect is also possible.

In terms of the phenomenological model discussed in the previous session, we can expand the free energy to incorporate the proper lattice and crystallographic properties into

$$P = D - eS - \varepsilon E - \alpha H \tag{11.14}$$

and

$$M = B - \mu H \tag{11.15}$$

where strain is introduced by

$$\sigma = c\varepsilon - e^{\mathrm{T}} E - c\varepsilon^{\mathrm{rms}} - \sigma_{\mathrm{s}} \tag{11.16}$$

Here, P and M are polarization and magnetization described previously. σ, S, D, and B are, respectively, the stress, strain, electric displacement, and magnetic induction, while c, ε, and μ are the stiffness, dielectric constant, and permeability. The parameter e is the piezoelectric coefficient and e^{T} is the transpose of tensor e. $\varepsilon^{\mathrm{rms}}$ is the magnetic field–induced strain produced by the field-dependent magnetostriction constants. The residual stress σ_{s} (or residual strain ε_{s}) is included for multiferroic heterojunctions. The breakdown of these tensor equations can be found in References [22,23,35].

The introduction of strain and stress brings complication to the general phenomenological model, which was demonstrated by Liu et al. using the Ginsburg–Landau theory [24]. This shows the complex nature of magnetoelectric heterojunctions, which will continue to serve as a playground for studying various coupling phenomena.

11.3 Conclusions and Outlook

The process of searching for new materials with multifunctionalities has led to the development of multiferroic heterostructures, where the specific material properties interact through an interface. While this has led to many challenges in and of itself, the benefit is that we gain a better understanding and control of magnetic and electronic states at the interface. In this chapter, we have presented a standard method for analyzing the interfacial coupling between polarization and magnetization with a brief introduction on effect of strain.

However, the complex interactions at the interface involving orbital reconstruction and hybridization, lattice strain, atomic defects and interdiffusion have placed an overwhelming challenge on researchers looking to enhance and control the magnetoelectric coupling effect. Future research should focus on deconvoluting these contributions. Furthermore, understanding domain walls and vortex formation as well as the effects of other order parameters like multipolar and orbital ordering may provide other avenues for device applications.

References

1. Dagotto, E. 2005. Complexity in strongly correlated electronic systems. *Science* 309:257–262.

2. Fiebig, M. 2005. Revival of the magnetoelectric effect. *Journal of Physics D Applied Physics* 38:R123–R152.

3. Spaldin, N. A., and M. Fiebig. 2005. The renaissance of magnetoelectric multiferroics. *Science* 309:391–392.

4. Lottermoser, T., T. Lonkai, U. Amann, D. Hohlwein, J. Ihringer, and M. Fiebig. 2004. Magnetic phase control by an electric field. *Nature* 430:541–544.

5. Balke, N., S. Choudhury, S. Jesse, M. Huijben, Y. H. Chu, A. P. Baddorf, L.-Q. Chen, R. Ramesh, and S. V. Kalinin. 2009. Deterministic control of ferroelastic switching in multiferroic materials. *Nature Nanotechnology* 4:868–875.

6. Cheong, S.-W., and M. Mostovoy. 2007. Multiferroics: a magnetic twist for ferroelectricity. *Nature Materials* 6:13–20.

7. Ramesh, R., and N. A. Spaldin. 2007. Multiferroics: progress and prospects in thin films. *Nature Materials* 6:21–29.

8. Martin, L., S. Crane, Y. Chu, M. Holcomb, M. Gajek, M. Huijben, C. Yang, N. Balke, and R. Ramesh. 2008. Multiferroics and magnetoelectrics: thin films and nanostructures. *Journal of Physics Condensed Matter* 20:434220.

9. Park, T.-J., G. C. Papaefthymiou, A. J. Viescas, A. R. Moodenbaugh, and S. S. Wong. 2007. Size-dependent magnetic properties of single-crystalline multiferroic BiFeO$_3$ nanoparticles. *Nano letters* 7:766–772.

10. Kharrazi, S., D. C. Kundaliya, S. Gosavi, S. Kulkarni, T. Venkatesan, S. Ogale, J. Urban, S. Park, and S. -W. Cheong. 2006. Multiferroic TbMnO$_3$ nanoparticles. *Solid State Communications* 138:395–398.

11. Bibes, M., J. E. Villegas, and A. Barthelemy. 2011. Ultrathin oxide films and interfaces for electronics and spintronics. *Advances in Physics* 60:5–84.

12. Béa, H., M. Gajek, M. Bibes, and A. Barthélémy. 2008. Spintronics with multiferroics. *Journal of Physics Condensed Matter* 20:434221.

13. Haraldsen, J., S. Trugman, and A. Balatsky. 2011. Induced polarization at a paraelectric/superconducting interface. *Physical Review B* 84:020103.

14. Hill, N. A. 2000. Why are there so few magnetic ferroelectrics? *The Journal of Physical Chemistry B* 104:6694–6709.

15. Neaton, J., C. Ederer, U. Waghmare, N. Spaldin, and K. Rabe. 2005. First-principles study of spontaneous polarization in multiferroic BiFeO$_3$. *Physical Review B* 71:014113.

16. Seshadri, R., and N. A. Hill. 2001. Visualizing the role of Bi 6s "lone pairs" in the off-center distortion in ferromagnetic BiMnO$_3$. *Chemistry of Materials* 13:2892–2899.

17. Ikeda, N., H. Ohsumi, K. Ohwada, K. Ishii, T. Inami, K. Kakurai, Y. Murakami, K. Yoshii, S. Mori, and Y. Horibe. 2005. Ferroelectricity from iron valence ordering in the charge-frustrated system LuFe$_2$O$_4$. *Nature* 436:1136–1138.

18. Katsura, H., N. Nagaosa, and A. V. Balatsky. 2005. Spin current and magnetoelectric effect in noncollinear magnets. *Physical Review Letters* 95:057205.

19. Sergienko, I. A., and E. Dagotto. 2006. Role of the Dzyaloshinskii-Moriya interaction in multiferroic perovskites. *Physical Review B* 73:094434.

20. Arima, T.-H. 2007. Ferroelectricity induced by proper-screw type magnetic order. *Journal of the Physical Society of Japan* 76:073702.

21. Haraldsen, J., and A. Balatsky. 2013. Effects of magnetoelectric ordering due to interfacial symmetry breaking. *Materials Research Letters* 1:39–44.

22. Nan, C.-W., G. Liu, Y. Lin, and H. Chen. 2005. Magnetic-field-induced electric polarization in multiferroic nanostructures. *Physical Review Letters* 94:197203.

23. Wang, Y., J. Hu, Y. Lin, and C.-W. Nan. 2010. Multiferroic magnetoelectric composite nanostructures. *NPG Asia Materials* 2:61–68.

24. Liu, G., C.-W. Nan, and J. Sun. 2006. Coupling interaction in nanostructured piezoelectric/magnetostrictive multiferroic complex films. *Acta Materialia* 54:917–925.

25. Zubko, P., S. Gariglio, M. Gabay, P. Ghosez, and J.-M. Triscone. 2011. Interface physics in complex oxide heterostructures. *Annual Review of Condensed Matter Physics* 2:141–165.

26. Kimura, T., T. Goto, H. Shintani, K. Ishizaka, T. Arima, and Y. Tokura. 2003. Magnetic control of ferroelectric polarization. *Nature* 426:55–58.

27. Duan, C.-G., S. S. Jaswal, and E. Y. Tsymbal. 2006. Predicted magnetoelectric effect in $Fe/BaTiO_3$ multilayers: ferroelectric control of magnetism. *Physical Review Letters* 97:047201.

28. Rondinelli, J. M., M. Stengel, and N. A. Spaldin. 2008. Carrier-mediated magnetoelectricity in complex oxide heterostructures. *Nature Nanotechnology* 3:46–50.

29. Eerenstein, W., N. Mathur, and J. F. Scott. 2006. Multiferroic and magnetoelectric materials. *Nature* 442:759–765.

30. Cai, T., S. Ju, J. Lee, N. Sai, A. A. Demkov, Q. Niu, Z. Li, J. Shi, and E. Wang. 2009. Magnetoelectric coupling and electric control of magnetization in ferromagnet/ferroelectric/normal-metal superlattices. *Physical Review B* 80:140415.

31. Jona, F., and G. Shirane. 1962. *Ferroelectric crystals.* Oxford: Pergamon.

32. Rabe, K. M., C. H. Ahn, and J. -M. Triscone. 2007. *Physics of ferroelectrics: a modern perspective.* Berlin: Springer.

33. Hu, J. -M., and C. Nan. 2009. Electric-field-induced magnetic easy-axis reorientation in ferromagnetic/ferroelectric layered heterostructures. *Physical Review B* 80:224416.

34. Ma, J., J. Hu, Z. Li, and C. W. Nan. 2011. Recent progress in multiferroic magnetoelectric composites: from bulk to thin films. *Advanced Materials* 23:1062–1087.

35. Nan, C.-W. 1994. Magnetoelectric effect in composites of piezoelectric and piezomagnetic phases. *Physical Review B* 50:6082–6088.

36. Thiele, C., K. Dörr, O. Bilani, J. Rödel, and L. Schultz. 2007. Influence of strain on the magnetization and magnetoelectric effect in $La_{0.7}A_{0.3}MnO_3/$ PMN–PT (001)(A= Sr, Ca). *Physical Review B* 75:054408.

SECTION IV

Emerging Topics in the Field

12

Topological Structures in Multiferroics
Domain Walls, Vortices, and Skyrmions

Jan Seidel

Contents

Multiferroic transition metal oxides exhibit a wide variety of properties that are related to the delicate balance between lattice, charge, spin, and orbital degrees of freedom [1]. Modern thin-film synthesis methods such as pulsed laser deposition (PLD), molecular beam epitaxy (MBE), and metal–organic chemical vapor deposition (MOCVD), among others, enable the engineering of interfaces between complex transition metal oxides with atomic-scale precision. Interfaces introduce a variety of local changes, for example, they break the symmetry of the material, induce stress, and lead to variation in the bonding between ions. These in turn give rise to changes in electronic structure such as bandwidth, orbital interactions, and energy-level degeneracy. Charge transfer in these correlated systems can induce changes in carrier densities at interfaces that result in physical properties that may be completely different from those of the parent bulk phases [2–6].

Physical phenomena involving topological structures, such as domain walls, vortices, and skyrmions, in multiferroics have recently received considerable attention. Nanoscale elements made from these materials are highly tunable through thin-film growth engineering or applied external strain, electric, and magnetic fields, offering unique possibilities for novel concepts in complex oxide nanoelectronics and spintronics.

12.1 Introduction

Many transition metal oxides are ferroics, that is, materials with a spontaneous, reversible ordering, below a critical temperature called the Curie temperature. In general, in the vicinity of a ferroic transition one or more macroscopic properties of the material associated with the order parameter can become large and very susceptible to external stimuli. Field-induced phase transitions around the Curie temperature are a common feature. Ferroic-phase transitions include ferroelectric, ferroelastic, and ferromagnetic transitions. These involve the emergence of spontaneous polarization, spontaneous strain, or spontaneous magnetization and are commonly referred to as primary or first-order ferroics. Many so-called smart materials and structures have at least one of these properties and are designed to change in a preconceived manner through the application of external fields. Multiferroics are defined as materials that have at least two of these ferroic properties: ferromagnetism, ferroelectricity, ferroelasticity, or ferrotoroidicity (or their antiferroic counterparts, e.g., antiferromagnetism).

The order parameter can have two or more distinct orientations. Spatial variations in the orientation of the order parameter give rise to distinct domain structures. Within each domain, the orientation of the order parameter is the same. Adjacent domains are separated by planar topological defects called domain walls. Domain walls are thus naturally occurring interfaces and topological boundaries in complex oxides as opposed to artificially created hetero-interfaces between different materials [7]. They have been investigated with respect to the mechanical and electrical behaviors in various studies involving ceramics, thin films, and single crystals. The fact that certain domain walls can be very conductive has sparked renewed interest and led to the discovery of unique properties associated with such domain walls [8,9]. First observations in $BiFeO_3$ have led to the investigation of other ferroelectrics and multiferroics, which also exhibit electronic conduction at domain walls, albeit in different forms. Similar phenomenon may also apply to structural phase boundaries [10–12], for example, morphotropic phase boundaries in ferroelectrics, including $BiFeO_3$ [13].

Although domain walls are ubiquitous in ferroic materials, it is not the only way in which spatially varying order parameters can be arranged. Alternatively, more complex patterns can develop in which the order parameter changes continuously over space without any clearly defined domain or domain wall structure. When combined with local defects or singularities, the number of possible geometrical patterns can be manifold, including vortex structures and skyrmions [14]. Which kind of micro- or nanostructure develops depends critically on the relative magnitudes of various energies such as exchange, crystallographic anisotropy, and adjacent surfaces. Physical dimensions of the specific material and its morphology are also important in determining the exact nature of the equilibrium ferroic patterns [15]. Tuning and utilizing the physical properties of such topological structures provides a new playground for research. In addition, it offers a novel platform for future nanotechnology [16].

12.2 Multiferroic Domain Walls

12.2.1 Local Structure Investigations of Multiferroic Domain Walls

Among the methods available for studying multiferroic materials at nano and atomic scales is high-resolution transmission electron microscopy (HRTEM) [17–21]. It allows for direct visualization of the lattice distortion across the domain wall by measuring the continuous deviation of a set of planes with respect to the undistorted lattice (exit-wave reconstruction), as shown in Figure 12.1. Current state-of-the-art techniques permit atomic-scale resolution of 0.5 Å, through aberration-corrected imaging. The exit-wave reconstruction approach eliminates the effects of objective-lens spherical aberrations, and images can be directly interpreted in terms of the projection of atomic columns.

Weak beam transmission electron microscopy has been used for a quantitative analysis of thickness fringes that appear on images of inclined domain walls. By fitting simulated fringe profiles to experimental ones, it is possible to extract the thickness of multiferroic domain walls in a quantitative manner. Regarding HRTEM images of domain walls, it has to be taken into account that the samples used are very thin (typically a few nm) so that surface pinning of the domain walls could play an important role. The atomic displacements across a typical wall are on the order of 0.02 nm, which makes direct imaging and interpretation still a challenge. HRTEM also offers the possibility of quantitatively measuring the local polarization and investigating the domain structure [22,23]. Elemental and electronic structure analysis by electron energy loss spectroscopy (EELS) has also been applied to the study of domain walls [24–26].

12.2.2 Conductivity at Multiferroic Domain Walls

Scanning probe microscopy (SPM) with its many variations, for example, conductive atomic force microscopy (c-AFM), piezoresponse force microscopy (PFM), are well suited for the characterization of prototype oxide structures at the nanoscale, including domain walls [27]. With c-AFM, one can

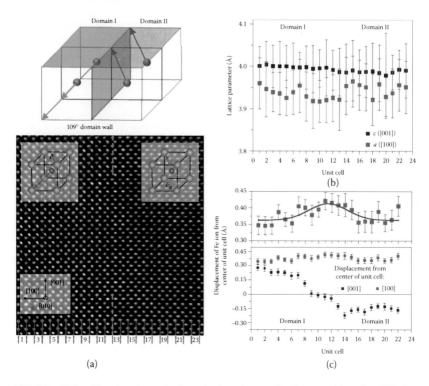

FIGURE 12.1 Structural analysis of domain walls in multiferroic $BiFeO_3$. (a) Schematic diagram of a 109° domain wall and exit-wave-reconstructed high-resolution transmission electron microscopy image of a 109° domain wall taken along the [010] zone axis. (b) Extracted a and c lattice parameters for each unit cell across the domain wall. (c) Extracted Fe-ion displacement relative to the Bi lattice for each unit cell across the domain wall. A close-up (upper panel) reveals an increase in the component of polarization perpendicular to the domain wall. (Adapted from Seidel, J. et al., *Nat. Mater.*, 8, 229–234, 2009.)

probe local conductivity at interfaces and topological defects. PFM has been further developed and now offers dynamic characterization of the switching process in stroboscopic PFM and PFM spectroscopy modes [28–31]. Ferroelectric thin films typically contain various structural defects such as cationic and anionic point defects, dislocations, and grain boundaries. The electric and stress fields around such defects are inhomogeneous. It can therefore be expected that the switching behavior near such a structural defect will be different from that found in the normal region. The recent spike of interest in local conductivity at domain walls arises from perspectives for nonvolatile electroresistive memory devices [32]. Such functionality is also interesting for the characterization of oxygen vacancy movement and ionic battery materials at the nanoscale [33–35]. In many cases, the presence of extended defects and oxygen vacancies makes the identification of polarization-modulated transport mechanisms difficult. The combination of local electromechanical and conductivity measurements has been an interesting approach and it has revealed the connection between local current and pinning at bicrystal grain boundaries in $BiFeO_3$ [36]. Electroresistance in ferroelectric structures has recently been reviewed by Watanabe [37]. The presence of extended defects and oxygen vacancy accumulation has also been shown to affect the transport and nucleation mechanisms at domain walls [38–40].

As another local characterization method, scanning tunneling microscopy (STM) can be used to probe directly the electronic structure at nanometer scale. For example, STM and scanning tunneling spectroscopy (STS) have been used to investigate the electronic structure of ferroelastic twin walls in $YBa_2Cu_3O_{7-\delta}$ [41], which play an important role in pinning the vortices in this type II superconductor and thereby enhance the currents that the material can support while remaining superconducting. In the case of insulators, STM/STS are considerably more difficult to conduct, primarily because of the small tunneling current and charging effect. The emergence of semiconducting ferroelectrics with smaller bandgaps and the possibility of conduction at domain walls, as described in the following, has stimulated renewed interest in exploring STM as a probe of local electronic structure in complex oxides. The emergence of combined AFM/STM or SEM/STM systems are especially interesting for exploring the electronic properties of domain walls in such insulating materials. Investigations using such combined tools have been demonstrated, including measurements on $BiFeO_3$ domain walls [42–45].

The changes in structure (and as a consequence electronic structure) that occur at ferroelectric (multiferroic) domain walls [46] can lead to changes in transport behavior (Figure 12.2). Indeed, domain wall conductivity has been shown in different ferroic materials, although with different transport behavior. The domain walls of $BiFeO_3$ were found to be more conductive than the domains [38,47,48], while those of $YMnO_3$ were more insulating or conductive depending on their orientation [49,50]. In multiferroic $YMnO_3$, a so-called improper ferroelectric where ferroelectricity is not the primary order parameter but rather induced by structural trimerization coexisting with magnetism, domain walls are found to be charged and stable and their conductivity depends on the specific orientation of the domain wall with respect to the polarization vector. The observed conduction suppression at domain walls at high voltages (still much less than the electric coercivity) is in striking contrast with what has been reported for $BiFeO_3$.

STM and STS measurements in cross-sectional samples have been used to directly investigate the nature of the local electronic conductivity at ferroelectric domain walls in multiferroic $BiFeO_3$ [45]. Samples with engineered stripe domain array were used to simplify the location of a single wall in the STM setup for local spectroscopy. It was found that in situ cleaved samples with ordered stripe arrays show decreases of the bandgap at the domain boundaries (Figure 12.3). In addition, a shift toward the Fermi level in the band edges of 109° and 71° domain walls has been observed. The approach demonstrated in this work serves as a model technique to investigate and understand electronic structure at complex oxide interfaces.

The electronic structure of domain walls in $BiFeO_3$ has been investigated theoretically by Lubk et al. [51]. Using density functional theory (DFT), the

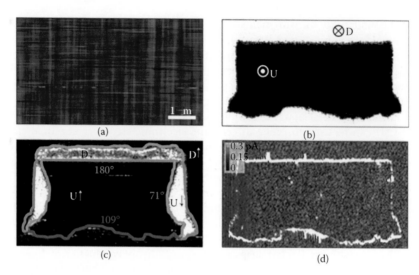

FIGURE 12.2 Three different types of domain walls in rhombohedral $BiFeO_3$. (a) AFM topography image, (b) out-of-plane piezoresponse force microscopy (PFM), (c) in-plane (PFM) image of a written domain pattern in a monodomain $BiFeO_3$ (110) thin film, and (d) the corresponding conductive atomic force microscopy (c-AFM) image showing conduction at both 109° and 180° domain walls (right). (Adapted from Seidel, J. et al., *Nat. Mater.*, 8, 229–234, 2009.)

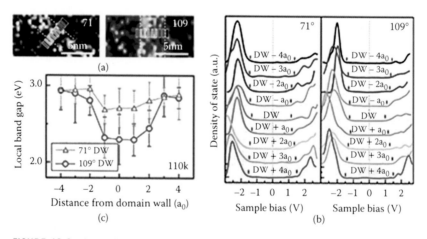

FIGURE 12.3 Layer-by-layer dI/dV measurements across 71° and 109° domain walls in $BiFeO_3$ acquired at 110 K. Bars in (a) denote positions where the electronic spectra are probed, and (b) shows the corresponding scanning tunneling spectroscopy (STS) spectra. The band edges are indicated by black tick marks in (b). (c) Extracted local bandgap across the domain walls. (Adapted from Chiu, Y.P. et al., *Adv. Mater.*, 23, 1530–1534, 2011.)

layer-by-layer densities of states were calculated and this shows that the domain walls have a significantly reduced bandgap compared to the R3c bulk. Structural changes at the walls lead to arrangements that approach the ideal cubic structure, in which the 180° Fe–O–Fe bond angle maximize the Fe 3d–O 2p hybridization and hence the bandwidth of the material. Note that in no case does the bandgap approach zero in the wall region. This study also gives insight into the changes in Fe–O–Fe bond angle in $BiFeO_3$. The rotations of the oxygen octahedra are found to remain in phase across the walls, that is, more favorable than that with rotation discontinuity (antiphase boundaries).

The same DFT calculations also revealed steps in the electrostatic potential for all domain wall types, which is important, as they also contribute to the electronic conductivity. Steps in the electrostatic potential at domain walls are correlated with (and caused by) small changes in the component of the polarization normal to the wall [38]. These changes in normal polarization are a consequence of the specific rotation of the polar vector across the domain wall, and are also found in other materials. Tetragonal $PbTiO_3$, for example, shows a similar effect for 90° domain walls [52]. Extended phase-field calculations for tetragonal $BaTiO_3$ also reveal intrinsic electrostatic potential drop across the 90° domain wall, regardless of the consideration of the ferroelectric as n-type semiconductor or dielectric [53]. This potential change creates a large local electric field that promotes asymmetric charge distribution around the walls, that is, they foster the accumulation of carriers and oxygen vacancies close to the wall. The increased charge density plays a role in increased conductivity at such walls.

As mentioned before, the semirigid rotation of the polar vector across a ferroelectric domain wall leads to an electrostatic potential that is screened by free charges, which enhance the local charge density and thus, presumably, the conductivity. Since this polar rotation [54] is not exclusive to $BiFeO_3$, other perovskite ferroelectrics should also be expected to display enhanced wall conductivity. This indeed has been found for other multiferroics such as hexagonal manganites ($YMnO_3$, $ErMnO_3$) [49,50]. For $BiFeO_3$, several other factors might be further helping the conductivity enhancement: First, magnetoelectric coupling between polarization and spin lattice is such that the magnetic sublattice rotates with the polarization [55,56]. Since spins rotate rigidly, they might favor a more rigid rotation of the polarization and hence a bigger electrostatic step at the wall (and of course, the polarization of $BiFeO_3$ is itself bigger than that of other known perovskite ferroelectrics, which means that all other things being equal a rigid polar rotation in $BiFeO_3$ will cause a larger electrostatic step). But perhaps the most obvious consideration is the fact that $BiFeO_3$ has intrinsically a smaller bandgap than other perovskite ferroelectrics (~2.7 eV instead of 3.5–4 eV). This means that the screening charges accumulated at the wall will be closer to the bottom of the conduction band and hence contribute more easily to the conductivity.

Apart from the undesirable effects associated with defects accumulation at domain walls, the intentional manipulation of electronic structure at walls by doping and strain in ferroelectric and ferroelastic oxides provides a way to effectively engineer and control nanoscale functionality in such materials. For the case of BiFeO$_3$, A-site doping with Ca and B-site substitution with Co or Ni, may prove viable to achieve new domain wall properties through manipulating the electronic structure, spin structure, and dipole moment [63]. Of obvious future interest is the question of what sets the limit to the current transport behavior at walls: can one "design" the domain wall structure to controllably induce electronic phase transitions within the wall? Is it possible to trigger an Anderson transition by doping or straining the domain walls? Domain wall (super)conductivity has been shown before by Aird and Salje [57]. Exposing WO$_3$ to sodium vapor, they observed preferential doping along the ferroelastic domain walls. Transport measurements showed superconductivity with a critical temperature of 3 K, while magnetic measurements did not, suggesting that superconductivity was confined to the domain walls only, which provided a percolating superconductive path while occupying a very small volume fraction of the crystal.

Recently, the tunable electronic conductivity at domain walls in La-doped BiFeO$_3$ was linked to oxygen vacancies (Figure 12.4) [39]. Specific growth conditions have been used to introduce varying amounts of vacancies in thin films with ordered stripe domain patterns [58]. The conductivity at 109° walls in such samples was shown to be thermally activated with activation energies of 0.24–0.5 eV. These results are a first step toward inducing an insulator–metal transition locally within the domain walls through careful design of the electronic structure, the state of strain, and chemical compositions. For actual device applications, it would be desirable to increase the magnitude of the wall currents from currently achieved values. The choice of the right shallow-level dopants and host material might be the key factor.

This observation may help merging magnetoelectrics and magnetoelectronics at room temperature by combining electronic conduction with electric and magnetic degrees of freedom [59–61]. With this in mind, Ca-doping of domain walls and phase boundaries [62–64] in BiFeO$_3$ constitutes an interesting direction for future research.

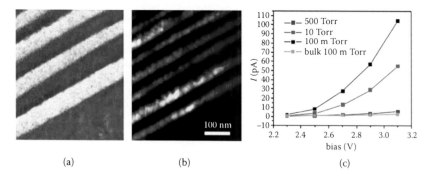

FIGURE 12.4 (a) PFM image showing ordered 109° stripe domains in BiFeO$_3$. (b) Simultaneously acquired c-AFM image of the same area showing that each 109° domain wall is electrically conductive. (c) Current levels for samples with different oxygen-cooling pressure and thus varying density of oxygen vacancies. (Adapted from Seidel, J. et al., *Phys. Rev. Lett.*, 105:197603, 2010.)

12.2.3 Light Interaction with Multiferroic Domain Walls

$BiFeO_3$ exhibits an anomalous photovoltaic effect in thin films arising from a unique mechanism, namely, structurally driven electrostatic potential steps at nanometer-scale domain walls [65–67]. In conventional solid-state photovoltaic cells, electron–hole pairs are created by light absorption in a semiconductor and separated by the electric field spanning a micrometer-thick depletion region. The maximum voltage these devices can produce is limited by the semiconductor bandgap. Interestingly, domain walls can give rise to a fundamentally different mechanism for photovoltaic charge separation, which operates over a distance of 1–2 nm and, when combined, produces voltages that are significantly higher than the bandgap. Recent investigations using c-AFM under illumination reveal these high photo-voltages at 71° and 109° domain walls in $BiFeO_3$ (see Figure 12.5) [67]. The charge separation happens at previously unobserved nanoscale electrostatic potential steps that naturally occur at ferroelectric domain walls in $BiFeO_3$. Electric-field control of domain structure allows the photovoltaic effect to be reversed in polarity or turned off.

Currently, the overall efficiency of this photovoltaic effect in devices is limited by the conductivity of the bulk material. Methods to increase the carrier mobility, as well as inducing the spatially periodic potential in an adjacent material with a lower gap than $BiFeO_3$ are possible routes to achieve larger current densities under visible light. Low bandgap semiconductors with asymmetric electron and hole mobilities are possible candidates to show such an effect. In addition, photoelectrochemical effects at domain walls are other interesting properties to explore for applications such as water splitting [68,69].

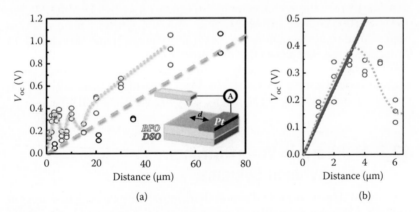

(a) (b)

FIGURE 12.5 Local measurements of open circuit voltage V_{OC} at domain walls in $BiFeO_3$. (a) 109° domain walls (blue) show different, oscillating behavior of Voc with distance compared with 71° domain walls (black). Inset: Schematic photovoltaic measurement setup using a c-AFM tip as a variable-distance counter electrode for $I–V$ characterization. (b) Initial large slope indicating large PV effect at 109° walls. (Adapted from Seidel, J. et al., *Ferroelectrics* 433:123–126, 2012.)

12.2.4 Magnetic Properties of Multiferroic Domain Walls

Over the last decade, many interesting phenomena have been observed in epitaxial heterostructures such as metallic and superconducting behavior at the interface between two insulators [3,70], metals and insulators [71], or gated oxide surfaces [72], promoted by electronic reconstruction due to charge transfer or carrier redistribution. Charge transfer has been found to trigger novel magnetic phases at the manganite–ruthenate interfaces [73,74]. Unusual phenomena such as inducing magnetism in a superconductor and the rearrangement of the magnetic domains in superconductor/ferromagnet heterostructures due to orbital rearrangement and strong hybridization [75] as well as other ways of combining superconductors with metallic ferromagnets [76] have been reported.

An important aspect of the multiferroic domain walls in BiFeO$_3$ concerns their true state of magnetism. Temperature-dependent transport measurements have been performed and are a possible route to clarify the actual spin structure and whether it exhibits a glasslike or ordered ferromagnetic state [7]. Of additional interest is the effect of extra carriers introduced into the system, for example, by doping or electric gating, on magnetism. This includes doping with magnetic ions like Co or Ni.

Is there a way to change the magnetic interaction from superexchange to double exchange in BiFeO$_3$, maybe locally at the walls? The strength of the coupling between the ferroelectric and antiferromagnetic walls in BiFeO$_3$ is an issue that still needs to be resolved from both theoretical and experimental perspectives. The role of dimensionality on electrical and magnetoelectrical transport properties needs to be elucidated and put into perspective with regard to known systems, such as manganites [77,78]. Recently, a notable magnetoelectric effect has been observed due to the coexistence of different spiral-spin domains and the motion of multiferroic domain walls in DyMnO$_3$ [79,80]. The interaction between ferroelectric and antiferromagnetic domain walls has been studied in model multiferroics such as YMnO$_3$ [81] and BiFeO$_3$ [82]. In both cases, it has been shown that the antiferromagnetic domain walls are significantly wider (by approximately one to two orders of magnitude) compared to the ferroelectric walls. This is also in agreement with the phenomenological predictions of Daraktchiev et al. [83] for coupling-mediated wall broadening.

12.3 Multiferroic Vortex Structures and Skyrmion Systems

Topological structures beyond simple domain walls are exciting new areas of research. A detailed discussion on vortex structures is presented in Chapter 13 of this book. Here we just give a brief discussion of this topic.

Ferroelectric nanowires have been shown to contain switchable quadrupoles (Figure 12.6a), which makes them interesting for nanodevice applications [84]. The vortex cores in BiFeO$_3$ have been shown to be controllable by the coupled response of polarization and mobile vacancy subsystems to

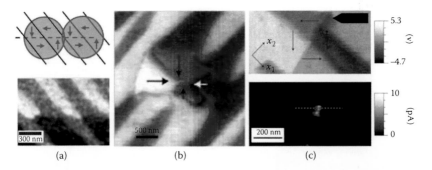

FIGURE 12.6 Vortex-like states can be created by direct writing using an SPM tip. (a) "Vortex quadrupole" chains [87]. (b) "Center" domain patterns [88]. (c) Enhanced conductivity has been observed at the core of flux closure domain patterns [85] in $BiFeO_3$.

external bias (Figure 12.6c) [85]. The details of these complex structures are of high importance. Indeed, a range of interesting dipole patterns surrounding domain walls have been investigated at a very fine scale using an AFM tip. Using a scanning probe to "write" specific domain topologies [86] was further developed in $BiFeO_3$ (Figure 12.6b). The ability to create new topological structures, rather than just observing their properties holds promise for future technologies.

Nelson et al. [90] in their HRTEM work on cross-sections of $BiFeO_3$ films, showed that the domain patterns that form close to interfaces depend on whether or not the interface concerned can sustain charge compensation. At metallic interfaces stripe domains formed, while at insulating interfaces small triangular domains were observed. This study shows that the local dipole reorientations at noncompensating interfaces reduce the depolarization field and exemplifies the continuous polar rotations.

Complex spin orders in magnetic materials can often be stabilized in non-centrosymmetric crystal lattices due to the Dzyaloshinsky–Moriya (DM) interaction. Recently, novel whirl-like spin patterns called skyrmions at length scales of 10–100 nm have been observed in chiral magnets [89,91,92]. The term skyrmion was originally introduced by Skyrme as a model in nuclear physics to describe a localized, particle-like configuration in field theory [86]. The typical two-dimensional spin pattern corresponding to a single skyrmion is shown in Figure 12.7a. It is characterized by a so-called winding number and is topologically stable. After several theoretical proposals, the experimental observation of magnetic skyrmions was first achieved in metallic alloys such as MnSi, FeGe, and $Fe_xCo_{1-x}Si$ by neutron scattering, Lorentz transmission electron microscopy (LTEM), and magnetic force microscopy (MFM) [15].

Interestingly, skyrmions have recently been observed in an insulating multiferroic material, namely, Cu_2OSeO_3 by Seki et al. [92]. The application of magnetic field H leads to the formation of a hexagonal lattice of skyrmions within a plane normal to H in this material only in a narrow H–T region just below T_c. In thin films, it is stabilized over a wider H–T range. Similar

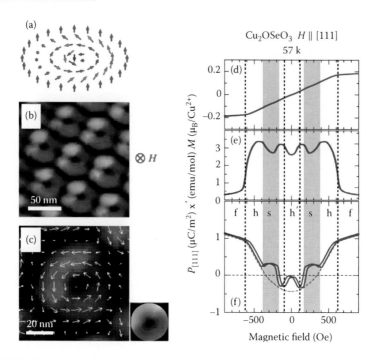

FIGURE 12.7 Skyrmions in multiferroic Cu_2OSeO_3 after Seki et al. [92]. (a) Schematic vector representation of single skyrmion. (b) and (c) Lateral magnetization distribution map for the (110) plane of a thin-film (~100 nm thick) sample of Cu_2OSeO_3 obtained by Lorentz transmission electron microscopy (LTEM) at 5 K, H ~ 800 Oe (white arrows represent the magnetization, color wheel shows the direction (hue), and magnitude (brightness) of the lateral magnetization). (d–f) Magnetic field ($H \parallel [111]$) dependence of magnetization M, ac magnetic susceptibility (χ'), and [111] component of electric polarization ($P[111]$) at 57 K. Red dashed line is the numerical fit for the single-domain helimagnetic state. Letters f, h, h', and s denote ferrimagnetic, helimagnetic (single q domain), helimagnetic (multiple q domains), and skyrmion crystal states, respectively. (Adapted from Seki, S. et al., *Science*, 336, 198-201, 2012.)

magnetic phase diagrams have been reported for the metallic alloys that show skyrmions. The skyrmion spin pattern in Cu_2OSeO_3 can lead to spin-driven electric polarization, indicating magnetoelectric coupling in this material.

Figure 12.7d through f shows the magnetic field dependence of magnetization, ac magnetic susceptibility, and electric polarization along [111] for bulk Cu_2OSeO_3. When the hexagonal skyrmion lattice is formed in the material with a small applied magnetic field, the system becomes polar along the magnetic field direction. The local magnetoelectric coupling in this case leads to skyrmions carrying an electric dipole or quadrupole [93]. Several studies have shown that skyrmions in insulators can be manipulated by electric fields. Small-angle, neutron-scattering experiments show that the application of electric fields can cause a slight rotation of the skyrmion crystal [94]. The hexagonal skyrmion crystal also shows unique magnetic resonance properties under an oscillating magnetic field in the GHz range [95,96]. Local magnetoelectric coupling allows for the activation of these excitations by alternating electric fields (i.e., electromagnon excitation) [97,98]. Thus, magnetoelectric skyrmions may contribute to the design of novel low-dissipation spintronic devices.

12.4 Conclusions and Outlook

Nanoscale domain walls, vortices, and skyrmions in complex oxides form an exciting and growing field of interest in materials science. There are many remaining questions to be answered and new areas to be explored. For example, the investigation of dynamic conductivity at domain walls is an exciting topic [99]. This addresses important factors: a possible electric field–induced distortion of the polarization structure at the domain wall; the dependence of conductivity on the degree of distortion; and weak-pinning scenarios of the distorted wall. The domain wall is very likely not a rigid electronic conductor, instead offering a quasi-continuous spectrum of voltage-tunable electronic states [80]. This aspect has been explored only rudimentarily, while ferroelectric domains were shown to exhibit concrete conductance levels [17,18]. The intrinsic dynamics of domain walls and other topological defects are expected not only to influence future theoretical and experimental interpretations of the electronic phenomena, but also offer a possibility to find unique properties of multiferroic domain walls, for example, magnetization and magnetoresistance within an insulating antiferromagnetic matrix, influenced by order parameter coupling and localized secondary order parameters [52]. Of obvious future interest is the question of what sets the limits to the current transport behavior at walls. It might be possible to "design" the topological structure of the domain wall to controllably induce electronic phase transitions within the wall arising from the correlated electron nature. The observation of superconductivity in ferroelastic walls of WO_3 is an indication of various exciting and unexplored areas of domain boundary physics [57].

Several applications have been proposed to make use of domain walls in multiferroic materials based on their additional functionalities as well as their effects on existing devices [8]. Uses that have been mentioned are as a local strain sensor incorporated on an AFM probe, reconfigurable nanocircuits [9], or a multilevel resistance-state device that is written by an electrical current [99]. Other possibilities include nonvolatile memories, piezoelectric actuators, ultrasound transducers, surface acoustic wave devices, and optical applications. For existing devices, the discovery of conducting domain walls indicates the possibility to engineer the right domain walls that could prevent leakage. Of strong interest are also the dynamic and field tunable properties in these materials [83,100,101]. Artificially engineered topological structures, including domain walls, may pave the way to novel states of matter with a wide range of electronic and magnetoelectric properties. Domain-wall electronics, particularly with multiferroics, may become interesting for future nanotechnology [102,103].

References

1. Imada, M., A. Fujimori, and Y. Tokura. 1998. Metal–insulator transitions. *Reviews of Modern Physics* 70:1039–1263.
2. Ogale, S. B. 2006. *Thin Films and Heterostructures for Oxide Electronics*. Berlin, Germany: Springer.
3. Ohtomo, A., D. Muller, J. Grazul, and H. Y. Hwang. 2002. Artificial charge-modulation in atomic-scale perovskite titanate superlattices. *Nature* 419:378–380.
4. Dagotto, E. 2007. When oxides meet face to face. *Science* 318:1076–1077.
5. Mannhart, J., and D. Schlom. 2010. Oxide interfaces—An opportunity for electronics. *Science* 327:1607–1611.
6. Yamada, H., Y. Ogawa, Y. Ishii, H. Sato, M. Kawasaki, H. Akoh, and Y. Tokura. 2004. Engineered interface of magnetic oxides. *Science* 305:646–648.
7. Heber, J. 2009. Materials science: Enter the oxides. *Nature News* 459:28–30.
8. Seidel, J. 2012. Domain walls as nanoscale functional elements. *The Journal of Physical Chemistry Letters* 3:2905–2909.
9. Seidel, J. 2015. Ferroelectric nanostructures: Domain walls in motion. *Nature Nanotechnology* 10:109–110.
10. Kim, K.-E., B.-K. Jang, Y. Heo, J. H. Lee, M. Jeong, J. Y. Lee, J. Seidel, and C.-H. Yang. 2014. Electric control of straight stripe conductive mixed-phase nanostructures in La-doped $BiFeO_3$. *NPG Asia Materials* 6:e81.
11. Heo, Y., B. K. Jang, S. J. Kim, C. H. Yang, and J. Seidel. 2014. Nanoscale mechanical softening of morphotropic $BiFeO_3$. *Advanced Materials* 26:7568–7572.
12. Lee, J. H., K. Chu, A. A. Ünal, S. Valencia, F. Kronast, S. Kowarik, J. Seidel, and C.-H. Yang. 2014. Phase separation and electrical switching between two isosymmetric multiferroic phases in tensile strained $BiFeO_3$ thin films. *Physical Review B* 89:140101.
13. Zhang, J., B. Xiang, Q. He, J. Seidel, R. Zeches, P. Yu, S. Yang, C. Wang, Y. Chu, and L. Martin. 2011. Large field-induced strains in a lead-free piezoelectric material. *Nature Nanotechnology* 6:98–102.
14. Mermin, N. D. 1979. The topological theory of defects in ordered media. *Reviews of Modern Physics* 51:591–648.
15. Milde, P., D. Köhler, J. Seidel, L. Eng, A. Bauer, A. Chacon, J. Kindervater, S. Mühlbauer, C. Pfleiderer, and S. Buhrandt. 2013. Unwinding of a skyrmion lattice by magnetic monopoles. *Science* 340:1076–1080.
16. Zubko, P., S. Gariglio, M. Gabay, P. Ghosez, and J.-M. Triscone. 2011. Interface physics in complex oxide heterostructures. *Annual Review of Condensed Matter Physics* 2:141–165.
17. Goo, E., R. Mishra, and G. Thomas. 1981. Electron microscopy study of the ferroelectric domains and domain wall structure in $PbZr_{0.52}Ti_{0.48}O_3$. *Journal of Applied Physics* 52:2940–2943.
18. Bursill, L., P. J. Lin, and F. Duan. 1983. HREM study of {100} ferroelectric domain walls in potassium niobate. *Philosophical Magazine A* 48:953–964.
19. Stemmer, S., S. Streiffer, F. Ernst, and M. Rüuhle. 1995. Atomistic structure of 90° domain walls in ferroelectric $PbTiO_3$ thin films. *Philosophical Magazine A* 71:713–724.
20. Lichte, H., M. Reibold, K. Brand, and M. Lehmann. 2002. Ferroelectric electron holography. *Ultramicroscopy* 93:199–212.
21. Jia, C., M. Lentzen, and K. Urban. 2003. Atomic-resolution imaging of oxygen in perovskite ceramics. *Science* 299:870–873.

22. Jia, C.-L., S.-B. Mi, K. Urban, I. Vrejoiu, M. Alexe, and D. Hesse. 2008. Atomic-scale study of electric dipoles near charged and uncharged domain walls in ferroelectric films. *Nature Materials* 7:57–61.

23. Lubk, A., M. Rossell, J. Seidel, Y. Chu, R. Ramesh, M. Hÿtch, and E. Snoeck. 2013. Electromechanical coupling among edge dislocations, domain walls, and nanodomains in BiFeO₃ revealed by unit-cell-wise strain and polarization maps. *Nano Letters* 13:1410–1415.

24. Jia, C., and K. Urban. 2004. Atomic-resolution measurement of oxygen concentration in oxide materials. *Science* 303:2001–2004.

25. Urban, K. 2008. *Advances in Imaging and Electron Physics*. New York: Elsevier.

26. Seidel, J., M. Trassin, Y. Zhang, P. Maksymovych, T. Uhlig, P. Milde, D. Köhler, A. P. Baddorf, S. V. Kalinin, and L. M. Eng. 2014. Electronic properties of iso-symmetric phase boundaries in highly strained Ca-doped BiFeO₃. *Advanced Materials* 26:4376–4380.

27. Eng, L. M. 1999. Nanoscale domain engineering and characterization of ferroelectric domains. *Nanotechnology* 10:405–411.

28. Kalinin, S. V., A. N. Morozovska, L. Q. Chen, and B. J. Rodriguez. 2010. Local polarization dynamics in ferroelectric materials. *Reports on Progress in Physics* 73:056502.

29. Gruverman, A., B. J. Rodriguez, C. Dehoff, J. Waldrep, A. Kingon, R. Nemanich, and J. Cross. 2005. Direct studies of domain switching dynamics in thin film ferroelectric capacitors. *Applied Physics Letters* 87:082902.

30. Rodriguez, B. J., S. Jesse, M. Alexe, and S. V. Kalinin. 2008. Spatially resolved mapping of polarization switching behavior in nanoscale ferroelectrics. *Advanced Materials* 20:109–114.

31. Jungk, T., Á. Hoffmann, and E. Soergel. 2007. Impact of elasticity on the piezoresponse of adjacent ferroelectric domains investigated by scanning force microscopy. *Journal of Applied Physics* 102:084102.

32. Yang, C.-H., J. Seidel, S. Kim, P. Rossen, P. Yu, M. Gajek, Y.-H. Chu, L. W. Martin, M. Holcomb, and Q. He. 2009. Electric modulation of conduction in multiferroic Ca-doped BiFeO₃ films. *Nature Materials* 8:485–493.

33. Seidel, J., W. Luo, S. Suresha, P.-K. Nguyen, A. Lee, S.-Y. Kim, C.-H. Yang, S. J. Pennycook, S. T. Pantelides, and J. Scott. 2012. Prominent electrochromism through vacancy-order melting in a complex oxide. *Nature Communications* 3:799.

34. Balke, N., S. Jesse, A. Morozovska, E. Eliseev, D. Chung, Y. Kim, L. Adamczyk, R. Garcia, N. Dudney, and S. Kalinin. 2010. Nanoscale mapping of ion diffusion in a lithium-ion battery cathode. *Nature Nanotechnology* 5:749–754.

35. Kumar, A., F. Ciucci, A. N. Morozovska, S. V. Kalinin, and S. Jesse. 2011. Measuring oxygen reduction/evolution reactions on the nanoscale. *Nature Chemistry* 3:707–713.

36. Rodriguez, B. J., Y. Chu, R. Ramesh, and S. V. Kalinin. 2008. Ferroelectric domain wall pinning at a bicrystal grain boundary in bismuth ferrite. *Applied Physics Letters* 93:142901.

37. Watanabe, Y. 2007. Review of resistance switching of ferroelectrics and oxides in quest for unconventional electronic mechanisms. *Ferroelectrics* 349:190–209.

38. Seidel, J., L. W. Martin, Q. He, Q. Zhan, Y.-H. Chu, A. Rother, M. Hawkridge, P. Maksymovych, P. Yu, and M. Gajek. 2009. Conduction at domain walls in oxide multiferroics. *Nature Materials* 8:229–234.

39. Seidel, J., P. Maksymovych, Y. Batra, A. Katan, S.-Y. Yang, Q. He, A. P. Baddorf, S. V. Kalinin, C.-H. Yang, and J.-C. Yang. 2010. Domain wall conductivity in La-doped BiFeO₃. *Physical Review Letters* 105:197603.

40. Fan, W., J. Cao, J. Seidel, Y. Gu, J. Yim, C. Barrett, K. Yu, J. Ji, R. Ramesh, and L. Chen. 2011. Large kinetic asymmetry in the metal–insulator transition nucleated at localized and extended defects. *Physical Review B* 83:235102.

41. Maggio-Aprile, I., C. Renner, A. Erb, E. Walker, and Ø. Fischer. 1997. Critical currents approaching the depairing limit at a twin boundary in YBa₂Cu₃O₇₋δ. *Nature* 390:487–490.

42. Wiessner, A., J. Kirschner, G. Schäfer, and T. Berghaus. 1997. Design considerations and performance of a combined scanning tunneling and scanning electron microscope. *Review of Scientific Instruments* 68:3790–3798.

43. Yang, B., N. Park, B. Seo, Y. Oh, S. Kim, S. Hong, S. Lee, and Y. Park. 2005. Nanoscale imaging of grain orientations and ferroelectric domains in (Bi₁₋ₓLaₓ)₄Ti₃O₁₂ films for ferroelectric memories. *Applied Physics Letters* 87:2902.

44. García, R. E., B. D. Huey, and J. E. Blendell. 2006. Virtual piezoforce microscopy of polycrystalline ferroelectric films. *Journal of Applied Physics* 100:064105.

45. Chiu, Y. P., Y. T. Chen, B. C. Huang, M. C. Shih, J. C. Yang, Q. He, C. W. Liang, J. Seidel, Y. C. Chen, and R. Ramesh. 2011. Atomic-scale evolution of local electronic structure across multiferroic domain walls. *Advanced Materials* 23:1530–1534.

46. Lubk, A., M. Rossell, J. Seidel, Q. He, S. Yang, Y. Chu, R. Ramesh, M. Hytch, and E. Snoeck. 2012. Evidence of sharp and diffuse domain walls in BiFeO₃ by means of unit-cell-wise strain and polarization maps obtained with high resolution scanning transmission electron microscopy. *Physical Review Letters* 109:047601.

47. Farokhipoor, S., and B. Noheda. 2011. Conduction through 71 domain walls in BiFeO₃ thin films. *Physical Review Letters* 107:127601.

48. Seidel, J., G. Singh-Bhalla, Q. He, S.-Y. Yang, Y.-H. Chu, and R. Ramesh. 2013. Domain wall functionality in BiFeO₃. *Phase Transitions* 86:53–66.

49. Choi, T., Y. Horibe, H. Yi, Y. Choi, W. Wu, and S.-W. Cheong. 2010. Insulating interlocked ferroelectric and structural antiphase domain walls in multiferroic YMnO₃. *Nature Materials* 9:253–258.

50. Meier, D., J. Seidel, A. Cano, K. Delaney, Y. Kumagai, M. Mostovoy, N. A. Spaldin, R. Ramesh, and M. Fiebig. 2012. Anisotropic conductance at improper ferroelectric domain walls. *Nature Materials* 11:284–288.

51. Lubk, A., S. Gemming, and N. Spaldin. 2009. First-principles study of ferroelectric domain walls in multiferroic bismuth ferrite. *Physical Review B* 80:104110.

52. Meyer, B., and D. Vanderbilt. 2002. Ab initio study of ferroelectric domain walls in PbTiO₃. *Physical Review B* 65:104111.

53. Hong, L., A. Soh, Q. Du, and J. Li. 2008. Interaction of O vacancies and domain structures in single crystal BaTiO₃: Two-dimensional ferroelectric model. *Physical Review B* 77:094104.

54. Borisevich, A., O. S. Ovchinnikov, H. J. Chang, M. P. Oxley, P. Yu, J. Seidel, E. A. Eliseev, A. N. Morozovska, R. Ramesh, and S. J. Pennycook. 2010. Mapping octahedral tilts and polarization across a domain wall in BiFeO₃ from Z-contrast scanning transmission electron microscopy image atomic column shape analysis. *ACS Nano* 4:6071–6079.

55. Zhao, T., A. Scholl, F. Zavaliche, K. Lee, M. Barry, A. Doran, M. Cruz, Y. Chu, C. Ederer, and N. Spaldin. 2006. Electrical control of antiferromagnetic domains in multiferroic BiFeO₃ films at room temperature. *Nature Materials* 5:823–829.

56. Lebeugle, D., D. Colson, A. Forget, M. Viret, A. Bataille, and A. Gukasov. 2008. Electric-field-induced spin flop in $BiFeO_3$ single crystals at room temperature. *Physical Review Letters* 100:227602.

57. Aird, A., and E. K. Salje. 1998. Sheet superconductivity in twin walls: Experimental evidence of. *Journal of Physics: Condensed Matter* 10:L377–L380.

58. Scullin, M. L., J. Ravichandran, C. Yu, M. Huijben, J. Seidel, A. Majumdar, and R. Ramesh. 2010. Pulsed laser deposition-induced reduction of $SrTiO_3$ crystals. *Acta Materialia* 58:457–463.

59. Ko, K.-T., M. H. Jung, Q. He, J. H. Lee, C. S. Woo, K. Chu, J. Seidel, B.-G. Jeon, Y. S. Oh, and K. H. Kim. 2011. Concurrent transition of ferroelectric and magnetic ordering near room temperature. *Nature Communications* 2:567.

60. Ramirez, M., A. Kumar, S. Denev, Y. Chu, J. Seidel, L. Martin, S.-Y. Yang, R. Rai, X. Xue, and J. Ihlefeld. 2009. Spin-charge-lattice coupling through resonant multimagnon excitations in multiferroic $BiFeO_3$. *Applied Physics Letters* 94:161905.

61. Ramirez, M. O., M. Krishnamurthi, S. Denev, A. Kumar, S.-Y. Yang, Y.-H. Chu, E. Saiz, J. Seidel, A. P. Pyatakov, A. Bush, D. Viehland, J. Orenstein, R. Ramesh, and V. Gopalan. 2008. Two-phonon coupling to the antiferromagnetic phase transition in multiferroic $BiFeO_3$. *Applied Physics Letters* 92:022511.

62. Zhou, J., M. Trassin, Q. He, N. Tamura, M. Kunz, C. Cheng, J. Zhang, W.-I. Liang, J. Seidel, and C.-L. Hsin. 2012. Directed assembly of nano-scale phase variants in highly strained $BiFeO_3$ thin films. *Journal of Applied Physics* 112:064102.

63. Yang, C.-H., D. Kan, I. Takeuchi, V. Nagarajan, and J. Seidel. 2012. Doping $BiFeO_3$: Approaches and enhanced functionality. *Physical Chemistry Chemical Physics* 14:15953–15962.

64. SooáLim, J. 2014. Investigation of continuous changes in the electric-field-induced electronic state in $Bi_{1-x} Ca_x FeO_{3-\delta}$. *Physical Chemistry Chemical Physics* 16:17412–17416.

65. Yang, S. Y., J. Seidel, S. J. Byrnes, P. Shafer, C. H. Yang, M. D. Rossell, P. Yu, Y. H. Chu, J. F. Scott, J. W. Ager, L. W. Martin, and R. Ramesh. 2010. Above-bandgap voltages from ferroelectric photovoltaic devices. *Nature Nanotechnology* 5:143–147.

66. Seidel, J., D. Fu, S.-Y. Yang, E. Alarcón-Lladó, J. Wu, R. Ramesh, and J. W. Ager, III. 2011. Efficient photovoltaic current generation at ferroelectric domain walls. *Physical Review Letters* 107:126805.

67. Seidel, J., S.-Y. Yang, E. Alarcòn-Lladò, J. Ager, III, and R. Ramesh. 2012. Nanoscale probing of high photovoltages at 109° domain walls. *Ferroelectrics* 433:123–126.

68. Kudo, A., and Y. Miseki. 2009. Heterogeneous photocatalyst materials for water splitting. *Chemical Society Reviews* 38:253–278.

69. Seidel, J., and L. M. Eng. 2014. Shedding light on nanoscale ferroelectrics. *Current Applied Physics* 14:1083–1091.

70. Reyren, N., S. Thiel, A. Caviglia, L. F. Kourkoutis, G. Hammerl, C. Richter, C. Schneider, T. Kopp, A.-S. Rüetschi, and D. Jaccard. 2007. Superconducting interfaces between insulating oxides. *Science* 317:1196–1199.

71. Logvenov, G., A. Gozar, and I. Bozovic. 2009. High temperature superconductivity in a single copper-oxygen plane. *Science* 326:699–702.

72. Ye, J., S. Inoue, K. Kobayashi, Y. Kasahara, H. Yuan, H. Shimotani, and Y. Iwasa. 2010. Liquid-gated interface superconductivity on an atomically flat film. *Nature Materials* 9:125–128.

73. Takahashi, K., M. Kawasaki, and Y. Tokura. 2001. Interface ferromagnetism in oxide superlattices of $CaMnO_3$/$CaRuO_3$. *Applied Physics Letters* 79:1324–1326.

74. Koida, T., M. Lippmaa, T. Fukumura, K. Itaka, Y. Matsumoto, M. Kawasaki, and H. Koinuma. 2002. Effect of A-site cation ordering on the magnetoelectric properties in $[(LaMnO_3)_m/(SrMnO_3)_m]_n$ artificial superlattices. *Physical Review B* 66:144418.

75. Chakhalian, J., J. Freeland, H.-U. Habermeier, G. Cristiani, G. Khaliullin, M. Van Veenendaal, and B. Keimer. 2007. Orbital reconstruction and covalent bonding at an oxide interface. *Science* 318:1114–1117.

76. Lyuksyutov, I., and V. Pokrovsky. 2005. Ferromagnet–superconductor hybrids. *Advances in Physics* 54:67–136.

77. Dagotto, E. 2013. *Nanoscale Phase Separation and Colossal Magnetoresistance: The Physics of Manganites and Related Compounds*. Berlin, Germany: Springer.

78. Salafranca, J., R. Yu, and E. Dagotto. 2010. Conducting Jahn-Teller domain walls in undoped manganites. *Physical Review B* 81:245122.

79. Kagawa, F., M. Mochizuki, Y. Onose, H. Murakawa, Y. Kaneko, N. Furukawa, and Y. Tokura. 2009. Dynamics of multiferroic domain wall in spin-cycloidal ferroelectric $DyMnO_3$. *Physical Review Letters* 102:057604.

80. Kagawa, F., Y. Onose, Y. Kaneko, and Y. Tokura. 2011. Relaxation dynamics of multiferroic domain walls in $DyMnO_3$ with cycloidal spin order. *Physical Review B* 83:054413.

81. Goltsev, A., R. Pisarev, T. Lottermoser, and M. Fiebig. 2003. Structure and interaction of antiferromagnetic domain walls in hexagonal $YMnO_3$. *Physical Review Letters* 90:177204.

82. Gareeva, Z. V., and A. K. Zvezdin. 2011 (private communication).

83. Daraktchiev, M., G. Catalan, and J. F. Scott. 2010. Landau theory of domain wall magnetoelectricity. *Physical Review B* 81:224118.

84. Hong, J., G. Catalan, D. Fang, E. Artacho, and J. Scott. 2010. Topology of the polarization field in ferroelectric nanowires from first principles. *Physical Review B* 81:172101.

85. Balke, N., B. Winchester, W. Ren, Y. H. Chu, A. N. Morozovska, E. A. Eliseev, M. Huijben, R. K. Vasudevan, P. Maksymovych, and J. Britson. 2012. Enhanced electric conductivity at ferroelectric vortex cores in $BiFeO_3$. *Nature Physics* 8:81–88.

86. Skyrme, T. H. R. 1962. A unified field theory of mesons and baryons. *Nuclear Physics* 31:556–569.

87. Balke, N., S. Choudhury, S. Jesse, M. Huijben, Y. H. Chu, A. P. Baddorf, L.-Q. Chen, R. Ramesh, and S. V. Kalinin. 2009. Deterministic control of ferroelastic switching in multiferroic materials. *Nature Nanotechnology* 4:868–875.

88. Vasudevan, R. K., Y.-C. Chen, H.-H. Tai, N. Balke, P. Wu, S. Bhattacharya, L.-Q. Chen, Y.-H. Chu, I.-N. Lin, and S. V. Kalinin. 2011. Exploring topological defects in epitaxial $BiFeO_3$ thin films. *ACS Nano* 5:879–887.

89. Mühlbauer, S., B. Binz, F. Jonietz, C. Pfleiderer, A. Rosch, A. Neubauer, R. Georgii, and P. Böni. 2009. Skyrmion lattice in a chiral magnet. *Science* 323:915–919.

90. Nelson, C. T., B. Winchester, Y. Zhang, S.-J. Kim, A. Melville, C. Adamo, C. M. Folkman, S.-H. Baek, C.-B. Eom, and D. G. Schlom. 2011. Spontaneous vortex nanodomain arrays at ferroelectric heterointerfaces. *Nano Letters* 11:828–834.

91. Yu, X., Y. Onose, N. Kanazawa, J. Park, J. Han, Y. Matsui, N. Nagaosa, and Y. Tokura. 2010. Real-space observation of a two-dimensional skyrmion crystal. *Nature* 465:901–904.

92. Seki, S., X. Yu, S. Ishiwata, and Y. Tokura. 2012. Observation of skyrmions in a multiferroic material. *Science* 336:198–201.
93. Seki, S., S. Ishiwata, and Y. Tokura. 2012. Magnetoelectric nature of skyrmions in a chiral magnetic insulator Cu_2OSeO_3. *Physical Review B* 86:060403.
94. White, J. S., I. Levatić, A. Omrani, N. Egetenmeyer, K. Prša, I. Živković, J. Gavilano, J. Kohlbrecher, M. Bartkowiak, and H. Berger. 2012. Electric field control of the skyrmion lattice in Cu_2OSeO_3. *Journal of Physics: Condensed Matter* 24:432201.
95. Mochizuki, M. 2012. Spin-wave modes and their intense excitation effects in skyrmion crystals. *Physical Review Letters* 108:017601.
96. Onose, Y., Y. Okamura, S. Seki, S. Ishiwata, and Y. Tokura. 2012. Observation of magnetic excitations of skyrmion crystal in a helimagnetic insulator Cu_2OSeO_3. *Physical Review Letters* 109:037603.
97. Mochizuki, M., and S. Seki. 2013. Magnetoelectric resonances and predicted microwave diode effect of the skyrmion crystal in a multiferroic chiral-lattice magnet. *Physical Review B* 87:134403.
98. Okamura, Y., F. Kagawa, M. Mochizuki, M. Kubota, S. Seki, S. Ishiwata, M. Kawasaki, Y. Onose, and Y. Tokura. 2013. Microwave magnetoelectric effect via skyrmion resonance modes in a helimagnetic multiferroic. *Nature Communications* 4:2391.
99. Maksymovych, P., J. Seidel, Y. H. Chu, P. Wu, A. P. Baddorf, L.-Q. Chen, S. V. Kalinin, and R. Ramesh. 2011. Dynamic conductivity of ferroelectric domain walls in $BiFeO_3$. *Nano Letters* 11:1906–1912.
100. Béa, H., and P. Paruch. 2009. Multiferroics: A way forward along domain walls. *Nature Materials* 8:168–169.
101. Goltsev, A., R. Pisarev, T. Lottermoser, and M. Fiebig. 2003. Structure and interaction of antiferromagnetic domain walls in hexagonal YMnO3. *Physical Review Letters* 90:177204.
102. Skumryev, V., V. Laukhin, I. Fina, X. Martí, F. Sánchez, M. Gospodinov, and J. Fontcuberta. 2011. Magnetization reversal by electric-field decoupling of magnetic and ferroelectric domain walls in multiferroic-based heterostructures. *Physical Review Letters* 106:057206.
103. Catalan, G., J. Seidel, R. Ramesh, and J. F. Scott. 2012. Domain wall nanoelectronics. *Reviews of Modern Physics* 84:119.

Topological Vortex Defects in an Improper Ferroelectric*

Seung Chul Chae and San-Wook Cheong

Contents

13.1 Introduction

In general, a matter of the fluid state with homogeneity and isotropy at high temperature exhibits a phase transition into the condensed state at the transition temperature with symmetry breaking. This symmetry breaking indicates that the system at the ground state has a partial symmetry of the Hamiltonian. Unless the temperature is 0 K, the system will excite in order to recover its symmetry. It has been known that Goldstone mode and topological singularity are possible for the system with broken continuous symmetry during the

---

* This chapter is a compilation of several publications written by Prof. San-Wook Cheong, Department of Physics and Astronomy, Rutgers University.

excitation. On the other hand, in the system with broken discrete symmetry, the allowed excitation is limited to the topological defect like domain wall.

For Goldstone mode, spin wave, phonon, and second sound can be considered examples for the conserving of rotational symmetry in a ferromagnetic, translation symmetry in the lattice structure of a condensed matter, and gauge symmetry of a superfluid, respectively. On the other hand, for topological singularity, a vortex in a two-dimensional (2D) XY spin system with U(1) symmetry can be considered the same. In terms of dimensionality of a system, the Heisenberg model, which has SU(2) or O(3) symmetry, can have monopole and/or vacancy as topological defects.

In this chapter, which is the compilation of 10 publications [1–10] written by Prof. S.-W. Cheong at the Department of Physics and Astronomy, Rutgers University, we introduce the hexagonal rare earth manganites (h-REMnO$_3$), which exhibit the rare topological defects in the condensed matter as shown in Figure 13.1. Hexagonal REMnO$_3$ (RE = Ho–Lu, Y, and Sc) are improper ferroelectrics where the size mismatch between RE and Mn ions induces a trimerization-type structural phase transition, and this structural transition leads to three structural domains, namely, α, β, and γ, each of which can support two directions of ferroelectric polarization. The TEM image as shown in Figure 13.1 demonstrates that one vertex consists of six distinguishable domains emerging from one point. Each vertex can be identified as a vortex or an antivortex according to its domain vorticity (α^+–β^-–γ^+–α^-–β^+–γ^- or

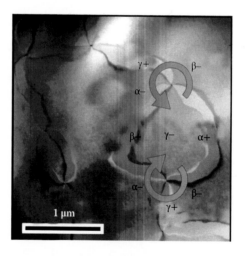

FIGURE 13.1 Configurations of multiferroic vortices and antivortices observed in a dark-field transmission electron microscope image of an h-YMnO$_3$ crystal. Six distinguishable domains merge at one point. Vorticity of the domain configuration (α^+–β^-–γ^+–α^-–β^+–γ^- or α^+–γ^-–β^+–α^-–γ^+–β^-) characterizes vortex versus antivortex. (Reprinted with permission from Chae, S.C. et al., Self-organization, condensation, and annihilation of topological vortices and antivortices in a multiferroic. *Proc. Natl. Acad. Sci.* U.S.A., 107, 21366–21370. Copyright 2010 National Academy of Sciences, U.S.A.)

$\alpha^+-\gamma^--\beta^+-\alpha^--\gamma^+-\beta^-$). Topological vortex and antivortex composed of these interlocked structural antiphase and ferroelectric domains have subsequently been reported to be paired and form the irregular network of a zoo of multiferroic vortices and antivortices. As reviewed in this chapter, the zoo of topology can be neatly analyzed in terms of graph theory. Not only its mathematical beauty but also their spontaneous symmetry breaking in relation to novel phenomena such as the early stage of universe after big bang has been reported in terms of Kibble–Zurek mechanism. The electric and magnetic properties of vortex domains will also be covered.

13.2 Analysis of Topology of Multiferroic Vortex Domain Based on Graph Theory

Since the famous problem of Seven Bridges of Königsberg solved by L. Euler in 1736, graph theory has become a good analysis method to reveal the connectivity in many systems ranging from humanity to natural science, and extended its scope of applications to various areas with the help of computer science [11]. Any configuration of two sets, a nonempty set of objects (vertices) and a set of the connections (edges) among the objects, can be considered graph. In analyzing the nature of underlying connectivity, graph theory has been a mathematical frame to understand many real configurational problems. It is widely used in science, engineering, and economics, covering phenomena ranging from atomic bonding in chemistry [12] to grander applications such as connectivity of neuron networks of cerebral cortexes [13] and even in sociology like the connectivity of the World Wide Web [14].

New topological defects with simultaneous ferroelectric and magnetic orders have been identified in h-REMnO$_3$ [1]—a multiferroic where ferroelectricity and magnetism coexist and are intriguingly coupled to each other [15–17]. Ferroelectricity in h-REMnO$_3$ is driven by a structural instability due to the size mismatch of RE–O and Mn–O layers [18–20]. It turns out that a structural trimerization via shrinking of the Mn–O cage is also induced by the structural transition, and three types (namely, α, β, and γ phases) of structural antiphase domains can result from the trimerization [21]. The topological defect in h-REMnO$_3$ is composed of six neighboring domains that converge with a sense of rotation of both structural antiphase and ferroelectric domains. The important ingredient for the formation of this topological structure is interlocking of the structural antiphase and ferroelectric domain walls. Because it has been known that ferroelectric domain walls pin antiferromagnetic domain walls in h-REMnO$_3$ [16], the topological defect is expected to accompany six antiferromagnetic domains with a sense of rotation at low temperatures. Thus, it is legitimate to call it a multiferroic vortex [22]. Chae et al. [2] discovered the self-organization of topological vortices and antivortices in h-YMnO$_3$ and its rich physics. Furthermore, the authors found that the connectivity of the self-organized complex network, rather than metric properties such as length or size, exhibits definite regularities and can be explicitly analyzed using mathematical graph theory.

Using an optical microscope, a network-like domain pattern on a chemically etched surface of a h-YMnO$_3$ single crystal was observed as shown in Figure 13.2a [23]. At first glance, the domain pattern seems to be in a random and irregular manner. However, upon careful inspection, a landscape with clear topological order and regularity is revealed, where every closed connected region (face) is surrounded by an "even" number of vertices connected by edges, exactly three of which are incident on each vertex. The coloring of maps without having adjacent regions in the same color has been considered for a long time, for example, four-color theorem for the coloring of a world map, and this coloring has played a key role of graph theory. In the famous four-color theorem, all regions of every 2D map can be colored with only four colors in a way that no two adjacent regions have the same color [24]. Mathematical statement of this four-color theorem is "all faces of every planar map can be four-proper-colorable." When a map satisfies certain regularities or rules, the number of colors for the proper coloring on the graph can be less than four. This coloring of regions in a pattern has inspired us to understand the mathematical connectivity in the organization of our domain pattern.

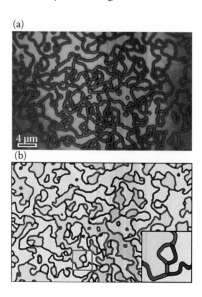

(a)

(b)

FIGURE 13.2 Ferroelectric domain pattern and coloring of the pattern in h-YMnO$_3$. (a) Ferroelectric domain pattern on the surface of an h-YMnO$_3$ crystal, imaged using an optical microscope. The upward-polarization and downward-polarization domains of the crystal were etched selectively utilizing phosphoric acid. (b) Proper coloring of faces (edges) with three colors: light red (red), light blue (blue), and light green (green). All adjacent edges or faces have different colors. The inset shows an enlarged area, denoted with a pink rectangle, for the clarification of the colors of adjacent edges and faces. (Reprinted with permission from Chae, S.C. et al., Self-organization, condensation, and annihilation of topological vortices and antivortices in a multiferroic. *Proc. Natl. Acad. Sci.* U.S.A., 107, 21366–21370. Copyright 2010, National Academy of Sciences, U.S.A.)

Chae et al. [2] found that every face in Figure 13.2a can be properly colored using three different colors (light red, light blue, and light green) as shown in Figure 13.2b. Additionally, focusing only on edges, Chae et al. found that every edge can also be properly colored with three colors (red, blue, and green) as clearly depicted in the inset of Figure 13.2b. To describe a graph mathematically, it should be defined in terms of the number of edges, faces, and vertices. A face with N vertices on its boundary is called an N-gon. A K-valent vertex denotes that K edges are incident to the vertex, and a K-valent graph means that all vertices in the graph are K-valent. Then it has been mathematically proven that all faces of a three-valent graph are three-proper-colorable if and only if all faces are composed of even-gons (Table 13.1) [25]. This theorem is certainly consistent with the fact that all faces in Figure 13.2b are colored with three colors without having adjacent faces in the same color. In addition, all edges of a three-valent graph with even-gons are three-proper-colorable [26], which is also consistent with the fact that all edges in Figure 13.2b are colored with three colors without having adjacent edges in the same color.

It turns out that the presence of "narrow" + domains (upward-polarization) on the surfaces of h-YMnO$_3$ crystals, displayed in Figure 13.2a, is due to a self-poling effect, probably due to oxygen off-stoichiometry near the surfaces [27]. In general, h-REMnO$_3$ system exhibits hole-type charge conduction, probably due to cation deficiency or excess oxygen [28]. Chae et al. [2] found that both flat surfaces of as-grown plate-like h-YMnO$_3$ crystals show narrow domains with upward-polarization near the surfaces. On the other hand, when the crystals were annealed in oxygen atmosphere at ~900°C, the + domains and – domains (downward-polarization) on the surfaces show a roughly 50:50% distribution in the entire area of both flat surfaces. These behaviors can be seen in Figure 13.3, where Chae et al. [2] display 3D illustrations with real atomic force microscopy (AFM) images, optical images, and the schematics corresponding to the optical images. In as-prepared crystals, the surfaces, compared with the bulk, tend to have more oxygen vacancies or less oxygen excess, and this reduced oxygen content near surfaces favors the presence of the tail of polarization near surfaces, inducing broad – and narrow + domains near surfaces. This kind of self-poling induced by oxygen off-stoichiometry is well documented in other ferroelectrics [27]. Note that the narrow + (broad –) domains on the surfaces of as-grown

TABLE 13.1 Number of Colors for Proper Coloring of Three- and Six-Valent Planar Graphs Whose Faces (or Domains) Are All Even Gons

Valence	Vertex Coloring	Edge Coloring	Face Coloring
3	2	3	3
6	2	6	2

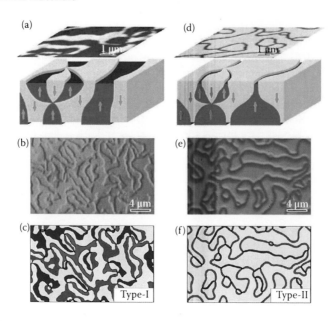

FIGURE 13.3 Two types of the configuration of copious topological vortices and antivortices in h-YMnO$_3$, demonstrating the condensation of numerous vortex–antivortex pairs. (a) and (d) Three-dimensional illustrations of distinct ferroelectric domain structures with real AFM images for O$_2$-annealed and as-grown h-YMnO$_3$ crystals, respectively. This difference originates from a chemistry-driven self-poling effect due to the low oxygen content on the surface of as-grown crystals. This self-poling of as-grown crystals favors energetically the presence of wide downward-polarization domains near surfaces. Note that the cartoons also depict vortex or antivortex cores curved along the *c*-axis. (b) and (c) Optical microscope image and a schematic of the surface of an O$_2$-annealed h-YMnO$_3$ crystal after chemical etching, respectively. These pictures show a type-I pattern with a roughly equal distribution of upward- and downward-polarization domains. (e) and (f) Optical microscopy image and a schematic of the surface of a self-poled h-YMnO$_3$ crystal after chemical etching, respectively. These pictures display a type-II pattern with narrow + and broad − domains. The difference between type-I and type-II patterns reflects the condensation of vortex–antivortex pairs. (Reprinted with permission from Chae, S.C. et al., Self-organization, condensation, and annihilation of topological vortices and antivortices in a multiferroic. *Proc. Natl. Acad. Sci.* U.S.A., 107, 21366–21370. Copyright 2010, National Academy of Sciences, U.S.A.)

crystals remain to be narrow (broad) after electric poling even though the overall domain patterns on the surfaces can change as discussed earlier, and the dominant electric poling appears to occur in the interior of crystals. Lastly, Chae et al. [2] emphasize that as the schematic in Figure 13.3c demonstrates, the roughly 50:50% mixture pattern also shows a six-valent graph with even-gons. However, in this graph, both + and − domains do have various even-gons. On the contrary, the patterns in Figure 13.2a correspond to six-valent graphs where − broad domains show various even-gons, but narrow + domains are always two-gons.

The domain patterns in h-YMnO$_3$, where ferroelectricity, magnetism, and structural distortions intertwined to each other, are found to be two types as depicted in Figure13.3: patterns with the roughly equal distribution of + and − domains (type I), and patterns with mixtures of narrow + and broad − domains (type II). The formation of type-II patterns stems from chemistry-driven self-poling, and Chae et al. [2] were able to obtain much larger-range and better-resolution images of type-II patterns than those of type-I patterns, so they have analyzed type-II patterns in great detail, as discussed earlier. Graph theoretical comparison between type-I and type-II patterns indicates the interesting possibility of the condensation of vortex–antivortex pairs, induced by chemistry-driven self-poling. As evident in Figure 13.3c, type-I patterns form six-valent graphs with even-gons (all possible + even-gons and also all possible − even-gons), the vertices and faces of which are two-proper-colorable. These patterns do have symmetry under the change of + and − signs. On the other hand, type-II patterns correspond to six-valent graphs with all possible − even-gons, but narrow + two-gons only. Self-poling induces the transition from type-I to type-II patterns, and thus the symmetry under the + and − sign change is broken after self-poling. Furthermore, if we consider the narrow two-gons connecting vortices and antivortices as edges, then "the six-valent graph with narrow + two-gons and broad − all-even-gons" can be readily considered "a three-valent graph with all-even-gons." This compactification of valence can occur only in type-II patterns, not in type-I patterns. High-gons are possibly related to large energy associated with, for example, strain, so conceptually they can be considered excited states, relative to the ground state of two-gons. Unlike type-I patterns, all vortices and antivortices in type-II patterns are paired, that is, linked by one or two narrow + two-gons, which may result in the lowest total energy.

Therefore, the symmetry breaking under the + and − sign change by self-poling can be considered "the condensation of topological vortex–antivortex pairs." Note that this condensation, induced by self-poling, can occur through continuous merging of domain walls without changing the number of vortices and antivortices. On the other hand, the electric poling of type-II patterns can induce the vortex–antivortex annihilation as discussed earlier. Chae et al. [2] found that electric poling of type-I patterns is difficult due to a drastic increase of coercivity and wall pinning after oxygen annealing. Finally, we noted that the transition from "+ all even-gons with a power-law distribution of even-gons" to "+ two-gons only" induced by self-poling resembles collapsing a sandpile, in the sense that both may involve a process from a self-organized critical state to the ground state induced by external stimulus.

13.3 Conduction Behavior of the Multiferroic Vortex Domain Walls

Domain walls are kink solitons that divide domains with different orientations of ferroic order. They may host novel properties or emergent phases like local conduction in an insulating media [29–31]. In typical ferroelectrics, most domain walls are known to be neutral, while charged domain walls are hardly observed, which is likely due to unfavorable electrostatic and/or strain energy cost [32]. Experimentally observed charged domain walls are often associated with defects or needle-shaped domains during polarization reversal [33,34]. However, it has been predicted that charged domain walls may be stabilized by charged defects or free charge carriers [35–37]. In contrast to conventional ferroelectrics, the formation of charged domain walls in multiferroic h-REMnO$_3$ is topologically inevitable because of the presence of highly curved vortex cores [2,38,39]. Recently, Wu et al. [5] reported the observation of nanoscale conduction characteristics of charged ferroelectric domain walls in h-HoMnO$_3$ (a p-type semiconductor) using in situ conductive atomic force microscopy (c-AFM), piezoresponse force microscopy (PFM), and Kelvin-probe force microscopy (KPFM) at low temperatures. Local conduction spectra indicate that the conduction at tail-to-tail (TT) domain walls is significantly (slightly) enhanced at high forward (reverse) bias compared with that of the domains themselves, probably stemming from the accumulation of hole-like carriers. In contrast, the conduction of head-to-head (HH) domain walls shows no enhancement at high forward bias and even suppression at high reverse bias, probably due to the depletion of hole-like carriers.

Figure 13.4a shows typical room temperature TEM dark-field images of h-HoMnO$_3$ taken with the electron beam either close to parallel (top view) or perpendicular (side view) to the crystallographic c-axis. The observation of the vortex pattern on the side view image demonstrates that curved vortex cores force the associated domain walls to intercept the polarization direction (the c-axis), as illustrated in the cartoon in Figure 13.4b. Therefore, these domain walls carry bound charges due to antagonistic polarizations as illustrated in Figure 13.4c. In this sense, the charged domain walls in h-REMnO$_3$ are "protected" by the formation of vortices instead of being randomly pinned by extrinsic defects. In conventional ferroelectrics, charged domain walls are thermodynamically unstable because of extra electrostatic or strain energy cost [40]. However, h-REMnO$_3$ are improper ferroelectrics where the ferroelectricity is a by-product of primary structural trimerization. Previous first-principles calculations revealed that the ferroelectric order parameter (polarization) is a

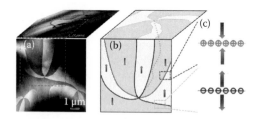

FIGURE 13.4 (a) TEM dark-field images of top and side views of vortex domains in h-HoMnO$_3$. (b) A cartoon sketch of the 3D profile of a curved vortex in the boxed area in (a). (c) Zoom-in cartoon of HH (TT) domain walls with positive (negative) bound charges. (Reprinted with permission from Wu, W. et al., Conduction of topologically protected charged ferroelectric domain walls. *Physical Review Letters*, 108, 077203. Copyright 2012, American Physical Society.)

cubic function of the primary order near T_c [19]. Therefore, the magnitude of the ferroelectric polarization is negligibly small in the proximity of T_c. In other words, the ferroelectric polarization and the associated electrostatic energy cost are irrelevant to the formation of the vortex–antivortex network near T_c. In addition, thermally excited charge carriers can effectively screen charged domain walls because h-REMnO$_3$ are small bandgap semiconductors [41,42]. From the thermodynamic point of view, the proliferation of vortex–antivortex pairs and highly curved vortex cores are more favorable because of the domination of the entropy contribution to the free energy at high temperature [43].

How vortices propagate along the c-axis can also be revealed by in-plane PFM with the c-axis in the surface plane [38,39]. As seen in the topography image shown in Figure 13.5a, Wu et al. [5] were able to obtain an atomically flat (110) surface of a HoMnO$_3$ single crystal by mechanical cleaving. Figure 13.5b shows the PFM images taken at room temperature to reveal the ferroelectric domain pattern. Wu et al. [5] aligned the orientation of the conductive cantilever so that it was parallel to the c-axis, which is along the slow scan axis (vertical direction). In this configuration, the PFM signal (vertical deflection) originates from the buckling of the cantilever caused by the shear deformation of in-plane ferroelectric domains in the presence of an out-of-plane electric field [44,39]. The dark and bright contrasts represent in-plane up and down ferroelectric domains, respectively. Figure 13.5c shows the derivative image of PFM (b) that highlights the HH (dark) and TT (bright) domain walls, respectively.

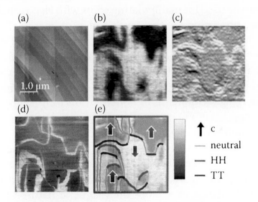

FIGURE 13.5 (a) Topographic and (b) PFM ($V_{ex} = 22$ V, $f = 21$ kHz) images taken simultaneously at 300 K. (c) Derivative map of the PFM image along the c-axis (vertical) where bright (dark) lines are TT (HH) domain walls. (d) cAFM image taken at the same location as PFM at 300 K with $V_{tip} = -10$ V. (e) A cartoon sketch of ferroelectric domain walls according to PFM (overlaid) and its derivative images. Red, blue, and gray lines represent HH, TT, and neutral domain walls, respectively. The arrows indicate in-plane polarization orientation determined from the phase of the PFM signal. Color scales are 4.5 nm, 8 pm, and 0.4 nA for topography, PFM, and cAFM images, respectively. (Reprinted with permission from Wu, W. et al., Conduction of topologically protected charged ferroelectric domain walls. *Physical Review Letters*, 108, 077203. Copyright 2012, American Physical Society.)

In cAFM images with $V_{tip} = -10$ V, there is no conduction (current) contrast between different domains, as shown in Figure 13.5d. In contrast, there are line features with significant extra current. More interestingly, these lines overlap with the ferroelectric domain walls observed by PFM at the same location. By correlating PFM and cAFM images at the same location, we can identify two vortices in this area, as illustrated in the cartoon sketch in Figure 13.5e. Note that the fast rastering of the cantilever artificially broadens the apparent width of some conduction lines. In the conductance image with very slow rastering, the observed width of conduction peaks is ~80–100 nm, which is comparable with the diameter (≤ 100 nm) of the conductive tip.

13.4 Magnetic Properties of Mutliferroic Vortex

Topological defects, such as domain walls and vortices, are very common in complex matter such as superfluids, liquid crystals, the Earth's atmosphere, and the early universe [43,45]. Topological defects have been fruitful playgrounds for emergent phenomena [46,31]. Recently, vortex-like topological defects, called magnetic skyrmions, were observed in helical magnets with broken inversion symmetry [47]. The interplay between the topological spin texture of skyrmions and the spins of conduction electrons may lead to novel spintronic applications [48]. Multiferroics are materials with coexisting magnetic and ferroelectric orders, where inversion symmetry is also broken [17]. The cross-coupling between two ferroic orders can result in strong magnetoelectric coupling. This coupling can be used for manipulating spins with electric fields. Therefore, multiferroics are promising for the development of energy-efficient memory and sensor applications [15,49–51]. Because the formation of domains is the hallmark of any ferroic order [52], it is of both fundamental and technological interests to visualize cross-coupled domains or walls in multiferroics. However, most multiferroics are antiferromagnets with vanishing magnetic moments, which makes imaging the domains or domain walls technically challenging.

Previous second harmonic generation (SHG) studies have suggested that ferroelectric domain walls in millimeter size h-YMnO$_3$ always pin 180° antiferromagnetic domain walls. However, free 180° antiferromagnetic domain walls also exist [16]. To date, SHG has been unable to resolve vortex domain structure because of the spatial resolution limitation (~10 μm) [53]. Thus, it is of fundamental interest to explore the magnetic nature of cross-coupled structural antiphase-ferroelectric domain walls (60° or 120° due to their antiphase relationship) with resolved vortex domain structure [54]. Visualizing antiferromagnetic domains or domain walls (especially in h-REMnO$_3$) has been an experimental challenge, particularly due to the lack of suitable high-resolution imaging techniques. Using a homemade low-temperature magnetic force microscope (MFM) [55], Geng et al. [6] observed and reported remarkable alternating net magnetic moments at interlocked antiphase ferroelectric domain walls around vortex cores in multiferroic h-ErMnO$_3$. Their results suggest that the intriguing domain wall magnetism originates from uncompensated Er^{3+} spins polarized by the Mn^{3+} antiferromagnetic order via anisotropic exchange interactions [56,57]. More interestingly, the alternating domain wall net moments correlate over the entire vortex network and can be controlled by cooling through T_N in a magnetic field.

Figure 13.6a shows a room temperature PFM image taken on the (001) surface of a single-crystal h-ErMnO$_3$, where six alternating up (red) and down (blue) ferroelectric domains merge at a vortex core. Figure 13.6b and c shows an MFM image measured at 5.5 K in a 0.2 T out-of-plane magnetic field after −0.2 T and +0.2 T field cooling from 100 K ($>T_N \approx$ 80 K), respectively. The MFM images were taken at the same location as the PFM image (Figure 13.6a). Clearly there are line features with alternating bright and dark colors in the MFM images (Figure 13.6b and c), correlating with the antiphase-ferroelectric domain walls around the vortex core in the PFM image (Figure 13.6a). Figure 13.6d shows the MFM measurement setup where a 50-nm Au film was deposited on the sample surface (after PFM measurements), to eliminate electrostatic stray fields from out-of-plane ferroelectric domains. The MFM tip moment is normal to the sample surface, so that the MFM signal (cantilever

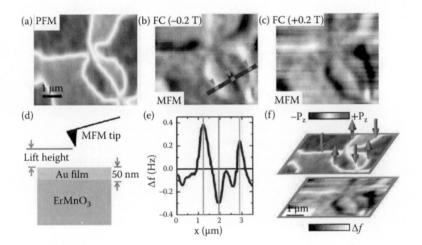

FIGURE 13.6 Coupled antiphase-ferroelectric and antiferromagnetic domain walls with alternating magnetic moments around multiferroic vortex cores. (a) Room temperature PFM image of the (001) surface of a single crystal h-ErMnO$_3$. The red and blue colors correspond to the up and down ferroelectric domains, respectively. (b) and (c) MFM images measured at 5.5 K in a 0.2 T out-of-plane magnetic field after −0.2 T (+0.2 T) field cooling from 100 ($>T_N$) to 5.5 K. The MFM images were taken at the same location as the PFM image. The color scale (Δf) is 0.8 Hz, and the lift height is 50 nm. (d) A cartoon sketch shows the setup of the MFM experiment. (e) The line profile of the MFM signal along the blue line in (b), vertical green lines note the position of the domain walls as indicated by the green arrows in (b). (f) A perspective view of PFM (a) and MFM (c) images with arrows representing the orientation of the uncompensated magnetic moments at structural antiphase ferroelectric domain walls. (Reprinted with permission from Geng, Y. et al., Collective magnetism at multiferroic vortex domain walls, *Nano Letters*, 12, 6055–6059. Copyright 2012, American Chemical Society.)

resonant frequency shift $\Delta f \propto$ force gradient) is due to the out-of-plane stray magnetic field gradient from the sample [58]. Note that the contrast inversion between Figure 13.6b and b is due to the reversal of local net moments because the orientation of the MFM tip moment is determined by the external magnetic field (0.2 T) $\mu_0 H_c \approx 0.02$ T). Reversing the tip moment by external magnetic fields at low temperatures merely inverted the MFM image contrast.

The MFM signal along a line drawn in Figure 13.6b is shown in Figure 13.6e. The single peak profile (width ~400 nm) of the MFM signal suggests that the local magnetization at domain walls is parallel to the c-axis. The MFM contrast of domain walls is essentially constant for 0.02–0.7 T, which excludes the possibility of local susceptibility differences as its origin [59]. Therefore, the MFM signal at domain walls originates from local net moments along the c-axis. Geng et al. [6] focused on the MFM results in low fields (<1 T) because $ErMnO_3$ undergoes a metamagnetic phase transition under ~1 T field. Since no net moment is expected inside the antiferromagnetic domains, the net moments at antiphase-ferroelectric domain walls must come from the uncompensated moments at the antiferromagnetic domain walls, which are coupled to the antiphase-ferroelectric DWs. Interestingly, there is negligible MFM contrast at the vortex core, indicating that alternating net moments from the 6 domain walls cancel each other at the core. The fact that the cooling magnetic field determines the magnetic state of alternating vortex domain walls in h-$ErMnO_3$ (shown in Figure 13.6b and c) suggests that it is likely a correlated phenomenon, tied to the antiferromagnetic order of the Mn^{3+} spins. Therefore, the results provide compelling evidence that the six-state vortices in h-$ErMnO_3$ are truly multiferroic. As summarized by the perspective view of the PFM and MFM images in Figure 13.6f, the uncompensated magnetic moments at domain walls around a vortex core are parallel to the c-axis with alternating orientation, similar to the alternating ferroelectric polarization of domains around the core. Therefore, there are only two types of magnetic domain walls, denoted as DW_I and DW_{II}. The cross-coupling, between uncompensated magnetic moments and antiphase ferroelectric domain walls, indicates the possibility of manipulating the net magnetic moments with electric fields, which will be further explored in future experimental studies.

To determine whether the alternating pattern of domain wall net magnetic moments is a local property of individual vortices or a global property of the vortex network, Geng et al. [6] obtained room temperature PFM and low-temperature MFM images (Figure 13.7b and c) on a larger area (16 μm × 16 μm), using topographic features (Figure 13.7a) as alignment marks. Figure 13.7d illustrates the pattern of domain wall net moments based on the vortex connectivity in the PFM and the magnetic pattern of one vortex. Clearly, the signals in the MFM image are in excellent agreement with that in the illustration, suggesting that the alternating domain wall net moments correlate over the entire field of view (and possibly over the whole sample), thus representing a collective phenomenon. The correlation of alternating domain wall moments was confirmed at multiple locations (>5) on two different crystals.

Previous micromagnetic analysis of 180° antiferromagnetic domain walls in h-h-$YMnO_3$ suggested that oscillatory uncompensated Mn^{3+} spins rotate

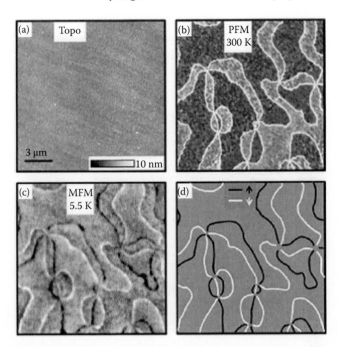

FIGURE 13.7 Correlation of domain wall magnetism over the vortex network. (a) Topography, (b) room-temperature PFM image, and (c) MFM image (5.5 K, 0.1 T, lift height: 40 nm) taken at the same location on the (001) surface of an h-ErMnO$_3$ single crystal (a different location from that in Figure 13.6). The color scale for the topography and MFM (Δf) are 10 nm and 0.15 Hz, respectively. The red and blue colors in the PFM image correspond to up and down ferroelectric domains, respectively. (d) A cartoon sketch of the domain wall net magnetic moments over the entire field of view based on MFM data of one vortex and the connectivity of vortices in (b) (PFM). Black (white) lines represent up (down) net moments. (Reprinted with permission from Geng, Y. et al., Collective magnetism at multiferroic vortex domain walls, *Nano Letters*, 12, 6055–6059. Copyright 2012, American Chemical Society.)

across the antiferromagnetic–ferroelectric domain walls in the a–b-plane due to in-plane anisotropy [60]. Therefore, the observed out-of-plane uncompensated moments, in h-ErMnO$_3$, likely come from Er^{3+} spins because RE^{3+} ions in h-REMnO$_3$ have out-of-plane anisotropy [61–64]. To understand the origin of the domain wall moments, Geng et al. studied the temperature dependence of the domain wall magnetic signal at the same location as that of Figure 13.6. The results are shown in Figure 13.8. The domain wall contrast, defined as the difference between bright and dark walls, decreases sharply for $T < 10$ K, then more slowly at higher temperature as shown in Figure 13.8g, which resembles the Curie–Weiss behavior, but is inconsistent with the temperature dependence of the Mn^{3+} order parameter (the green dashed line in Figure 13.8g) [65]. Assuming that the MFM signal is proportional to the net moments, Geng et al.

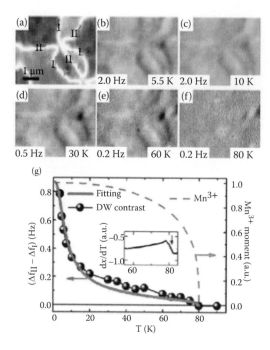

FIGURE 13.8 Temperature dependence of the net magnetic moments at anti-phase-ferroelectric domain walls. (a) PFM image at the same location as the MFM images, the two domains are labeled by "I" ("II") (DW$_I$ and DW$_{II}$). (b–f) selected MFM images at various temperatures (warming) in 0.2 T magnetic field. The color scales are noted at the left bottom of each image. (g) Temperature dependence of the domain wall contrast ($\Delta f_{II}-\Delta f_I$) of two walls noted by green arrows in (a). The red curve is a fitting curve based on a phenomenological doublet model. The inset shows the derivative of the dc susceptibility with respect to temperature, $d\chi_{DC}/dT$. The blue arrow indicates the anomaly at T_N. The green dashed line is the temperature dependence of the Mn^{3+} order parameter moment according to neutron scattering [35]. (Reprinted with permission from Geng, Y. et al., Collective magnetism at multiferroic vortex domain walls, *Nano Letters*, 12, 6055–6059. Copyright 2012, American Chemical Society.)

obtained a good fit (the red solid line in Figure 13.8g) of the temperature dependence of the domain wall contrast by using a phenomenological doublet model. In this model, the effective doublet ground state of Er^{3+} ions is split by the exchange fields from neighboring Mn^{3+} spins [61]. Indeed, the domain wall contrast of the MFM data essentially disappears above 80 K, in excellent agreement with $T_N \approx 80$ K, which can be inferred from the bulk susceptibility data (inset of Figure 13.8g). The doublet model has been successful in explaining the bulk (i.e., domains) partial RE^{3+} ordering in other h-REMnO$_3$ materials (RE = Ho, Tm, Yb) as has been revealed by x-ray magnetic resonant scattering, neutron diffraction, and Mössbauer spectroscopy [61–63]. The good agreement Geng et al. [6] found between the doublet model and the

MFM data suggest that this model is also applicable for the domain wall magnetism. Therefore, Geng et al. [6] proposed that it comes from uncompensated Er^{3+} spins polarized by the exchange fields from neighboring Mn^{3+} spins.

It is believed that the effective exchange fields originate from anisotropic exchange interactions, for which the antisymmetric components are the well-known Dzyaloshinskii–Moriya (DM) interactions [56,57]. The presence of DM interactions is the key ingredient for noncollinear spin orders, which induce ferroelectricity by breaking inversion symmetry and become multiferroic [17,66]. In multiferroic h-REMnO$_3$, the DM interactions between RE^{3+} and Mn^{3+} spins are likely responsible for inducing the partial RE^{3+} antiferromagnetic order [61–63]. The dipolar interactions between RE^{3+} and Mn^{3+} spins are of the order of 1 K and thus are too weak to account for the observed coupling strength ($\sim$10 K) [61]. However, dipolar interactions between RE^{3+} ions may be responsible for the additional RE^{3+} ordering below 5 K [61,64]. In zero or low-magnetic field (< 1 T), the magnetic space group of 120° order of Mn^{3+} spins in ErMnO$_3$ is P6$_3$cm, that is, B$_2$ (or Γ_4) using a 1D irreducible representation. Indeed, the Er^{3+} moments inside the domains that are polarized by DM exchange fields also respect B$_2$.

To provide a simple physical insight of the domain wall, net moments without invoking elaborate theoretical analyses, Geng et al. [6] proposed a model by making a few very simple assumptions. A realistic model of magnetic domain walls would require proper consideration of the exchange and anisotropy energies and the symmetries of the order parameters [54]. It has been suggested that there are two types of interlocked structural antiphase ferroelectric domain walls that alternate around vortex cores, which may be the structural origin of the two types of magnetic domain walls [1,67]. Assuming atomically sharp Mn^{3+} spin variation at the domain walls, a single Mn^{3+} spin configuration in domains across the walls, and abrupt structural distortion variation across the two types of walls, we find opposite uncompensated Er^{3+} moments polarized by DM exchange fields at the two types of domain walls. The results of the simple model are shown in Figure 13.9b and c. A simplified form is shown in Figure 13.9d. Using the aforementioned caveats, the MFM results provide additional evidence of the existence of the two types of antiphase-ferroelectric domain walls in h-REMnO$_3$.

Furthermore, MFM studies also reveal that there are two distinct domain wall states [(DW$_I$, DW$_{II}$) = ($\uparrow$, $\downarrow$) or ($\downarrow$, $\uparrow$)] in the B$_2$ phase that can be controlled by field cooling the sample through T_N with different magnetic field orientations (Figure 13.9b and c). Geng et al. [6] defined domain walls with dark (bright) color in the MFM image, which have uncompensated moments parallel (antiparallel) to the cooling magnetic field as DW$_I$ (DW$_{II}$). It is possible that the DW$_I$ has a larger uncompensated magnetization than DW$_{II}$ near T_N so that it is always parallel to the cooling field. Unfortunately, MFM results are not sufficient to pin down this possibility. Note that there are two degenerate Mn^{3+} spin states of B$_2$ symmetry, which are related to each other by time-reversal symmetry. Based on this simple model, the two degenerate Mn^{3+} spin states provide a natural origin of the two magnetic domain wall states [(DW$_I$, DW$_{II}$) = ($\uparrow$, $\downarrow$) or ($\downarrow$, $\uparrow$)] because the domain wall exchange field switches sign

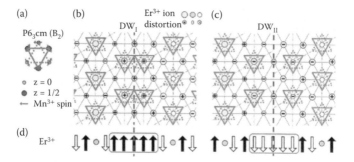

FIGURE 13.9 A simple mode of uncompensated Er^{3+} moments at domain walls in B_2 phase. (a) Mn^{3+} spins configuration of B_2 magnetic symmetry. Solid (dotted) green triangles represent Mn^{3+} trimers at $z = c/2$ ($z = 0$) layer. Mn^{3+} ions and spins are not shown in the rest of the cartoons for clarity. (b) and (c) Local distortion and spin configuration of Er^{3+} ions in type I and type II domain walls in $ErMnO_3$, respectively. Yellow (white) circle corresponds to an Er^{3+} ion distorting out of (into) paper. "+" ("−") denotes the induced Er^{3+} moment is out of (into) paper. (d) Illustration of spin configuration of the moments of each Er^{3+} atomic plane in b and c near DW_I and DW_{II} domain walls; the arrows inside the red boxes represent uncompensated moments at the domain walls. (Reprinted with permission from Geng, Y. et al., Collective magnetism at multiferroic vortex domain walls, *Nano Letters*, 12, 6055–6059. Copyright 2012, American Chemical Society.)

when the orientation of the neighboring Mn^{3+} spins reverse [56,57]. Therefore, the correlation of uncompensated moments at domain walls, shown in Figure 13.9, indicates a single Mn^{3+} spin state over the scanned area (16 μm × 16 μm) and possibly even over the entire sample. Whether a single Mn^{3+} spin state could be controlled by field cooling needs to be confirmed by future investigation, for example, neutron scattering.

With uniquely correlating ambient PFM and low-temperature MFM images, Geng et al. [6] identified unprecedented correlated domain-wall magnetism, that is, alternating uncompensated Er^{3+} moments at antiphase-ferroelectric domain walls interconnected through self-organized vortex networks in multiferroic h-$ErMnO_3$. Furthermore, results demonstrate that the state of these cross-coupled domain walls can be controlled by field cooling through T_N. The direct observation of uncompensated moments at coupled antiferromagnetic and ferroelectric domain walls not only represents a major advancement in both magnetism and ferroelectrics, but also paves the way for potential multifunctional applications of cross-coupled domain walls [51,68,69].

13.5 Dynamics of Multiferroic Vortex Domains

Chae et al. [4] have found that the ferroelectric domain configurations at both the original surfaces of an h-$ErMnO_3$ (with upward polarization favored near the top a–b surface, EMO-A) sample were type-II, but became type-I in the interior of the crystal as shown in Figure 13.10. Note that the two parallel

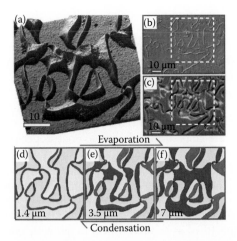

FIGURE 13.10 Depth profiles using sequential chemical etching and $Z_2 \times Z_3$ coloring. (a) Three-dimensional atomic force microscope (AFM) image of the top (001) surface of the ErMnO$_3$ sample after 150 minutes of chemical etching. (b) and (c) Optical microscope images of the top (001) surface of ErMnO$_3$ after 30 and 150 min of chemical etching, respectively. Dashed rectangles in (b) and (c) correspond to the AFM scanned region of (a). (d) and (f) Schematics of the white-dashed-line rectangle region in (b) and (c) with $Z_2 \times Z_3$ coloring, respectively. The depth noted in each schematic was estimated from the etching rate. (e) Schematic of a domain pattern estimated from the mid-height contour plot of (a), corresponding to an intermediate domain pattern between (d) and (f). The depth was also estimated from the mid-height contour plot. (Reprinted with permission from Chae, S.C. et al., *Phys. Rev. Lett.*, 110, 167601. Copyright 2013, American Physical Society.)

surfaces of the crystal favor opposite polarization domains. The differential chemical etching between upward and downward polarization domains resulted in etched surfaces containing shapes of mountain ridges and valley floors as shown in Figure 13.10a. Chae et al. [4] emphasized that both surfaces show the similar structure with narrow mountain ridges and broad valley floors. Figure 13.10b and c shows the optical microscope images of the top surface after chemical etching of ~1.4 and 7 μm, respectively. These images demonstrate that a type-II pattern with narrow downward polarization domains near the top surface evolves into a type-I pattern with increasing depth: the ridges of the mountains in Figure 13.10a reflect the narrow downward polarization domains near the top surface, and the valley floors in Figure 13.10a exhibit the upward polarization domains inside the crystal. The corresponding schematics of ferroelectric domain configurations and their evolution are displayed in Figure 13.10d through f (Figure 13.10d and f is the schematic of Figure 13.10b and c, respectively. Figure 13.10e is drawn from the mid-height contour plot of Figure 13.10a.). As demonstrated in Figure 13.10d, the ferroelectric domain patterns near the original surfaces are type-II, but the patterns are type-I inside the crystal as Figure 13.10f shows.

Graph theory is useful to understand the seemingly irregular patterns of ferroelectric domains in h-REMnO$_3$ [2]. For example, Figure 13.10f can be considered a six-valent graph, where six domain walls always merge at one vortex or antivortex core, and each domain is surrounded by an even number of vortices and antivortices. Each domain can be called an even-gon graphically, since it is surrounded by an even number of vertices (vortices and antivortices). This type-I pattern is $Z_2 \times Z_3$ colorable in the sense that all domains can be colored with 2 (dark and light) $\times$ 3 (red, blue, green) colors in a way that adjacent domains are colored in different colors (proper-colorable), and, for example, a dark red domain is never surrounded by light red domains. These dark and light colors correspond to upward and downward polarizations. On the other hand, Figure 13.10d can be considered a three-valent graph where all domains with one of dark or light colors are always two-gons. When these two-gons are considered lines (or edges), then the six-valent graph with even-gons can be compactified as a "three-valent graph with even-gons," which is three proper-colorable. These three colors (red, blue, and green) correspond to the three structural antiphases.

The physical meaning of this $Z_2 \times Z_3$ coloring is that all domains of any ferroelectric domain pattern forming a six-valent graph with even-gons can be assigned with α^+, α^-, β^+, β^-, γ^+, and γ^- in a way that, for example, an α^+ domain is surrounded only by β^-, and γ^- domains. The type-I patterns exhibit $Z_2 \times Z_3$ symmetry in the sense that the topology of the patterns remains intact with respect to the exchange of (+, −) or (α, β, γ) indices, and the symmetry between + and − is broken in the type-II patterns. In other words, the type-II patterns, which can be considered three-valent graphs with even-gons after compactification, show only Z_3-symmetry with broken Z_2-symmetry. All color schemes in the schematics of Figure 13.10d through f are consistent with the $Z_2 \times Z_3$ coloring, where the dark and bright contrasts correspond to the initial 2-proper-coloring and the blue-red-green colors correspond to the second 3-proper-coloring.

Interesting systematics emerge when the $Z_2 \times Z_3$ colors in the schematics of Figure 13.10d and f are compared. First, the switching from Figure 13.10f to Figure 13.10d through e can be considered a topological condensation through the breaking of Z_2-symmetry in the sense that all dark downward-polarization domains become two-gons, with each two-gon connecting one vortex and one antivortex. Then, one can consider the opposite process as topological evaporation through the restoration of Z_2-symmetry. Naturally, the opposite processes between a type-I pattern and a type-II pattern with two-gons of bright upward-polarization can be considered topological anticondensation and antievaporation. Chae et al. [4] noted that during topological (anti)condensation and (anti)evaporation, most of the cores of vortices and antivortices are hardly influenced since their locations are nearly fixed.

Choi et al. [1] have observed a remarkable dependence of local conduction of h-YMnO$_3$ on the magnitude of applied voltage. In particular, the observed conduction suppression at ferroelectric domain boundaries (FEBs) at high voltages (still much less than the electric coercivity) is in striking contrast with what has been reported in BiFeO$_3$ [31]. A series of c-AFM current images with negative bias voltages down to −5 V and also a positive bias voltage (+10 V) were obtained on the same 6 μm × 6 μm region-data and shown in Figure 13.11a. Naturally, the overall current increases with increasing voltage. However, one distinct feature is that as the voltage increases, significant conduction starts to occur in the middle of the initially least conducting regions (dark domains in c-AFM images). In addition, at high voltages, the current is significantly suppressed at the simultaneous antiphase boundary (APB) + FEB. These conduction variations with increasing voltage are clearer in the conduction profiles (Figure 13.11b). Nine conduction profiles in Figure 13.11b were constructed from lateral scans along the dotted lines in Figure 13.11a. At low voltages of −2 V (black curve) and −3 V (red curve), bright and dark domains show plateau in the conduction profiles. As the voltage increases, the high-conduction plateau (bright domains) change into broad peaks because there is relatively less conduction at domain boundaries, and broad peaks also grow at the lower regions of the conduction profiles (dark domains). At high voltages, the conduction in the middle of the initially dark domain becomes comparable to that of the initially bright domain and the suppression of conduction at domain boundaries becomes evident. These conduction behaviors can also be seen in the same-location scans (Figure 13.11c) with varying voltages along the vertical dotted lines in Figure 13.11b. The conduction in the middle of the initially bright domain (red scan) is highest at all voltages, but the conduction in the middle of the initially dark domain (blue scan) becomes comparable to that of the bright domain at high voltages. Domain boundary (black) scans always show the lowest conduction. In contrast, ferroelectric domain walls in $BiFeO_3$, particularly 180° walls, do have significant electric conduction owing to the reduced bandgap [31]. This unexpected insulating nature of domain walls in h-$YMnO_3$ seems consistent with the surprising resistivity behavior. First, the resistivity data show that the ferroelectric state is more conducting than the paraelectric state, in contrast with the typical behavior in other ferroelectrics; usually band gap and/or resistivity increases when ferroelectric distortions occur [70,71]. This insulating paraelectric state is consistent with the insulating nature of ferroelectric domain walls, as the domain walls can be considered in the paraelectric state in terms of structural distortion. When the specimen is quenched from 1,100 K, the resistivity below Curie temperature is significantly larger than that of the slowly cooled case, as the quenched specimen tends to have a larger number of FEBs. At the ferroelectric transition of h-$YMnO_3$, there is a significant length shortening of the Y–O bonds along the c-axis [72] by as much as 0.54 Å (from 2.85 to 2.31 Å). This may enhance conduction along the c-axis in the ferroelectric phase. The absence or reduction of this bond shortening at ferroelectric domain walls may be responsible for the reduction of local conduction at simultaneous APB + FEB.

It is noteworthy that there exists a significant diode behavior between positive and negative bias in addition to the earlier discussed results of local conduction change with varying negative bias. There exists little conduction for +10 V (Figure 13.11a). The local I–V curve measured by c-AFM demonstrates a clear rectification effect for a highly conductive domain (Figure 13.11d). The corresponding rectification ratio, defined as the ratio of the forward bias

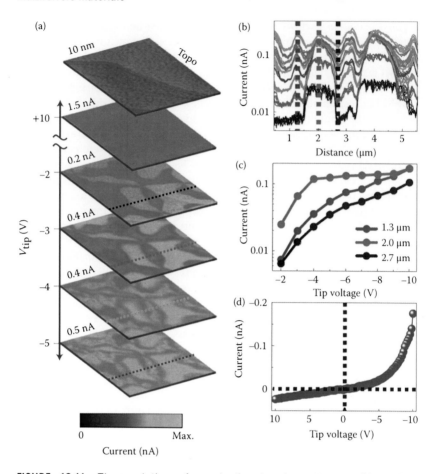

FIGURE 13.11 The evolution of conductive-domain patterns with various bias voltages, and conduction profiles of ferroelectric domains and domain boundaries. (a) Topography after repeated c-AFM scanning over the same area and a series of c-AFM current images with various forward bias voltages in the range of −2 to −5V and with a reverse bias of +10 V. The maximum current for the full color scale for each bias is denoted in each panel. (b) Conduction profiles acquired at the dotted lines on the current images (in [a] and similar images for up to −10 V). (c) Current–voltage characteristics of various locations extracted from the vertical dotted lines in (b). The blue and red lines correspond to the middle of upward and downward polarization domains, respectively. The black line represents a scan of a domain wall. (d) Local *I–V* curve of a highly conductive domain through c-AFM, demonstrating a rectification effect. (Reprinted by permission from Macmillan Publishers Ltd. *Nat. Mater.* Choi, T. et al., Insulating interlocked ferroelectric and structural antiphase domain walls in multiferroic h-YMnO3, 9:253–258, Copyright 2010.)

current at -10 V divided by the reverse bias current at $+10$ V, is 7.30 at room temperature.

To understand the conduction mechanism of ferroelectric domains in h-YMnO$_3$, Choi et al. [1] considered both interface-limited charge injection and bulk-limited charge transport. Note that the charge carriers in h-YMnO$_3$ have been reported to be the hole type [73]. The polarization-orientation dependence of conduction, especially at low voltages, may be understood in terms of the Schottky barrier at a metallic tip–semiconductor-rectifying junction [74]. The sheet of surface charge owing to ferroelectric polarization results in a reduction of the Schottky barrier height for downward polarization and an enhancement for upward polarization. The difference in barrier height, $\triangle\Phi \sim 44$ meV, can be estimated from the maximum current ratio (that is, $\sim$5.5 at -4 V), defined as the maximum ratio of the current for the downward-polarization domain and that for the upward-polarization domain. In addition, the overall rectification effect of ferroelectric domains can be explained by the effect of forward (large current) and reverse (small current) bias voltages on the Schottky barrier junction. The difference in conduction between the two different ferroelectric domains reaches a maximum at -4 V and then gradually decreases with increasing bias voltage. This suggests that the electronic transport mechanism changes from Schottky emission at low bias voltages to bulk-limited conduction such as space-charge-limited conduction [75] at high voltages.

Figure 13.12c shows a low-magnification optical microscope image of the surface of the h-ErMnO$_3$ crystal (with upward polarization favored near the top a-b surface, EMO-A) after applying strain at high temperatures [7]. White dashed lines indicate the position of the alumina rod exerting a downward force on the crystal. The yellow dashed lines show the location of the edges of the groove in the alumina plate. The tilted dark line between the two white lines indicates where the alumina rod touched the crystal. The rod was off centered in the vertical direction to make the force in the bottom triangular region larger than in the top region. Stripe-like domains along the alumina rod direction (perpendicular to the top edge of h-ErMnO$_3$) were observed near the alumina rod, whereas the remaining area showed vortex domains. The vortex-to-stripe transformation takes place near the boundary where vortices meet stripes, as shown in Figure 13.12d, displaying a high-magnification optical microscope image of the green-boxed area in Figure 13.12c. Vortices evidently were unfolded and became stripes. The opposite surface exhibits similar domain patterns. In order to clarify the mechanism for the vortex-to-stripe transformation, we performed experiments on two more crystals: the rectangular-shaped h-ErMnO$_3$ (EMO-B) with a centered alumina rod (Figure 13.12e) and another triangular-shaped h-ErMnO$_3$ (EMO-C) with an off-centered rod. Surprisingly, EMO-B showed only vortices with no hint of stripes, even in the region right under the alumina rod (see Figure 13.12f), while EMO-C exhibited the vortex-to-stripe transformation, similar to that observed in EMO-A. Evidently, the crystal shape is crucial for the vortex-to-stripe transformation. Figure 13.12g shows schematically the in-plane strain distribution in triangular-shape and rectangular-shape specimens to illustrate what may be happening in these

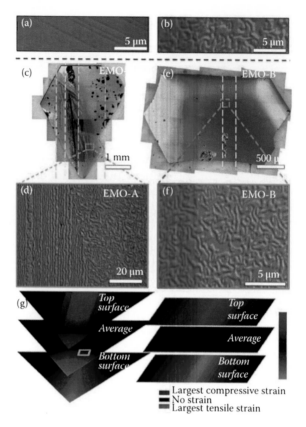

FIGURE 13.12 Effect of the crystal shape for annealing under strain (triangle vs. rectangle). (a) and (b) Optical images of chemically etched h-ErMnO₃ crystals, which indicate two distinct domain patterns: stripes and vortices. (c) Collaged optical microscope image of EMO-A after chemical etching. (d) Enlarged optical image of the green-box area in (c) showing the vortex-to-stripe transformation. (e) Optical images of triangular EMO-B. (f) Enlarged image of the green-boxed area in € shows only vortices. (g) Schematics of in-plane strain on the top surface, average (or middle region), and bottom surface. Blue, red, and black colors indicate compressive, tensile, and no strain, respectively. (Reprinted with permission from Wang, X. et al., *Phys. Rev. Lett.*, 112, 247601. Copyright 2014 by the American Physical Society.)

different-shaped crystals (the out-of-plane strain is not shown, as it does not couple to the trimerization phase). In rectangular EMO-B, the top surface is compressed under the weight of the alumina rod, while the bottom surface is stretched. The top compressive and the bottom tensile strains cancel each other on average, that is, in the middle of the crystal. The triangular corner of EMO-A and EMO-C exposed to an additional strain. To amplify this effect, we intentionally shifted the center of mass of the alumina rod closer to the triangular corner, which produced an additional shear strain with a large in-plane gradient in the corner. This average shear strain with a large gradient induces the vortex-to-stripe transformation.

Han et al. directly observed switching dynamics near the topological defect (vortex) by applying external electric fields in-situ along the c-axis (Figure 13.13) [8]. They employed the dark-field TEM imaging method with a large objective aperture including the 020-, 030-, 022-, and 032-spots during switching experiment, and found that this dark-field imaging optimally visualized the domain walls as lines when the sample was thick. Additionally, the images showed several thickness fringes that are extraneous to the domain wall observations. In Figure 13.13, Han et al. [8] drew lines for domain walls observed in dark-field images. They carried out a series of switching experiments, denoted in alphabetical order, and correspondingly illustrated in Figure 13.13a through m. Domains with parallel polarization to the applied electric field expand, while those with antiparallel polarization shrink, as one can predict for typical ferroelectric domain switching. By measuring the area

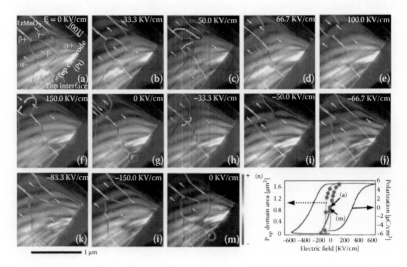

FIGURE 13.13 Switching dynamics around a vortex. (a–m) Dark-field images showing the switching sequence, denoted alphabetically, with an applied field along the [001] direction. Yellow arrows indicate the polarization direction for each domain. The vortex core is denoted by green dots. Electrostatic charges associated with the domain walls are indicated in red (positive) and blue (negative). The abrupt changes in the domain wall's position from 50 to 66.7 kV/cm, from 150 to 0 kV/cm, and from −33.3 to −50 kV/cm are shown by white arrows. Note that three 0 kV/cm states have similar configurations of the surface domain, indicated by the red circles in (a), (g), and (m). A hysteresis loop (n) was obtained by measuring the P_{up} domain areas for each biased condition represented by red dots. Significant back switching, indicated with the blue arrow [from l through m], is visible. For comparison, a P–E loop electrically measured from a bulk LuMnO$_3$ crystal is also shown. (Han, M.-G. et al.: Ferroelectric switching dynamics of topological vortex domains in a hHexagonal manganite. *Adv. Mater.* 25. 2415–2421. 2013. Copyright Wiley-VCH Verlag GmbH & Co. KGaA. Reprinted with permission.)

of P_{up} domains (polarization pointing toward the surface of the sample, or along the c-axis), a hysteresis behavior is observed (Figure 13.13n). For comparison, a polarization (P)–electric field (E) loop electrically measured from a bulk $LuMnO_3$ crystal is shown in Figure 13.13n, which shows larger coercive fields. In fact, it is consistent in that a larger field is typically required to achieve a global poling of a bulk $LuMnO_3$ crystal, while a smaller field is enough to achieve a local poling of a few micron size TEM sample. Han et al. note that the three 0 V states (Figure 13.13a, g, and m) exhibit a strong preference of P_{up} domains near the surface, which thus suppress the P_{down}-dominant remnant state. It indicates the presence of an internal electric field near the surface, locally lowering the energy of the P_{up} domain with respect to that of P_{down} domain. The internal electric field near the surface resulted in significant back switching when the negative external field was removed, as can be seen in the domain structure change from Figure 13.13l through m, shifting the P-E loop in Figure 13.13n toward negative voltage side. Han et al. [8] attribute this internal electric field to inhomogeneous oxygen vacancy or metallic impurity distributions along the c-axis near the surface [76,77]. Interestingly, we note that the position of paired walls at the top electrode interface is preserved for all three 0 V states, as depicted by red circles in Figure 13.13a, g, and m; this feature is indicative of the restoration of the configuration of the surface domain after removing the applied electric field.

In Figure 13.13, all TEM images show that the vortex core (marked with a green dot) was fixed during the entire switching process, revealing that its topology protected it. The vortex core where the three up domains and three down domains meet may be electrically neutral and is not influenced by applied electric fields. Also, vortex core can be pinned at defects [76], such as oxygen vacancy, and thus becomes immobile. In addition, the domain walls are closely paired with large electric fields, rather than pair-annihilated as often happens in typical ferroelectric crystals without accompanying antiphase boundaries; examples are $PbTiO_3$ and $BaTiO_3$, where a single domain state is easily obtained by electrical poling. The absence of pair-annihilation here can be understood by the partial unit-cell-shift vectors across each domain wall. Around a vortex core, each domain wall carries a unit-cell-shift vector $\left(1/3[\bar{1}10], -\right)$. For two domain walls paired by an applied electric field, their vector sum becomes $\left(2/3[\bar{1}10], 0\right)$, that is, incommensurate with respect to the underlying lattice. The lattice cannot accommodate this partial unit-cell-shift, consequently prohibiting pair-annihilation.

13.6 Evolution of Topological Vortex Domains

Improper ferroelectric h-$YMnO_3$ with the size mismatch between Y–O and Mn–O layers shows a trimerization-type structural phase transition, which leads to three antiphase domains (α, β, γ), each of which can have two directions (+, −) of ferroelectric polarization [16,21,20]. The antiphase and ferroelectric domains of h-$YMnO_3$ form the vortex structure with six domain configurations where each domain neighbors with other domains of different polarization as well as structural antiphase. Occurring in pairs, the vortices can be assigned as vortices or antivortices, according to the domain configurations [1]. In addition, the analogy between 2D six-state clock model and the three antiphase domains with two polarization directions (+, –) of h-YMnO$_3$ remind us of three different phases, that is, Ising-type long-range-ordered ground phase, intermediate Kosterlitz–Thouless (KT) phase, and high-temperature disordered phase [78].

Interestingly, KT-like phases with vortex–antivortex domain patterns rather than the Ising-type long-range-ordered phase have been observed in h-YMnO$_3$ at room temperature as shown in Figure 13.14c. Chae et al. [3] studied systematically the thermal evolution of domain configurations in a series of h-REMnO$_3$ crystals in order to unveil the origin of this inconsistency. The traditional method of preferential chemical etching followed by

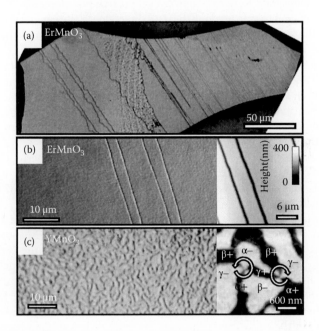

FIGURE 13.14 Two distinct domain patterns of h-REMnO$_3$ crystals; stripe vs. vortex patterns. (a) Low-magnification optical microscope image of the entire surface of a chemically etched ErMnO$_3$ crystal. (b) High-magnification optical microscope image of a chemically etched ErMnO$_3$ crystal surface. The inset shows the AFM image of a stripe domain pattern after chemical etching. These domain patterns for an Ising-type long-range-ordered phase are also observed in HoMnO$_3$, TmMnO$_3$, YbMnO$_3$, and LuMnO$_3$ crystals. (c) Optical microscope image of a chemically etched h-YMnO$_3$ crystal surface. The inset shows the AFM image of an h-YMnO$_3$ surface after chemical etching, showing a vortex–antivortex domain pattern. Vortex and antivortex are distinguished by the arrangement of trimerization antiphase and ferroelectric domains with the opposite sense of rotation around a core. Note that the entire surfaces of all other as-grown crystals h-REMnO$_3$ (RE = Ho, Er, Tm, Yb, Lu) exhibit no hint of the presence of any vortices or antivortices. (Reproduced with permission from Chae, S.C. et al., *Physical Review Letters*, 108, 167603. Copyright 2012 by the American Physical Society.)

optical microscopy as well as HRTEM was used to reveal domain patterns of thin plate-like crystals. The preferential etching of surface areas with upward polarization makes it possible to observe the ferroelectric domain patterns on a crystal surface using optical microscope or AFM.

Unlike in h-YMnO$_3$, long-range-ordered domain pattern, that is, stripe patterns with large downward (−) domains are observed in most of h-REMnO$_3$ crystals (RE = Ho, Er, Tm, Yb, Lu) [3]. The optical image shown in Figure 13.14a shows no sign of domain merging at one point for the formation of vortex on one entire surface of an ErMnO$_3$ crystal. Using AFM, the dark stripes in the optical images were identified as narrow trenches of ~500 nm in depth resulting from the selective chemical etching as shown in the inset of Figure 13.14b. These trenches correspond to narrow upward (+) polarization domains [2]. In addition, the trenches tend to be along the [110] direction (the hexagonal P6$_3$cm notation). In the vortex–antivortex domain pattern in h-YMnO$_3$, six trimerization antiphase (α, β, γ) and ferroelectric (+,−) domains merge to the vortex, and each vortex is paired with other vortices with opposite vorticity in terms of structural antiphase and ferroelectric relationship as shown in the inset of Figure 13.14c [2]. However, stripe domain pattern as shown in Figure 13.14a spans over the entire crystal surface in all other h-REMnO$_3$ crystals. Chae et al. [3] elucidated that the different domain patterns depend on whether the growth temperature is above the structural transition temperature (T_c) or not.

To investigate the thermal evolution of the stripe domain patterns, Chae et al. quenched crystals from various temperatures and checked the domain pattern of crystals etched chemically at 130°C. Figure 13.15a shows the AFM image of a chemically etched ErMnO$_3$ surface after quenching from 1120°C. Although the annealing temperature is very high, the Ising-type stripe domain pattern remained. (Note that the domain patterns were observed by AFM at room temperature after chemical etching.) However, when the temperature is raised by only 20°C, the pattern changes significantly, and exhibits highly curved lines with the appearance of many curved closed loops, as shown in Figure 13.15b. Dark stripe lines, conserved robustly up to 1120°C, start to wiggle heavily at 1140°C, but never cross to each other; that is, there is no hint of the presence of vortices in the entire crystal surface. The more-or-less straight parts of dark stripe lines are indicated with white dashed lines in the upper region of Figure 13.15b. A TEM image of closed loop domains is shown in Figure 13.15c, and the corresponding possible schematic is shown in Figure 13.15d. Therefore, when the system approaches T_c from below, thermal fluctuations induce roughening of the stripe domain walls and the appearance of a large number of loop domains.

When the temperature is further raised by 30°C, more complicated patterns with crossing lines and a large number of vortices appeared as shown in Figure 13.15e. Finally, we can observe a more vortex–antivortex domain pattern formation after the crystal is quenched from 1200°C, as shown in Figure 13.15f. Chae et al. [3] have determined the critical temperatures (T_c) of the other h-REMnO$_3$ crystal (the temperature at which stripe domains change into vortex–antivortex domains). The results, including that of h-YMnO$_3$, are plotted in the inset of Figure 13.15f [79]. The linear dependence suggests the correlation between ferroelectric transition and structural-phase transition in

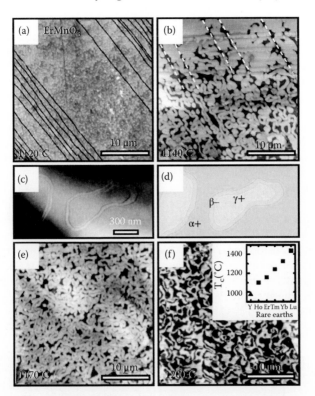

FIGURE 13.15 The thermal evolution of stripe domain patterns in $ErMnO_3$ and the creation of dislocation loops and vortex–antivortex pairs. (a), (b), (e), and (f) The AFM images of chemically etched $ErMnO_3$ crystals quenched from 1120°C, 1140°C, 1170°C, and 1200°C, respectively. A large number of dislocation loops are evident in (b), but (e) and (f) display vortex–antivortex patterns. The inset displays T_c's of $REMnO_3$ estimated from the formation temperature of vortex–antivortex domain patterns. T_c of h-$YMnO_3$ is from Reference [79]. (c) and (d) TEM image of dislocation loops of $ErMnO_3$ quenched from 1140°C and a possible corresponding schematic, respectively. The dislocation-loop domains observed in the TEM image are assigned with three antiphase domains (α, β, γ). (Reprinted with permission from Chae, S.C. et al., *Physical Review Letters*, 108, 167603. Copyright 2012 by the American Physical Society.)

h-$REMnO_3$. T_c's of h-$REMnO_3$ are reliably determined for the first time and it increases with decreasing RE size, which is consistent with the fact that the structural transition is induced by the mismatch between small RE–O layers and large Mn–O layers. Since the growth temperature of h-$YMnO_3$ is higher than T_c, it can be concluded that stripe domain patterns form when the crystal growth temperature is below T_c, while vortex–antivortex domain patterns are realized when crystals are exposed to temperatures above T_c.

Chae et al. [3] also found that once vortex–antivortex domain patterns form after crossing T_c, the patterns spanning the entire crystal surface are conserved against various thermal treatments but the domain size or the distance between vortices and antivortices can vary in a systematic manner. The thermal evolution of vortex–antivortex domain patterns and its kinetics were investigated by varying the cooling rate around T_c from 0.5°C/h to 300°C/h. In addition, one specimen was cooled from 1220°C to 677°C at a cooling rate of 5°C/h and quenched to room temperature to investigate the abnormal transport result below 677°C. Figure 13.16a through d show the AFM images of etched $ErMnO_3$ crystals following these thermal treatments. Despite the large variation of cooling rate, the topology of vortex–antivortex domain patterns remains robust, but the domain size shows systematic change.

Looking at the vortex–antivortex domain patterns in Figure 13.16b and c, the similarity indicates that the cooling rate below 677°C has no influence on the domain patterns. This is important for the origin of the mysterious second transition near 600°C reported in many early publications [21,19,80,79,81–84]. This second transition near 600°C was suggested to be the ferroelectric transition from centrosymmetric $P6_3/mmc$ to low-temperature polar $P6_3cm$ structure via an intermediate $P6_c/mcm$ structure [83], whereas other results argued against the intermediate $P6_3/mcm$ state, but the presence of an isosymmetric phase transition with Y–O hybridization [19,79,21,81].

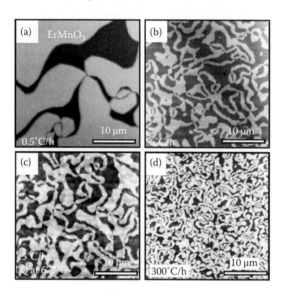

FIGURE 13.16 The evolution of vortex–antivortex domain patterns with varying cooling rate. The AFM images of chemically etched $ErMnO_3$ crystals: (a) cooled from 1220°C to 890°C with a rate of 0:5°C/h, followed by furnace cooling; (b) cooled from 1200°C to room temperature with a rate of 5°C/h; (c) cooled from 1200°C to 677°C with a rate of 5°C/h, followed by quenching; and (d) cooled from 1200°C to room temperature with a rate of 300°C/h. (Reprinted with permission from Chae, S.C. et al., *Physical Review Letters*, 108, 167603. Copyright 2012 by the American Physical Society.)

Li et al. [85] have modeled the ferroelectric loop domains with different shell structures (Figure 13.17), which match the experimental observation (SEM images, Figure 13.17b) very well [10]. Note that these ferroelectric loop domains can be indexed by their annular number n (i.e., the domain shells) and they are very similar to the magnetic bubble domains, which have been used in various electronic devices [86,87]. As the magnetic bubble domain can be manipulated by magnetic field, these ferroelectric bubbles might be manipulated by an electric field.

To understand the essential structural features of the annular domains, Li et al. [85] have conducted Monte Carlo simulations on the 2D six-state clock model for the triangular lattice, and the Hamiltonian used in their study is similar to that in References [3,88]. The six distinctive states in the model correspond to the six domains, that is, α^+, β^-, γ^+, α^-, β^+, and γ^-, respectively. Figure 13.17c shows a typical result for structural domains within the KT phase after 3500 steps per lattice at a temperature of $T = 0.1J/k$, where J is the coupling constant and k is the Boltzmann constant. In addition to vortex–antivortex domains, double-wall annular domains appear during the simulation (Figure 13.17c), which is consistent with our experimental observations. As the transition from long-range order to KT order is accompanied by changes of correlation function and free energy, a large quantity of annular domains condense while the vortex pairs are protected by their logarithmic entropy. However, a few residual annular domains coexist with vortex pairs in the KT phase, so the so-called KT phase is a little different from the low-temperature phase in the XY model in which only vortex-type topological excitations exist. This conclusion is supported by the experimental and simulated results of Li et al. The coexistence of circular-type and vortex-type excitations is also observed on the side surface parallel to the c-axis. It suggests a potential for experimental 3D six-state clock model study. Moreover, the structural configurations, as well as the poling features, for each domain can be directly read out from this simulated image as clearly illustrated in the inset images. According to our theoretical data, the double-wall structure could have the configurations of $(\alpha+/-\beta-/+\gamma+/-)$ or $(\alpha+/-\gamma-/+\beta+/-)$. On the other hand, it is also noted that the annular domains could commonly appear at room temperature in h-REMnO$_3$ crystals. This fact suggests that this kind of domain can also be arrested by slow kinetic and strong pinning features as discussed in Reference [87].

The restoration of continuous U(1) symmetry, like the superfluid to normal state transition of ^{4}He, can occur via proliferation of vortices. The potential of having emergent continuous symmetries in magnets or ferroelectrics with discrete microscopic symmetries opens the possibility of observing a similar proliferation of vortices in these materials. While the emergence of continuous symmetries from discrete variables is theoretically established, the same is not true at the experimental level. Among other reasons, it is always challenging to measure the critical exponents with the required resolution to distinguish between discrete and continuous symmetry breaking. Recently, the emergence of a continuous U(1) symmetry at the ferroelectric transition of h-REMnO$_3$ (R = Y, Ho, ... Lu, Sc) was suggested by directly measuring the vortices or disorder field, instead of addressing the order parameter field [9].

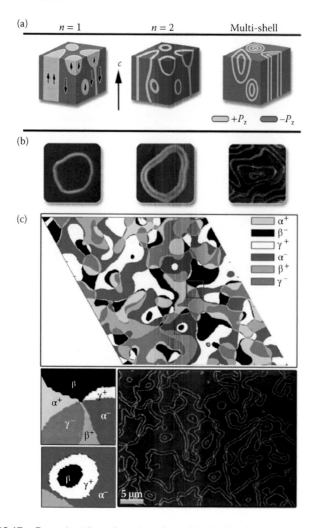

FIGURE 13.17 Ferroelectric onion domains with shells of $n = 1$, 2, and multi-shell. (a) Structural models for ferroelectric onion domains observed in h-EMnO$_3$. The $n = 1$ onion domain often shows poling features very similar to those of magnetic bubbles. Polarization in h-REMnO$_3$ occurs along the c-axis direction as indicated by the arrows. (b) SEM images showing the domain walls for ferroelectric domains with different annular numbers in the a–b plane. (c) The domain structures obtained from the six-state clock model in two dimensions through Monte Carlo simulations, showing the coexistence of cloverleaf vortices and annular domains. Lower panel: The left insets show the structural features and poling configurations for a vortex and an annular domain from the square marked areas. The right inset is an experimental image showing the coexistence of cloverleaf patterns and annular domains obtained from an annealed ErMnO$_3$ sample. (Reprinted with permission from J. Li et al. Physical Review B, vol 87, 094106. Copyright 2013, American Physical Society.)

An emergent U(1) symmetry implies that the critical point belongs to the XY universality class, which is the class of the superfluid transition of a neutral system such as ^{4}He. Therefore, analogous to the case of superfluid ^{4}He, the transition must be driven by proliferation of vortices spanning the whole system above T_c. Based on the disorder field of topological vortices instead of the order parameter field [89–91], the phase transition is described as a condensation of the disorder field, which is coupled to a gauge field [90,92]. Upon heating across T_c, the vortex condensation makes the gauge field massive via the Higgs mechanism. Consequently, the vortex–vortex interaction becomes screened above T_c, instead of the Biot–Savart interaction that characterizes the Coulomb phase below T_c. Figure 13.18 shows the same phase diagram from two different viewpoints based on the order and disorder fields.

The local order parameter of our problem is a complex field $\varphi_j = |\varphi_j| e^{i\phi}$, that takes six possible values ($Z_2 \times Z_3$) corresponding to the even times of a clock. By assuming that the local trimerization and dipole moments develop above T_c, we neglect the amplitude ($|\phi_j|$) fluctuations near T_c. The six orientations of ϕ_j are enforced by an effective potential, $V(\phi_j) = A \cos(6\varphi_j)$, which reflects the anisotropy of the underlying crystal lattice. $V(\phi_j)$ is dangerously irrelevant at T_c; that is, the coarse grained action near the ferroelectric transition becomes

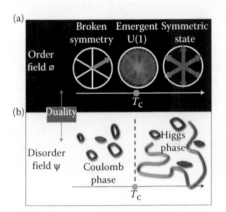

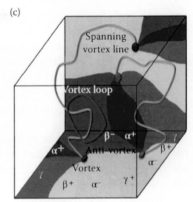

FIGURE 13.18 Dual description of a phase transition with $Z_2 \times Z_3$ symmetry. The phase transition can be described in terms of (a) order field Φ or (b) disorder field Ψ. The local order parameter Φ takes six values, represented by the even hours in the clock dials in (a). They correspond to the six multiferroic states or domains α^+ through γ^- distinguished by the polarization direction (+ or −) and the trimerization phase (α, β, γ), as described in the text. The multiferroic $Z_2 \times Z_3$ vortices are line defects where the six domains meet with each other, as shown in (c). Continuous U(1) symmetry emerges from $Z_2 \times Z_3$ order parameter at the critical temperature. The disordered phase above T_c can be described as a condensation of the disorder field Ψ signaled by the proliferation of vortex lines spanning the whole system (yellow lines). Only quickly fluctuating closed vortex loops (red lines) are present for $T < T_c$. (Reprinted by permission from Macmillan Publishers Ltd. *Nat. Phys.* Lin, S.-Z. et al., Topological defects as relics of emergent continuous symmetry and Higgs condensation of disorder in ferroelectrics, 10:970–977, copyright 2014.)

identical to the isotropic ϕ^4 action for the normal to superfluid transition of a neutral system like ^{4}He (see Figure 13.18a):

$$H_\varphi = m_\varphi^2 \varphi^2 + u_\varphi \varphi^4 + \left(\nabla \varphi\right)^2 \qquad (13.1)$$

The superfluid to normal transition occurs via proliferation of vortex lines at $T > T_c$. The problem admits a dual description, in which the proliferation of vortex lines spanning the whole system arises from a condensation of a dual or "disorder" field $\psi = |\psi| e^{i\theta}$ minimally coupled to an effective gauge field. Below T_c, the "photon" of this gauge field is the Goldstone mode of the superfluid field ϕ. This photon acquires a finite mass via the Higgs mechanism for $T > T_c$ (see Figure 13.18b). Consequently, the Biot–Savart (Coulomb) interaction between vortex segments for $T < T_c$ becomes screened (Yukawa) for $T > T_c$.

The dual description is obtained after a sequence of transformations. The original ϕ^4 theory (B1) for a neutral superfluid is first mapped into a gas of vortex loops coupled to a vector gauge field **A** generated by the smooth phase fluctuations of the original field ϕ. The fluctuating vortex loops are then described by a disorder $|\psi|^4$ field theory in which the vortex loops correspond to "supercurrents" of ψ, which remain minimally coupled to A [89–91]:

$$H_\psi = m_\psi^2 \psi^2 + u_\psi \psi^4 + \frac{1}{2t}\left|\left(\nabla - iq_{\text{eff}} A\right)\psi\right|^2 + \frac{1}{2}\left(\nabla \times A\right)^2 \qquad (13.2)$$

The constants t and q_{eff} are determined by nonuniversal parameters, such as the vortex core energy and the transition temperature [89–91]. Having direct experimental access to the vortex field, we can observe the Higgs condensation of ψ: the emergence of vortex lines that span the whole system above T_c implies that superfluid currents of the disorder field ψ connect opposite ends of the sample; that is, the disorder field has condensed into a "superfluid state" (see Figure 13.18b).

The dynamics of symmetry breaking in phase transitions is another fascinating phenomenon that can be tested in h-REMnO$_3$. Its salient features are captured by the Kibble–Zurek mechanism (KZM), which combines cosmological motivations with information about the near-critical behavior. Symmetry breaking is thought to be responsible for the emergence of the familiar fundamental interactions from the unified field theory at GUT temperatures of $\sim 10^{15}$ GeV in the Universe after Big Bang. As Kibble [94] noted, relativistic causality limits the size of domains that can coordinate the choice of broken symmetry in the nascent Universe. This results in a random selection of local broken symmetry, and can lead to creation of topological defects (e.g., monopoles or cosmic strings) that influence the evolution of the Universe.

Because phase transitions are ubiquitous, their dynamics can be investigated experimentally. Although relativistic causality is no longer a useful constraint in the laboratory, cosmological motivations can be combined with the scaling relations in the near-critical regime of second order phase transitions to estimate the density of topological defects as a function of the quenching rate [95]. This combination is the KZM [94,96].

The main difference between the cosmological and laboratory settings is that now the relaxation time and coherence length (and speed of the relevant sound rather than the speed of light) determine the sonic horizon—the linear size ξ of regions that can break symmetry in step. The basic idea [95] is to compare the relaxation time τ with the timescale of change of the key parameter [here, relative temperature $\varepsilon = (T - T_c)/T_c$]. We assume $= t/\tau_Q$, where τ_Q is the quench time. The relaxation time $\tau(\varepsilon) = \tau_0/|\varepsilon|^{vz}$ (where v and z are spatial and dynamical critical exponents, and τ_0 is a timescale set by microphysics) determines the reaction time of the order parameter. Relaxation characterized by $\tau(\varepsilon)$ is faster than $\varepsilon/\dot{\varepsilon} = t$ outside interval $\hat{t} = \left(\tau_0 \tau_Q^{zv}\right)^{1/(1+vz)}$ around the transition, so the system can quasi-adiabatically follow the change imposed by the quench. This instant is determined by the following equation [95]:

$$\tau(\varepsilon(\hat{t})) = \varepsilon/\dot{\varepsilon} = \hat{t} \tag{13.3}$$

The system will cease to keep up with the imposed change at time $\hat{t}$ before reaching the critical point, while its reflexes are recovered at time $\hat{t}$ (i.e., when $\hat{\varepsilon} = \left(\tau_0/\tau_Q\right)^{1/(1+vz)}$) after the transition. Thus, broken symmetry is chosen by fluctuations when their coherence length is [95]

$$\hat{\xi} = \frac{\xi_0}{|\hat{\varepsilon}|^v} = \xi_0 \left(\tau_Q/\tau_0\right)^{v/(1+vz)} \tag{13.4}$$

The choice of broken symmetry is random within the fluctuating domains of this size. Topological defects are then expected to form with the density of one defect fragment per domain. Thus, the scaling of $\hat{\xi}$ with quenching rate set by the universality class of the transition translates into the scaling of the defect density. This prediction has been verified for the 3D XY model [3]. Indeed, even the KZM predictions of the actual density (and not just its scaling) are close to those observed [97]. Our experimental results (see Supplementary Information, Section 4 of Reference [9]) for h-REMnO$_3$, as well as our simulations (for the simulation details, see Supplementary Information section 2 of Reference [9]) of H of Equation 13.1, confirm this prediction and corroborate KZM (see Figure 13.19). The obtained exponent of ~0.59 is very close to the value $\left(2v/(1+vz) \cong 0.57\right)$ that is expected for a 3D XY fixed point: $v = 0.67155(27)$ [31] and $z \cong 2$ [98]. We emphasize that the 3D XY fixed point is a consequence of the Z_6 symmetry of h-REMnO$_3$ compounds (the Z_6 anisotropy is dangerously irrelevant at $T = T_c$ [99,100]). Moreover, we verify the KZM for rapid quenches (where Reference [97] observed an unexpected decrease of defect density which they termed "anti-KZM").

By using the earlier estimation of one $\hat{\xi}$ defect fragment per volume of the domain of that linear size, one can estimate the defect density as functions of quenching rate and of τ_0 and ξ_0—2D constants that characterize the system—and its universality class given by the spatial and dynamical critical exponents v and z.

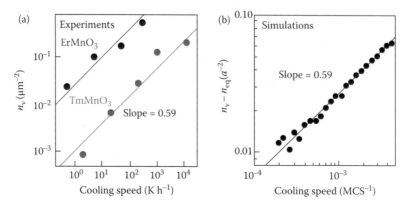

FIGURE 13.19 Dependence of the vortex density n_v on cooling rate. (a) Experimental vortex density in the final state as a function of the cooling rate. TmMnO$_3$ ($T_c \approx 1523$ K) samples with the cooling rates ranging from 2 to 12,000 K/h, and ErMnO$_3$ ($T_c \approx 1403$ K) with the cooling rates from 0.5 to 300 K/h were measured. The vortex density as a function of cooling rate is consistent with a power law dependence with exponent 0.59 (full lines), that is obtained from our Monte Carlo simulations shown in (b) for a final temperature of 0.92 T_c (black dots), as well as with the prediction of ~0.57 that follows from the KZM. In (b), the cooling speed is given in inverse Monte Carlo sweep (arbitrary units), n_{eq} is the density of the thermally excited vortices subtracted to reveal the KZM scaling, and a is the lattice parameter. (Reprinted by permission from Macmillan Publishers Ltd. *Nat. Phys.* Lin, S.-Z. et al., Topological defects as relics of emergent continuous symmetry and Higgs condensation of disorder in ferroelectrics, 10:970–977, copyright 2014.)

The essence of KZM is the randomness of the choices of broken symmetry in domains of size $\hat{\xi}$. This randomness—in addition to defect density—predicts [95] scaling of the winding number W subtended by a contour C. The winding number is the net topological charge: $W = n_+ - n_-$, the difference of the numbers n_+ and n_- of vortices and antivortices inside C. If these charges were assigned at random, typical net charge would be proportional to the square root of their total number, $n = n_+ + n_-$, inside C, so it would scale as a square root of the area A inside C. Therefore, for contours of a fixed shape, it would scale as the length of the contour, $W \propto \sqrt{A} \propto C$.

According to the KZM, W is set by the winding of the phase along C. In our clock model, broken symmetry phases correspond to even hours on the clock face. W is then the "number of days" elapsed along the contour C. As choices of even hours (phases) are random in $\hat{\xi}$-sized domains, the typical net winding number W scales as $\sqrt{C/\hat{\xi}}$—it is proportional to the square root of the number of steps.

This scaling with the square root of the circumference can be tested by finding the net charge of vortices inside C. The results are shown in Figure 13.20. The typical winding number (characterized either by the average absolute value $\langle |W| \rangle$ or the dispersion $\sqrt{\langle W^2 \rangle}$) indeed scales like $\sqrt{C/\hat{\xi}}$, as long as $C > \hat{\xi}$.

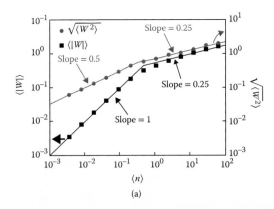

(a)

FIGURE 13.20 Winding numbers for KZM defects. Absolute winding numbers $\langle |W| \rangle$ and their dispersions $\sqrt{\langle W^2 \rangle}$ obtained from the coordinates of ~4100 defects in the h-YMnO$_3$ crystal, as functions of the average number of defects <n> inside the contour C. Typical winding numbers $W = n_+ - n_-$ inside C are predicted by KZM [95,101]. Their scaling follows from the key idea that local random choices of broken symmetry in domains of size ~$\hat{\xi}$ determine defect locations. For contours with circumference $C > \hat{\xi}$ that contain many defects (large average $n = n_+ + n_-$) KZM predicts that $\langle |W| \rangle$ and $\sqrt{\langle W^2 \rangle}$ depend only on $\sqrt{C/\hat{\xi}}$, and vary as $<n>^{1/2} \sim A^{1/2} \sim C$, independently of its shape, area A, or the average <n> ~ A inside C. This may seem surprising, for if defect charges were random, one would expect a winding number (the mismatch between vortices and antivortices inside C) to vary as $C \sim <n>^{1/2}$. KZM prediction is confirmed in (a) for randomly placed contours of a fixed shape (here squares, like the green one in Figure 13.4b). For a fixed shape <n> and large <n> typical winding number scaling $|W| \sim \sqrt{C/\hat{\xi}}$ result in $\langle |W| \rangle$ and $\sqrt{\langle W^2 \rangle}$ proportional to $<n>^{1/4}$. By contrast, when $<n> < 1$, there is usually at most one defect inside C, so W can be only 0, +1, or −1, so $\langle |W| \rangle$ is proportional to probability p of finding a defect. Moreover, p ~ A, so now $\langle |W| \rangle \sim A \sim <n>$—typical winding numbers depend on the area A inside C. However, dispersion is proportional to $p^{1/2}$, so $\sqrt{\langle W^2 \rangle} \sim C \sim <n>^{1/2}$. Thus, when $<n> < 1$ scaling of average $\langle |W| \rangle$ and $\sqrt{\langle W^2 \rangle}$ differ [101]. This is also seen in panel (a). Panel (b) shows $\langle |W| \rangle$ for contours of the same circumference, but with different shapes and, hence, areas that differ by a factor of ~3. As expected, $\langle |W| \rangle$ depends on $C^{1/2}$ for large <n>, but on $A \sim <n> \sim C^2$ for fractional <n>. Panel (c) shows the same data as (b) redrawn as a function of <n>. In all panels, solid lines show power laws predicted for $\langle |W| \rangle$ and $\sqrt{\langle W^2 \rangle}$ by KZM [101]. C is normalized so that a square of circumference $C = 4$ (A = 1) contains one defect on average. KZM prediction for W, based on randomness of broken symmetry choice in domains of size ~$\hat{\xi}$, is thus verified in all cases. (Reprinted by permission from Macmillan Publishers Ltd. *Nat. Phys.* Lin, S.-Z. et al., Topological defects as relics of emergent continuous symmetry and Higgs condensation of disorder in ferroelectrics, 10:970–977, copyright 2014.)

(Continued)

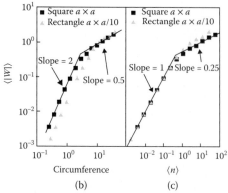

FIGURE 13.20 (*Continued*) Winding numbers for KZM defects. Absolute winding numbers $\langle |W| \rangle$ and their dispersions $\sqrt{\langle W^2 \rangle}$ obtained from the coordinates of ~4100 defects in the h-YMnO$_3$ crystal, as functions of the average number of defects <n> inside the contour C. Typical winding numbers $W = n_+ - n_-$ inside C are predicted by KZM [95,101]. Their scaling follows from the key idea that local random choices of broken symmetry in domains of size ~$\hat{\xi}$ determine defect locations. For contours with circumference $C > \hat{\xi}$ that contain many defects (large average $n = n_+ + n_-$) KZM predicts that $\langle |W| \rangle$ and $\sqrt{\langle W^2 \rangle}$ depend only on $\sqrt{C/\hat{\xi}}$, and vary as $<n>^{1/2} \sim A^{1/2} \sim C$, independently of its shape, area A, or the average $<n> \sim A$ inside C. This may seem surprising, for if defect charges were random, one would expect a winding number (the mismatch between vortices and antivortices inside C) to vary as $C \sim <n>^{1/2}$. KZM prediction is confirmed in (a) for randomly placed contours of a fixed shape (here squares, like the green one in Figure 13.4b). For a fixed shape $<n>$ and large $<n>$ typical winding number scaling $|W| \sim \sqrt{C/\hat{\xi}}$ result in $\langle |W| \rangle$ and $\sqrt{\langle W^2 \rangle}$ proportional to $<n>^{1/4}$. By contrast, when $<n> < 1$, there is usually at most one defect inside C, so W can be only 0, +1, or −1, so $\langle |W| \rangle$ is proportional to probability p of finding a defect. Moreover, $p \sim A$, so now $\langle |W| \rangle \sim A \sim <n>$—typical winding numbers depend on the area A inside C. However, dispersion is proportional to $p^{1/2}$, so $\sqrt{\langle W^2 \rangle} \sim C \sim <n>^{1/2}$. Thus, when $<n> < 1$ scaling of average $\langle |W| \rangle$ and $\sqrt{\langle W^2 \rangle}$ differ [101]. This is also seen in panel (a). Panel (b) shows $\langle |W| \rangle$ for contours of the same circumference, but with different shapes and, hence, areas that differ by a factor of ~3. As expected, $\langle |W| \rangle$ depends on $C^{1/2}$ for large $<n>$, but on $A \sim <n> \sim C^2$ for fractional $<n>$. Panel (c) shows the same data as (b) redrawn as a function of $<n>$. In all panels, solid lines show power laws predicted for $\langle |W| \rangle$ and $\sqrt{\langle W^2 \rangle}$ by KZM [101]. C is normalized so that a square of circumference C = 4 (A = 1) contains one defect on average. KZM prediction for W, based on randomness of broken symmetry choice in domains of size ~$\hat{\xi}$, is thus verified in all cases. (Reprinted by permission from Macmillan Publishers Ltd. *Nat. Phys.* Lin, S.-Z. et al., Topological defects as relics of emergent continuous symmetry and Higgs condensation of disorder in ferroelectrics, 10:970–977, copyright 2014.)

This scaling dependence changes when the magnitude of W falls below 1. Moreover, the scalings of $\langle|W|\rangle$ and of the dispersion $\sqrt{\langle W^2 \rangle}$ diverge in this regime. This may seem surprising, but it is actually predicted by the KZM [101]: $|W| < 1$ occurs when $C < \hat{\xi}$, that is, C normally contains a single defect or none. In this case, $\langle|W|\rangle \approx p_+ + p_- = p_{\text{DEFECT}}$, while $\sqrt{\langle W^2 \rangle} = p_{\text{DEFECT}}$ in terms of probabilities. Moreover, the probability p_{DEFECT} of finding a defect inside C is proportional to area A subtended by C, accounting for both the change and divergence of the scalings of $\langle|W|\rangle$ and $\sqrt{\langle W^2 \rangle}$ seen in Figure 13.20a.

Further evidence of the KZM is found in the scaling of $\langle|W|\rangle$ and $\sqrt{\langle W \rangle^2}$ with the deformation of the shape of the contour (and the consequent changes of the area A inside). Figure 13.20b and c show that, as long as the size of the contour is large compared to $\hat{\xi}$, the winding number depends only on its length, and not on the area enclosed by the loop. However, as expected, the area becomes important when the number of defects falls below 1 and the scalings of $\langle|W|\rangle$ and $\sqrt{\langle W \rangle^2}$ steepen and diverge.

13.7 Conclusions and Outlook

Intriguing patterns of the ferroelectric and structural antiphase domains in h-REMnO$_3$ were investigated in conjunction with macroscopic connectivity and topology. Graph theoretical consideration of the domain pattern evolution revealed the condensation and annihilation of topological vortices–antivortices pairs. Electric and magnetic characterizations of domain walls revealed an unprecedented conduction behavior and correlated magnetism through self-organized domain-wall networks in multiferroic h-REMnO$_3$. Topological symmetry exists in the seemingly irregular domain patterns: $Z_2 \times Z_3$ symmetry emerges through a ferroelectric phase transition and it remains intact against external stimulus. The appearance of topological vortices in h-REMnO$_3$ reflects the rich nature of a ferroelectric phase transition. Note that topological vortices are relevant to the formation of cosmic topological defects during the birth of our universe as well as the proliferation of quantum topological defects in type-II superconductors and superfluids. Since numerous topological vortices can be visualized readily in h-REMnO$_3$, they can be a test bed to learn the interaction among topological defects as well as dynamics during a phase transition of a matter with nontrivial topological symmetry. For example, the universality of KZM can be neatly tested in topological vortex patterns in h-REMnO$_3$. Finally, we would like to note the importance of realizing nontrivial real-space topology of domain patterns in complex materials. Often this macroscopic topology is responsible for emergent phenomena in functional materials such as the presence of conducting ferroelectric domain walls. Therefore, this exploration of macroscopic domain topology should be further pursued for functional complex materials.

Reference

1. Choi, T., Y. Horibe, H. T. Yi, Y. J. Choi, W. Wu, and S. W. Cheong. 2010. Insulating interlocked ferroelectric and structural antiphase domain walls in multiferroic h-YMnO$_3$. *Nature Materials* 9:253–258.

2. Chae, S. C., Y. Horibe, D. Y. Jeong, S. Rodan, N. Lee, and S. W. Cheong. 2010. Self-organization, condensation, and annihilation of topological vortices and antivortices in a multiferroic. *Proceedings of the National Academy of Sciences of the United States of America* 107:21366–21370.

3. Chae, S. C., N. Lee, Y. Horibe, M. Tanimura, S. Mori, B. Gao, S. Carr, and S. W. Cheong. 2012. Direct observation of the proliferation of ferroelectric loop domains and vortex-antivortex pairs. *Physical Review Letters* 108:167603.

4. Chae, S. C., Y. Horibe, D. Y. Jeong, N. Lee, K. Iida, M. Tanimura, and S. W. Cheong. 2013. Evolution of the domain topology in a ferroelectric. *Physical Review Letters* 110:167601.

5. Wu, W., Y. Horibe, N. Lee, S. W. Cheong, and J. R. Guest. 2012. Conduction of topologically protected charged ferroelectric domain walls. *Physical Review Letters* 108:077203.

6. Geng, Y., N. Lee, Y. J. Choi, S. W. Cheong, and W. Wu. 2012. Collective magnetism at multiferroic vortex domain walls. *Nano Letters* 12:6055–6059.

7. Wang, X., M. Mostovoy, M. G. Han, Y. Horibe, T. Aoki, Y. Zhu, and S. W. Cheong. 2014. Unfolding of vortices into topological stripes in a multiferroic material. *Physical Review Letters* 112:247601.

8. Han, M.-G., Y. Zhu, L. Wu, T. Aoki, V. Volkov, X. Wang, S. C. Chae, Y. S. Oh, and S.-W. Cheong. 2013. Ferroelectric switching dynamics of topological vortex domains in a hHexagonal manganite. *Advanced Materials* 25:2415–2421.

9. Lin, Q., A. Armin, R. C. R. Nagiri, P. L. Burn, and P. Meredith. 2014. Electro-optics of perovskite solar cells. *Nature Photonics* 9:106–112.

10. Burschka, J., N. Pellet, S.-J. Moon, R. Humphry-Baker, P. Gao, M. K. Nazeeruddin, and M. Grätzel. 2013. Sequential deposition as a route to high-performance Perovskite-sensitized solar cells. *Nature* 499:316–319.

11. Wilson, R. J. 1970. *An Introduction to Graph Theory*. India: Pearson Education India.

12. Balaban, A. T. 1985. Applications of graph-theory in chmistry. *Journal of Chemical Information and Computer Sciences* 25:334–343.

13. Hagmann, P., L. Cammoun, X. Gigandet, R. Meuli, C. J. Honey, V. J. Wedeen, and O. Sporns. 2008. Mapping the structural core of human cerebral cortex. *Plos Biology* 6:1479–1493.

14. Barabasi, A. L., and R. Albert. 1999. Emergence of scaling in random networks. *Science* 286:509–512.

15. Ramesh, R., and N. A. Spaldin. 2007. Multiferroics: Progress and prospects in thin films. *Nature Materials* 6:21–29.

16. Fiebig, M., T. Lottermoser, D. Frohlich, A. V. Goltsev, and R. V. Pisarev. 2002. Observation of coupled magnetic and electric domains. *Nature* 419:818–820.

17. Cheong, S.-W., and M. Mostovoy. 2007. Multiferroics: A magnetic twist for ferroelectricity. *Nature Materials* 6:13–20.

18. dela Cruz, C., F. Yen, B. Lorenz, Y. Q. Wang, Y. Y. Sun, M. M. Gospodinov, and C. W. Chu. 2005. Strong spin-lattice coupling in multiferroic HoMnO$_3$: Thermal expansion anomalies and pressure effect. *Physical Review B* 71:060407.

19. Fennie, C. J., and K. M. Rabe. 2005. Ferroelectric transition in h-YMnO$_3$ from first principles. *Physical Review B* 72:100103.

20. Van Aken, B. B., T. T. M. Palstra, A. Filippetti, and N. A. Spaldin. 2004. The origin of ferroelectricity in magnetoelectric h-YMnO3. *Nature Materials* 3:164–170.

21. Katsufuji, T., S. Mori, M. Masaki, Y. Moritomo, N. Yamamoto, and H. Takagi. 2001. Dielectric and magnetic anomalies and spin frustration in hexagonal $RMnO_3$ (R = Y, Yb, and Lu). *Physical Review B* 64:104419.

22. Mostovoy, M. 2010. Multifferoics A whirlwind of opportunities. *Nature Materials* 9:188–190.

23. Safranko.M, J. Fousek, and S. A. Kizaev. 1967. Domains in ferroelectric h-$YMnO_3$. *Czechoslovak Journal of Physics* 17:559–560.

24. Appel, K., and W. Haken. 1976. Every planar map is 4 colorable. *Bulletin of the American Mathematical Society* 82:711–712.

25. Wilson, R. A. 2002. *Graphs, Colourings and the Four-Colour Theorem*. Oxford: Oxford University Press.

26. König, D. 1916. Über graphen und ihre anwendung auf determinantentheorie und mengenlehre. *Mathematische Annalen* 77:453–465.

27. Wang, R. V., D. D. Fong, F. Jiang, M. J. Highland, P. H. Fuoss, C. Thompson, A. M. Kolpak, J. A. Eastman, S. K. Streiffer, A. M. Rappe, and G. B. Stephenson. 2009. Reversible chemical switching of a ferroelectric film. *Physical Review Letters* 102:047601.

28. Rao, G. V. S., B. M. Wanklyn, and C. N. R. Rao. 1971. Electrical transport in rare earth ortho chromites, ortho manganites and ortho ferrites. *Journal of Physics and Chemistry of Solids* 32:345.

29. Maksymovych, P., J. Seidel, Y. H. Chu, P. Wu, A. P. Baddorf, L.-Q. Chen, S. V. Kalinin, and R. Ramesh. 2011. Dynamic conductivity of ferroelectric domain walls in $BiFeO_3$. *Nano Letters* 11:1906–1912.

30. Seidel, J., P. Maksymovych, Y. Batra, A. Katan, S. Y. Yang, Q. He, A. P. Baddorf, S. V. Kalinin, C. H. Yang, J. C. Yang, Y. H. Chu, E. K. H. Salje, H. Wormeester, M. Salmeron, and R. Ramesh. 2010. Domain wall conductivity in La-doped $BiFeO_3$. *Physical Review Letters* 105:197603.

31. Seidel, J., L. W. Martin, Q. He, Q. Zhan, Y. H. Chu, A. Rother, M. E. Hawkridge, P. Maksymovych, P. Yu, M. Gajek, N. Balke, S. V. Kalinin, S. Gemming, F. Wang, G. Catalan, J. F. Scott, N. A. Spaldin, J. Orenstein, and R. Ramesh. 2009. Conduction at domain walls in oxide multiferroics. *Nature Materials* 8:229–234.

32. Zhong, W., R. D. Kingsmith, and D. Vanderbilt. 1994. Giant lo-to splitting sin perovskite ferroelectrics. *Physical Review Letters* 72:3618–3621.

33. Jia, C.-L., S.-B. Mi, K. Urban, I. Vrejoiu, M. Alexe, and D. Hesse. 2008. Atomic-scale study of electric dipoles near charged and uncharged domain walls in ferroelectric films. *Nature Materials* 7:57–61.

34. Shur, V. Y., E. L. Rumyantsev, E. V. Nikolaeva, and E. I. Shishkin. 2000. Formation and evolution of charged domain walls in congruent lithium niobate. *Applied Physics Letters* 77:3636–3638.

35. Gureev, M. Y., A. K. Tagantsev, and N. Setter. 2011. Head-to-head and tail-to-tail 180 degrees domain walls in an isolated ferroelectric. *Physical Review B* 83:184104.

36. Koehler, W. C., H. L. Yakel, E. O. Wollan, and J. W. Cable. 1964. A note on the magnetic structures of rare earth manganese oxides. *Physics Letters* 9:93–95.

37. Wu, X. F., and D. Vanderbilt. 2006. Theory of hypothetical ferroelectric superlattices incorporating head-to-head and tail-to-tail 180 degrees domain walls. *Physical Review B* 73:020103.

38. Jungk, T., A. Hoffmann, M. Fiebig, and E. Soergel. 2010. Electrostatic topology of ferroelectric domains in h-YMnO3. *Applied Physics Letters* 97:012904.

39. Lochocki, E. B., S. Park, N. Lee, S. W. Cheong, and W. Wu. 2011. Piezoresponse force microscopy of domains and walls in multiferroic $HoMnO_3$. *Applied Physics Letters* 99:232901.

40. Lines, M. E., and A. M. Glass. 1977. *Principles and Applications of Ferroelectrics and Related Materials.* Oxford: Oxford University Press.

41. Choi, W. S., S. J. Moon, S. S. A. Seo, D. Lee, J. H. Lee, P. Murugavel, T. W. Noh, and Y. S. Lee. 2008. Optical spectroscopic investigation on the coupling of electronic and magnetic structure in multiferroic hexagonal $RMnO_3$ (R = Gd, Tb, Dy, and Ho) thin films. *Physical Review B* 78:054440.

42. Souchkov, A. B., J. R. Simpson, M. Quijada, H. Ishibashi, N. Hur, J. S. Ahn, S. W. Cheong, A. J. Millis, and H. D. Drew. 2003. Exchange interaction effects on the optical properties of $LuMnO_3$. *Physical Review Letters* 91:027203.

43. Chaikin, P. M., and T. C. Lubensky. 2000. *Principles of Condensed Matter Physics.* Cambridge: Cambridge University Press.

44. Jungk, T., A. Hoffmann, and E. Soergel. 2009. Contrast mechanisms for the detection of ferroelectric domains with scanning force microscopy. *New Journal of Physics* 11:033029.

45. Fraisse, A. A., C. Ringeval, D. N. Spergel, and F. R. Bouchet. 2008. Small-angle CMB temperature anisotropies induced by cosmic strings. *Physical Review D* 78:043535.

46. Mesaros, A., K. Fujita, H. Eisaki, S. Uchida, J. C. Davis, S. Sachdev, J. Zaanen, M. J. Lawler, and E.-A. Kim. 2011. Topological defects coupling smectic modulations to intra-unit-cell nematicity in cuprates. *Science* 333:426–430.

47. Ye, J., S. Inoue, K. Kobayashi, Y. Kasahara, H. Yuan, H. Shimotani, and Y. Iwasa. 2010. Liquid-gated interface superconductivity on an atomically flat film. *Nature Materials* 9:125–128.

48. Pfleiderer, C., and A. Rosch. 2010. Condensed-matter physics single skyrmions spotted. *Nature* 465:880–881.

49. Eerenstein, W., N. D. Mathur, and J. F. Scott. 2006. Multiferroic and magneto-electric materials. *Nature* 442:759–765.

50. Scott, J. F. 2007. Data storage—Multiferroic memories. *Nature Materials* 6:256–257.

51. Spaldin, N. A., and M. Fiebig. 2005. The renaissance of magnetoelectric multi-ferroics. *Science* 309:391–392.

52. Abrahams, S. C. 2001. Ferroelectricity and structure in the h-$YMnO_3$ family. *Acta Crystallographica Section B-Structural Science* 57:485–490.

53. Fiebig, M., D. Frohlich, S. Leute, and R. V. Pisarev. 1998. Second harmonic spectroscopy and control of domain size in antiferromagnetic h-$YMnO_3$. *Journal of Applied Physics* 83:6560–6562.

54. Artyukhin, S., K. T. Delaney, N. A. Spaldin, and M. Mostovoy. 2014. Landau theory of topological defects in multiferroic hexagonal manganites. *Nature Materials* 13:42–49.

55. Park, S., P. Ryan, E. Karapetrova, J. W. Kim, J. X. Ma, J. Shi, J. W. Freeland, and W. Wu. 2009. Microscopic evidence of a strain-enhanced ferromagnetic state in $LaCoO_3$ thin films. *Applied Physics Letters* 95:072508.

56. Dzyaloshinskii, I. E. 1960. On the magneto-electrical effect in antiferromagnets. *Soviet Physics Jetp-Ussr* 10:628–629.

57. Moriya, T. 1960. Anisotropic superexchange interaction and weak ferromagnetism. *Physical Review* 120:91–98.

58. Rugar, D., H. J. Mamin, P. Guethner, S. E. Lambert, J. E. Stern, I. McFadyen, and T. Yogi. 1990. Magnetic force microscopy—general principles and application to longitudinal recording media. *Journal of Applied Physics* 68:1169–1183.

59. Israel, C., W. Wu, and A. de Lozanne. 2006. High-field magnetic force microscopy as susceptibility imaging. *Applied Physics Letters* 89:032502.

60. Goltsev, A. V., R. V. Pisarev, T. Lottermoser, and M. Fiebig. 2003. Structure and interaction of antiferromagnetic domain walls in hexagonal h-YMnO3. *Physical Review Letters* 90:177204.

61. Fabreges, X., I. Mirebeau, P. Bonville, S. Petit, G. Lebras-Jasmin, A. Forget, G. Andre, and S. Pailhes. 2008. Magnetic order in $YbMnO_3$ studied by neutron diffraction and Mossbauer spectroscopy. *Physical Review B* 78:214422.

62. Nandi, S., A. Kreyssig, L. Tan, J. W. Kim, J. Q. Yan, J. C. Lang, D. Haskel, R. J. McQueeney, and A. I. Goldman. 2008. Nature of Ho magnetism in multiferroic $HoMnO_3$. *Physical Review Letters* 100:217201.

63. Salama, H. A., and G. A. Stewart. 2009. Exchange-induced Tm magnetism in multiferroic h-$TmMnO_3$. *Journal of Physics Condensed Matter* 21:386001.

64. Yen, F., C. dela Cruz, B. Lorenz, E. Galstyan, Y. Y. Sun, M. Gospodinov, and C. W. Chu. 2007. Magnetic phase diagrams of multiferroic hexagonal $RMnO_3$ (R = Er, Yb, Tm, and Ho). *Journal of Materials Research* 22:2163–2173.

65. Lonkai, T., D. Hohlwein, J. Ihringer, and W. Prandl. 2002. The magnetic structures of h-$YMnO_3$-delta and $HoMnO_3$. *Applied Physics a-Materials Science and Processing* 74:S843–S845.

66. Spaldin, N. A., S.-W. Cheong, and R. Ramesh. 2010. Multiferroics: Past, present, and future. *Physics Today* 63:38–43.

67. Zhang, Q. H., L. J. Wang, X. K. Wei, R. C. Yu, L. Gu, A. Hirata, M. W. Chen, C. Q. Jin, Y. Yao, Y. G. Wang, and X. F. Duan. 2012. Direct observation of interlocked domain walls in hexagonal $RMnO_3$ (R = Tm, Lu). *Physical Review B* 85:020102.

68. Skumryev, V., V. Laukhin, I. Fina, X. Marti, F. Sanchez, M. Gospodinov, and J. Fontcuberta. 2011. Magnetization reversal by electric-field decoupling of magnetic and ferroelectric domain walls in multiferroic-based heterostructures. *Physical Review Letters* 106:057206.

69. Tokunaga, Y., N. Furukawa, H. Sakai, Y. Taguchi, T.-h. Arima, and Y. Tokura. 2009. Composite domain walls in a multiferroic perovskite ferrite. *Nature Materials* 8:558–562.

70. Palaniappan, L., F. D. Gnanam, and P. Ramasamy. 1986. Electrical characterization of sbsi grown from the melt by a temperature-fluctuation technique. *Journal of Materials Science Letters* 5:1007–1008.

71. Wemple, S. H. 1970. Polarization fluctuations and optical-absorption edge in $BaTiO_3$. *Physical Review B* 2:2679–2689.

72. Cho, D. Y., J. Y. Kim, B. G. Park, K. J. Rho, J. H. Park, H. J. Noh, B. J. Kim, S. J. Oh, H. M. Park, J. S. Ahn, H. Ishibashi, S. W. Cheong, J. H. Lee, P. Murugavel, T. W. Noh, A. Tanaka, and T. Jo. 2007. Ferroelectricity driven by Y d(0)-ness with rehybridization in h-YMnO3. *Physical Review Letters* 98:217601.

73. Wood, V. E., A. E. Austin, E. W. Collings, and K. C. Brog. 1973. Magnetic properties of heavy-rare-earth orthomanganites. *Journal of Physics and Chemistry of Solids* 34:859–868.

74. Pintilie, L., I. Vrejoiu, D. Hesse, G. LeRhun, and M. Alexe. 2007. Ferroelectric polarization-leakage current relation in high quality epitaxial $Pb(Zr, Ti)O_3$ films. *Physical Review B* 75:104103.

75. Kim, S. H., S. H. Lee, T. H. Kim, T. Zyung, T. H. Jeong, and M. S. Jang. 2000. Growth, ferroelectric properties, and phonon modes of h-YMnO₃ single crystal. *Crystal Research and Technology* 35:19–27.

76. Catalan, G., J. Seidel, R. Ramesh, and J. F. Scott. 2012. Domain wall nanoelectronics. *Reviews of Modern Physics* 84:119–156.

77. Dawber, M., K. M. Rabe, and J. F. Scott. 2005. Physics of thin-film ferroelectric oxides. *Reviews of Modern Physics* 77:1083–1130.

78. Einhorn, M. B., R. Savit, and E. Rabinovici. 1980. A physical picture for the phase-transitions in Zn symmetric models. *Nuclear Physics B* 170:16–31.

79. Gibbs, A. S., K. S. Knight, and P. Lightfoot. 2011. High-temperature phase transitions of hexagonal h-YMnO₃. *Physical Review B* 83:094111.

80. Fujimura, N., T. Ishida, T. Yoshimura, and T. Ito. 1996. Epitaxially grown h-YMnO₃ film: New candidate for nonvolatile memory devices. *Applied Physics Letters* 69:1011–1013.

81. Lonkai, T., D. G. Tomuta, U. Amann, J. Ihringer, R. W. A. Hendrikx, D. M. Tobbens, and J. A. Mydosh. 2004. Development of the high-temperature phase of hexagonal manganites. *Physical Review B* 69:134108.

82. Lukaszew.K, and Karutkal.J. 1974. X-ray investigations of crystal-structure and phase-transitions of h-YMnO₃. *Ferroelectrics* 7:81–82.

83. Nenert, G., M. Pollet, S. Marinel, G. R. Blake, A. Meetsma, and T. T. M. Palstra. 2007. Experimental evidence for an intermediate phase in the multiferroic h-YMnO₃. *Journal of Physics Condensed Matter* 19:466212.

84. Smolenskii, G. A., and V. A. Bokov. 1964. Coexistence of magnetic and electric ordering in crystals. *Journal of Applied Physics* 35:915–918.

85. Li, J., X. Yang, H. F. Tian, S.-W. Cheong, C. Ma, S. Zhang, Y. G. Zhao, and J. Q. Li. 2013. Ferroelectric annular domains in hexagonal manganites. *Physical Review B* 87:094106.

86. O'Dell, T. H. 1974. *Magnetic Bubbles*. New York: Halsted Press.

87. Cockayne, B., M. Chesswas, J. G. Plant, and A. W. Vere. 1969. Ferroelectric domains and growth striae in barium sodium niobate single crystals. *Journal of Materials Science* 4:565–569.

88. Grest, G. S., and D. J. Srolovitz. 1984. Vortex effects on domain growth—the clock model. *Physical Review B* 30:6535-6539.

89. Kiometzis, M., H. Kleinert, and A. M. J. Schakel. 1995. Dual description of the superconducting phase transition. *Fortschritte Der Physik Progress of Physics* 43:697–732.

90. Kleinert, H. 1989. *Gauge Fields in Condensed Matter*. Singapore: World Scientific.

91. Kleinert, H., R. Ruffini, and G. Vereshchagin. 2010. From Landau's order parameter to modern disorder fields. *AIP Conference Proceedings* 1205:103–107.

92. Kiometzis, M., H. Kleinert, and A. M. J. Schakel. 1994. Critical exponents of the superconducting phase-transition. *Physical Review Letters* 73:1975–1977.

93. Lin, S.-Z., X. Wang, Y. Kamiya, G-W. Chern, F. Fan, D. Fan, B. Casas, Y. Liu, V. Kiryukhin, W. H. Zurek, C. D. Batista, S.-W. Cheong. 2014. Topological defects as relics of emergent continuous symmetry and Higgs condensation of disorder in ferroelectrics. *Nature Physics*, 10:970–977.

94. Kibble, T. W. B. 1976. Topology of cosmic domains and strings. *Journal of Physics a-Mathematical and General* 9:1387–1398.

95. Zurek, W. H. 1985. Cosmological experiments in superfluid helium. *Nature* 317:505–508.

96. del Campo, A., and W. H. Zurek. 2014. Universality of phase transition dynamics: Topological defects from symmetry breaking. *International Journal of Modern Physics A* 29:1430018.

97. Griffin, S. M., M. Lilienblum, K. T. Delaney, Y. Kumagai, M. Fiebig, and N. A. Spaldin. 2012. Scaling behavior and beyond equilibrium in the hexagonal manganites. *Physical Review X* 2:041022.

98. Hohenberg, P. C., and B. I. Halperin. 1977. Theory of dynamic critical phenomena. *Reviews of Modern Physics* 49:435–479.

99. Liu, F., and G. F. Mazenko. 1992. Defect-defect correlation in the dynamics of 1st-order phase-transitions. *Physical Review B* 46:5963–5971.

100. Zurek, W. H. 1996. Cosmological experiments in condensed matter systems. *Physics Reports Review Section of Physics Letters* 276:177–221.

101. Zurek, W. H. 2013. Topological relics of symmetry breaking: winding numbers and scaling tilts from random vortex-antivortex pairs. *Journal of Physics Condensed Matter* 25:404209.

14

Multiferroics and Beyond

Lu You, Yang Zhou, and Junling Wang

Contents

The past decade has witnessed the renaissance of research in magnetoelectric and multiferroic materials, which has largely remained quiet after the pioneer work that is traced back to the 1960s. Supported by the advances in both material synthesis and first-principles calculation techniques, unprecedented achievements have been made in this field. General consensus has been reached in terms of the classification of multiferroics based on the origins of the electric and magnetic orders. As introduced in Chapter 1, type-I multiferroics refer to those with their ferroelectricity and magnetism coming from different sources, whereas in type-II multiferroics the ferroelectricity is induced by some type of magnetic order [1]. Coincidentally, groundbreaking progress was made on both type-I ($BiFeO_3$, 2003) [2] and type-II multiferroics ($TbMnO_3$ 2003 and $TbMn_2O_5$ 2004) [3,4] at about the same time, which led to the revival of this field. The experimental demonstrations of electric-field control of magnetic order and magnetic-field switching of the ferroelectric polarization substantiate the main characteristic of multiferroics—the cross-coupling between ferroic orders—and lay the cornerstone for the next-generation switches, sensors, and memories. However, after a decade of intensive research, major bottlenecks still limit the applications of multiferroics. For example, although type-II multiferroics usually show strong magnetoelectric coupling, their ferroic-ordering temperatures are way below the room temperature, and thus are not viable for practical applications. Type-I multiferroics, such as $BiFeO_3$, may possess both electric and magnetic long-range orders above the room temperature. Unfortunately, most of them are antiferromagnets, which exhibit no macroscopic magnetization. Therefore, heterostructures with a ferromagnetic layer have to be fabricated to indirectly control the ferromagnetism using an electric field [5,6]. However, the degree of control is not reliable due to the fragile interfacial exchange coupling [7]. It is fair to say that the main obstacle for this field is the lack of a true room-temperature multiferroic material with robust ferroelectric and ferromagnetic orders.

Unfortunately, it seems that ferroelectricity and ferromagnetism are mutually exclusive fundamentally [8]. To face this dilemma, an open mind is indispensible for an illuminating outlook and prospect. Future research on multiferroics is not necessary to be restricted to single-phase materials or composites, but should also include all "multiferroic phenomena," which are prevalent in various frontiers of condense matter physics. Here we try to outline some possible future directions, which, hopefully, will spark readers' imaginations to keep the momentum going in this fascinating field.

14.1 Interfaces Host New Multiferroics

"The interface is the device," as coined by Nobel laureate Herbert Kroemer [9], highlights the crucial role of heterointerface in semiconductor devices. Similarly, heterointerfaces can also host various emergent properties that are absent in the parent materials, even multiferroicity. Due to the symmetric breaking at the interface, reconstructions of the charge, orbital, spin, and lattice degrees of freedom are inevitable, which, in turn, generates new physics and exotic properties. Equipped with the state-of-the-art expitaxy techniques, researchers can grow various thin films layer-by-layer with atomic precision. Playing a "LEGO" toy at the atomic level is now within reach. In the recent years, notable examples of emergent phenomena arising from oxide interfaces include the orbital reconstruction at $(Y,Ca)Ba_2Cu_3O_7$ (superconductor)–$La_{2/3}Ca_{1/3}MnO_3$ (ferromagnetic half-metal) interface [10], 2D electron gas at $LaAlO_3$ (nonmagnetic insulator)–$SrTiO_3$ (nonmagnetic insulator) interface [11], exchange coupling at $BiFeO_3$ (antiferromagnetic ferroelectric)–$La_{2/3}Sr_{1/3}MnO_3$ (ferromagnetic half-metal) interface [12], and so on. Although not all constituent materials are ferroic in nature, multiferroicity does appear at some of these interfaces. Intrinsically, charge transfer is inevitable when two materials bond with each other, rendering the interface polar. Meanwhile, a nonmagnetic material could become magnetic because the loss or gain of electrons results in partially filled orbitals, for instance, ferromagnetism of the $LaAlO_3$–$SrTiO_3$ interface [13]. More importantly, it is possible to control the interface properties via an electric field, be it a true magnetoelectric coupling effect like in $BiFeO_3$ [14] or an electrochemical effect in other systems [15]. It is likely that heterointerfaces will become the new platform for future explorations on magnetoelectric coupling phenomena and other relevant effects. Last but not least, the numerous mix-and-match combinations make the guidance of high-throughput theoretical predictions indispensable. A "material genome" database can act as the blueprint for design strategies of functional heterointerfaces and superlattices [16].

14.2 Ionic Defects—An Additional Degree of Freedom

Ferroelasticity, ferroelectricity, and ferromagnetism are the three main order parameters in multiferroics. Their cross-couplings have been the main focus throughout the new wave of the multiferroic research. Recently, ionic defects have emerged as an additional degree of freedom in tuning materials' properties [17]. At first glance, this is counterintuitive as defects are usually detrimental to the material performance. However, more and more evidence has shown that these "undesirable" species may play a crucial role in determining materials' functionalities if they are properly engineered and manipulated. Microscopically, ionic point defects, such as oxygen vacancies ubiquitous in oxides, are charged particles. They interact with the ferroelectric polarization electrostatically and are highly active under electric field. Polarization screening by charged point defects has long been proposed to prevail on ferroelectric surfaces and domain walls, and recently verified experimentally with the assistance of high-resolution TEM [18]. Furthermore, ionic vacancies impose huge chemical strain to the lattice locally, leading to modifications of bond-length and bond-angle that directly affect the electronic and magnetic behaviors at the nanoscale [19]. The valence change of the magnetic ions in transition metal oxides induced by oxygen vacancies has an even more significant impact on the magnetic order parameter [20]. These charged point defects may even form certain ordered nanostructures with distinct properties [21]. All of these fascinating phenomena demonstrate that ionic defects are strongly entangled with the three primary ferroic order parameters. More importantly, owing to their high activity under external stimuli (strain, electric field, magnetic field, etc.), they may prove to be an additional tool in shaping the functionalities of multiferroic materials and heterostructures in the future.

14.3 Topological Defects (Domain Walls, Vortices, Skyrmion)

Topological defects, such as domain walls and vortices, constitute another burgeoning field in the multiferroic community [22]. Similar to the heterointerfaces, domain walls and vortices can be viewed as the homo-interfaces where symmetry breaking occurs intrinsically. The striking resembling in the scaling behavior of electric vortices emerging from a ferroelectric transition and the condensation of the cosmic strings in the early universe makes people think whether it is a simple coincidence, or whether a water drop can really reflect the whole universe [23,24]. Nevertheless, the attempt to connect two seemingly irrelevant physical phenomena is illuminating albeit audacious. Aside from the rich fundamental physics, multiferroic domain walls also host a vast array of emergent characters [25]. Specifically to the perovskite oxides are the complex rotation patterns of the oxygen octahedra, which directly modify the bond angles and lengths, leading to diverse electronic and magnetic properties at the domain walls [26–29]. Moreover, due to the structural and electrostatic discontinuities, the ionic defects discussed previously are also intimately coupled to the domain walls [30,31]. Further studies should be devoted to the interplay between them and deconvolute the intrinsic properties from extrinsic effects.

Magnetic skyrmion is yet another topological state that is closely linked to the magnetoelectric effect [32,33]. The vortex-like spin configuration is believed to originate from the Dzyaloshinskii–Moriya (DM) interaction, which happens to be responsible for the ferroelectric polarization in type-II multiferroics [34]. Similar flexomagnetoelectric interaction also exists in type-I multiferroic such as antiferromagnetic $BiFeO_3$, leading to the spin cycloid and strong coupling with the polar order. Furthermore, ferrotoroidicity has long been hypothesized to be the fourth primary ferroic orders, which violates both the time- and space-inversion symmetries. Although it has been observed experimentally [35,36], the ferrotoroidicity receives relatively less attention compared to other ferroic orders. With the groundbreaking progress in the magnetic vortices study, it may be a good opportunity to reexamine this topic in related multiferroics. The underlying physics behind these topological defects and the ability to manipulate their states warrants more intensive research in the future.

14.4 Multiferroics Go Hybrid and 2D

Lastly, we would like to draw your attention to some unconventional multiferroics beyond the inorganic oxides. Great achievements have been made in organic/polymer ferroelectrics [37], and we point out here that organic–inorganic hybrid materials, including the renowned metal–organic frameworks (MOFs), could be the next seedbed for novel multiferroics [38–40]. A stunning example is the organometallic lead halide perovskites that whip up a storm in the solar cell community with unprecedented improvement in power conversion efficiencies within 2–3 years, compared to conventional inorganic solar cells [41]. Chemists have demonstrated their capabilities to synthesize a broad range of hybrid multiferroics, whose ferroelectricity and magnetism usually comes separately from the organic and inorganic components, respectively [42–45]. However, a strong magnetoelectric coupling in them remains elusive, and most of the studies still stay on the theoretical stage [46–50]. This task should be followed up by the physicists in the future.

Another intriguing feature of these hybrid materials is that a number of them exhibit layered crystal structures. Individual atomic sheets are bonded weakly through hydrogen bonds or Van der Waals force, which can be easily broken down into two-dimensional (2D) atomic layers. Inspired by the discovery of graphene, 2D materials are another fast-growing research field that holds great potential for the next-generation atomic-level electronics [51–54]. Although the family of 2D materials is expanding rapidly, members with long-range electric and magnetic orders remain to be discovered. The interplay between quantum confinement effect and long-range ferroic orders would be an exciting topic to explore in the future. From the application point of view, 2D multiferroics will greatly enhance the versatility of the van der Waals heterostructures by adding new functionalities and making them nonvolatile.

The research on single-phase multiferroics may run aground due to the material limitations, but the study of multiferroic phenomena will certainly remain strong. The emergent multiferroic phenomena host new physics and possibly revolutionary technologies that wait for researchers to unveil.

References

1. Khomskii, D. 2009. Classifying multiferroics: Mechanisms and effects. *Physics* 2:20.
2. Wang, J., J. B. Neaton, H. Zheng, V. Nagarajan, S. B. Ogale, B. Liu, D. Viehland, V. Vaithyanathan, D. G. Schlom, U. V. Waghmare, N. A. Spaldin, K. M. Rabe, M. Wuttig, and R. Ramesh. 2003. Epitaxial $BiFeO_3$ multiferroic thin film heterostructures. *Science* 299:1719–1722.
3. Kimura, T., T. Goto, H. Shintani, K. Ishizaka, T. Arima, and Y. Tokura. 2003. Magnetic control of ferroelectric polarization. *Nature* 426:55–58.
4. Hur, N., S. Park, P. A. Sharma, J. S. Ahn, S. Guha, and S. W. Cheong. 2004. Electric polarization reversal and memory in a multiferroic material induced by magnetic fields. *Nature* 429:392–395.
5. Zhao, T., A. Scholl, F. Zavaliche, K. Lee, M. Barry, A. Doran, M. P. Cruz, Y. H. Chu, C. Ederer, N. A. Spaldin, R. R. Das, D. M. Kim, S. H. Baek, C. B. Eom, and R. Ramesh. 2006. Electrical control of antiferromagnetic domains in multiferroic $BiFeO_3$ films at room temperature. *Nature Materials* 5:823–829.
6. Chu, Y. H., L. W. Martin, M. B. Holcomb, M. Gajek, S. J. Han, Q. He, N. Balke, C. H. Yang, D. Lee, W. Hu, Q. Zhan, P. L. Yang, A. Fraile-Rodriguez, A. Scholl, S. X. Wang, and R. Ramesh. 2008. Electric-field control of local ferromagnetism using a magnetoelectric multiferroic. *Nature Materials* 7:478–482.
7. Heron, J. T., J. L. Bosse, Q. He, Y. Gao, M. Trassin, L. Ye, J. D. Clarkson, C. Wang, J. Liu, S. Salahuddin, D. C. Ralph, D. G. Schlom, J. Iniguez, B. D. Huey, and R. Ramesh. 2014. Deterministic switching of ferromagnetism at room temperature using an electric field. *Nature* 516:370–373.
8. Hill, N. A. 2000. Why are there so few magnetic ferroelectrics? *The Journal of Physical Chemistry B* 104:6694–6709.
9. Kroemer, H. 2001. Nobel lecture: Quasielectric fields and band offsets: Teaching electrons new tricks. *Reviews of Modern Physics* 73:783–793.

10. Chakhalian, J., J. W. Freeland, H.-U. Habermeier, G. Cristiani, G. Khaliullin, M. van Veenendaal, and B. Keimer. 2007. Orbital reconstruction and covalent bonding at an oxide interface. *Science* 318:1114–1117.

11. Ohtomo, A., and H. Y. Hwang. 2004. A high-mobility electron gas at the $LaAlO_3$/$SrTiO_3$ heterointerface. *Nature* 427:423–426.

12. Yu, P., J. S. Lee, S. Okamoto, M. D. Rossell, M. Huijben, C. H. Yang, Q. He, J. X. Zhang, S. Y. Yang, M. J. Lee, Q. M. Ramasse, R. Erni, Y. H. Chu, D. A. Arena, C. C. Kao, L. W. Martin, and R. Ramesh. 2010. Interface ferromagnetism and orbital reconstruction in $BiFeO_3$-$La_{0.7}Sr_{0.3}MnO_3$ heterostructures. *Physical Review Letters* 105:027201.

13. Brinkman, A., M. Huijben, M. van Zalk, J. Huijben, U. Zeitler, J. C. Maan, W. G. van der Wiel, G. Rijnders, D. H. A. Blank, and H. Hilgenkamp. 2007. Magnetic effects at the interface between non-magnetic oxides. *Nature Materials* 6:493–496.

14. Wu, S. M., S. A. Cybart, P. Yu, M. D. Rossell, J. X. Zhang, R. Ramesh, and R. C. Dynes. 2010. Reversible electric control of exchange bias in a multiferroic field-effect device. *Nature Materials* 9:756–761.

15. Cen, C., S. Thiel, G. Hammerl, C. W. Schneider, K. E. Andersen, C. S. Hellberg, J. Mannhart, and J. Levy. 2008. Nanoscale control of an interfacial metal-insulator transition at room temperature. *Nature Materials* 7:298–302.

16. Curtarolo, S., G. L. W. Hart, M. B. Nardelli, N. Mingo, S. Sanvito, and O. Levy. 2013. The high-throughput highway to computational materials design. *Nature Materials* 12:191–201.

17. Kalinin, S. V., and N. A. Spaldin. 2013. Functional ion defects in transition metal oxides. *Science* 341:858–859.

18. Kim, Y.-M., A. Morozovska, E. Eliseev, M. P. Oxley, R. Mishra, S. M. Selbach, T. Grande, S. T. Pantelides, S. V. Kalinin, and A. Y. Borisevich. 2014. Direct observation of ferroelectric field effect and vacancy-controlled screening at the $BiFeO_3$/$LaxSr_{1-x}MnO_3$ interface. *Nature Materials* 13:1019–1025.

19. Becher, C., L. Maurel, U. Aschauer, M. Lilienblum, C. Magén, D. Meier, E. Langenberg, M. Trassin, J. Blasco, I. P. Krug, P. A. Algarabel, N. A. Spaldin, J. A. Pardo, and M. Fiebig. 2015. Strain-induced coupling of electrical polarization and structural defects in $SrMnO_3$ films. *Nature Nanotechnology* 10:661–665.

20. Bauer, U., L. Yao, A. J. Tan, P. Agrawal, S. Emori, H. L. Tuller, S. van Dijken, and G. S. D. Beach. 2015. Magneto-ionic control of interfacial magnetism. *Nature Materials* 14:174–181.

21. Farokhipoor, S., C. Magen, S. Venkatesan, J. Iniguez, C. J. M. Daumont, D. Rubi, E. Snoeck, M. Mostovoy, C. de Graaf, A. Muller, M. Doblinger, C. Scheu, and B. Noheda. 2014. Artificial chemical and magnetic structure at the domain walls of an epitaxial oxide. *Nature* 515:379–383.

22. Artyukhin, S., K. T. Delaney, N. A. Spaldin, and M. Mostovoy. 2014. Landau theory of topological defects in multiferroic hexagonal manganites. *Nature Materials* 13:42–49.

23. Griffin, S. M., M. Lilienblum, K. T. Delaney, Y. Kumagai, M. Fiebig, and N. A. Spaldin. 2012. Scaling behavior and beyond equilibrium in the hexagonal manganites. *Physical Review X* 2:041022.

24. Lin, S.-Z., X. Wang, Y. Kamiya, G.-W. Chern, F. Fan, D. Fan, B. Casas, Y. Liu, V. Kiryukhin, W. H. Zurek, C. D. Batista, and S.-W. Cheong. 2014. Topological defects as relics of emergent continuous symmetry and Higgs condensation of disorder in ferroelectrics. *Nature Physics* 10:970–977.

25. Catalan, G., J. Seidel, R. Ramesh, and J. F. Scott. 2012. Domain wall nanoelectronics. *Reviews of Modern Physics* 84:119–156.

26. Seidel, J., L. W. Martin, Q. He, Q. Zhan, Y. H. Chu, A. Rother, M. E. Hawkridge, P. Maksymovych, P. Yu, M. Gajek, N. Balke, S. V. Kalinin, S. Gemming, F. Wang, G. Catalan, J. F. Scott, N. A. Spaldin, J. Orenstein, and R. Ramesh. 2009. Conduction at domain walls in oxide multiferroics. *Nature Materials* 8:229–234.

27. Yang, S. Y., J. Seidel, S. J. Byrnes, P. Shafer, C. H. Yang, M. D. Rossell, P. Yu, Y. H. Chu, J. F. Scott, J. W. Ager, L. W. Martin, and R. Ramesh. 2010. Above-bandgap voltages from ferroelectric photovoltaic devices. *Nature Nanotechnology* 5:143–147.

28. Balke, N., B. Winchester, W. Ren, Y. H. Chu, A. N. Morozovska, E. A. Eliseev, M. Huijben, R. K. Vasudevan, P. Maksymovych, J. Britson, S. Jesse, I. Kornev, R. Ramesh, L. Bellaiche, L. Q. Chen, and S. V. Kalinin. 2012. Enhanced electric conductivity at ferroelectric vortex cores in BiFeO$_3$. *Nature Physics* 8:81–88.

29. He, Q., C. H. Yeh, J. C. Yang, G. Singh-Bhalla, C. W. Liang, P. W. Chiu, G. Catalan, L. W. Martin, Y. H. Chu, J. F. Scott, and R. Ramesh. 2012. Magnetotransport at domain walls in BiFeO$_3$. *Physical Review Letters* 108:067203.

30. Seidel, J., P. Maksymovych, Y. Batra, A. Katan, S. Y. Yang, Q. He, A. P. Baddorf, S. V. Kalinin, C. H. Yang, J. C. Yang, Y. H. Chu, E. K. H. Salje, H. Wormeester, M. Salmeron, and R. Ramesh. 2010. Domain wall conductivity in La-doped BiFeO$_3$. *Physical Review Letters* 105:197603.

31. Farokhipoor, S., and B. Noheda. 2011. Conduction through 71 domain walls in BiFeO$_3$ thin films. *Physical Review Letters* 107:127601.

32. Yu, X. Z., Y. Onose, N. Kanazawa, J. H. Park, J. H. Han, Y. Matsui, N. Nagaosa, and Y. Tokura. 2010. Real-space observation of a two-dimensional skyrmion crystal. *Nature* 465:901–904.

33. Seki, S., X. Z. Yu, S. Ishiwata, and Y. Tokura. 2012. Observation of skyrmions in a multiferroic material. *Science* 336:198–201.

34. Nagaosa, N., and Y. Tokura. 2013. Topological properties and dynamics of magnetic skyrmions. *Nature Nanotechnology* 8:899–911.

35. Van Aken, B. B., J.-P. Rivera, H. Schmid, and M. Fiebig. 2007. Observation of ferrotoroidic domains. *Nature* 449:702–705.

36. Zimmermann, A. S., D. Meier, and M. Fiebig. 2014. Ferroic nature of magnetic toroidal order. *Nature Communications* 5.

37. Horiuchi, S., and Y. Tokura. 2008. Organic ferroelectrics. *Nature Materials* 7:357–366.

38. Ramesh, R. 2009. Materials science: Emerging routes to multiferroics. *Nature* 461:1218–1219.

39. Rogez, G., N. Viart, and M. Drillon. 2010. Multiferroic materials: The attractive approach of metal–organic frameworks (MOFs). *Angewandte Chemie International Edition* 49:1921–1923.

40. Zhang, W., and R.-G. Xiong. 2012. Ferroelectric metal–organic frameworks. *Chemical Reviews* 112:1163–1195.

41. Green, M. A., K. Emery, Y. Hishikawa, W. Warta, and E. D. Dunlop. 2015. Solar cell efficiency tables (Version 45). *Progress in Photovoltaics: Research and Applications* 23:1–9.

42. Jain, P., V. Ramachandran, R. J. Clark, H. D. Zhou, B. H. Toby, N. S. Dalal, H. W. Kroto, and A. K. Cheetham. 2009. Multiferroic behavior associated with an order–disorder hydrogen bonding transition in metal–organic frameworks (MOFs) with the perovskite ABX$_3$ architecture. *Journal of the American Chemical Society* 131:13625–13627.

43. Kundys, B., A. Lappas, M. Viret, V. Kapustianyk, V. Rudyk, S. Semak, C. Simon, and I. Bakaimi. 2010. Multiferroicity and hydrogen-bond ordering in $(C_2H_5NH_3)_2CuCl_4$ featuring dominant ferromagnetic interactions. *Physical Review B* 81:224434.

44. Leblanc, N., N. Mercier, L. Zorina, S. Simonov, P. Auban-Senzier, and C. Pasquier. 2011. Large spontaneous polarization and clear hysteresis loop of a room-temperature hybrid ferroelectric based on mixed-halide [BiI$_3$Cl$_2$] polar chains and methylviologen dication. *Journal of the American Chemical Society* 133:14924–14927.

45. Polyakov, A. O., A. H. Arkenbout, J. Baas, G. R. Blake, A. Meetsma, A. Caretta, P. H. M. van Loosdrecht, and T. T. M. Palstra. 2012. Coexisting ferromagnetic and ferroelectric order in a CuCl$_4$-based organic–inorganic hybrid. *Chemistry of Materials* 24:133–139.

46. Stroppa, A., P. Jain, P. Barone, M. Marsman, J. M. Perez-Mato, A. K. Cheetham, H. W. Kroto, and S. Picozzi. 2011. Electric control of magnetization and interplay between orbital ordering and ferroelectricity in a multiferroic metal-organic framework. *Angewandte Chemie* 123:5969–5972.

47. Di Sante, D., A. Stroppa, P. Jain, and S. Picozzi. 2013. Tuning the ferroelectric polarization in a multiferroic metal–organic framework. *Journal of the American Chemical Society* 135:18126–18130.

48. Stroppa, A., P. Barone, P. Jain, J. M. Perez-Mato, and S. Picozzi. 2013. Hybrid improper ferroelectricity in a multiferroic and magnetoelectric metal-organic framework. *Advanced Materials* 25:2284–2290.

49. Wang, W., L. Q. Yan, J. Z. Cong, Y. L. Zhao, F. Wang, S. P. Shen, T. Zou, D. Zhang, S. G. Wang, X. F. Han, and Y. Sun. 2013. Magnetoelectric coupling in the paramagnetic state of a metal-organic framework. *Scientific Reports* 3:2024.

50. Tian, Y., A. Stroppa, Y. Chai, L. Yan, S. Wang, P. Barone, S. Picozzi, and Y. Sun. 2014. Cross coupling between electric and magnetic orders in a multiferroic metal-organic framework. *Scientific Reports* 4:6062.

51. Novoselov, K. S., V. I. Falko, L. Colombo, P. R. Gellert, M. G. Schwab, and K. Kim. 2012. A roadmap for graphene. *Nature* 490:192–200.

52. Geim, A. K., and I. V. Grigorieva. 2013. Van der Waals heterostructures. *Nature* 499:419–425.

53. Ferrari, A. C., F. Bonaccorso, V. Fal'ko, K. S. Novoselov, S. Roche, P. Boggild, S. Borini, F. H. L. Koppens, V. Palermo, N. Pugno, J. A. Garrido, R. Sordan, A. Bianco, L. Ballerini, M. Prato, E. Lidorikis, J. Kivioja, C. Marinelli, T. Ryhanen, A. Morpurgo, J. N. Coleman, V. Nicolosi, L. Colombo, A. Fert, M. Garcia-Hernandez, A. Bachtold, G. F. Schneider, F. Guinea, C. Dekker, M. Barbone, Z. Sun, C. Galiotis, A. N. Grigorenko, G. Konstantatos, A. Kis, M. Katsnelson, L. Vandersypen, A. Loiseau, V. Morandi, D. Neumaier, E. Treossi, V. Pellegrini, M. Polini, A. Tredicucci, G. M. Williams, B. Hee Hong, J.-H. Ahn, J. Min Kim, H. Zirath, B. J. van Wees, H. van der Zant, L. Occhipinti, A. Di Matteo, I. A. Kinloch, T. Seyller, E. Quesnel, X. Feng, K. Teo, N. Rupesinghe, P. Hakonen, S. R. T. Neil, Q. Tannock, T. Lofwander, and J. Kinaret. 2015. Science and technology roadmap for graphene, related two-dimensional crystals, and hybrid systems. *Nanoscale* 7:4598–4810.

54. Kim, S. J., K. Choi, B. Lee, Y. Kim, and B. H. Hong. 2015. Materials for flexible, stretchable electronics: Graphene and 2D materials. *Annual Review of Materials Research* 45:63–84.

Index

Index

Index